Fundamentals of Convective Heat Transfer

Fundamentals of Convective Heat Transfer

Gautam Biswas

Amaresh Dalal

Vijay K. Dhir

CRC Press
Taylor & Francis Group
Boca Raton London New York

CRC Press is an imprint of the
Taylor & Francis Group, an **informa** business

CRC Press
Taylor & Francis Group
6000 Broken Sound Parkway NW, Suite 300
Boca Raton, FL 33487-2742

First issued in paperback 2021

Version Date: 20190418

ISBN 13: 978-1-03-224198-2 (pbk)
ISBN 13: 978-1-138-10329-0 (hbk)

DOI: 10.1201/9780429198724

Visit the Taylor & Francis Web site at
http://www.taylorandfrancis.com

and the CRC Press Web site at
http://www.crcpress.com

Contents

Contents

Authors

Gautam Biswas is presently a professor of mechanical engineering at the Indian Institute of Technology (IIT) Kanpur. Previously, he was director of IIT Guwahati, and director of the CSIR-Central Mechanical Engineering Research Institute at Durgapur. He was the G.D. and V.M. Mehta Endowed Chair professor, and dean of academic affairs at IIT Kanpur. He served as a guest professor at the University of Erlangen-Nuremberg, Germany in 2002.

Prof. Biswas was a Humboldt Fellow in Germany in 1987 and 1988 and an Invited Fellow by the Japan Society for the Promotion of Science in 1994. He is a fellow of the three major science academies in his country: the Indian National Science Academy, New Delhi, the Indian Academy of Sciences, Bangalore, and the National Academy of Sciences of India, Allahabad. He is a fellow of the Indian National Academy of Engineering, New Delhi and Institution of Engineers India, Kolkata.

Prof. Biswas is a fellow of the American Society of Mechanical Engineers (ASME) and served as the associate editor of its *Journal of Heat Transfer*. He delivered the prestigious Prof. C.N.R. Rao Lecture in 2010. The Department of Science and Technology, India, awarded him the esteemed J.C. Bose National Fellowship in 2011. He is an associate editor of the journal *Computers & Fluids*. He has been conferred Honorary Doctorate by the Aristotle University of Thessaloniki, Greece in 2018.

Amaresh Dalal is an associate professor in the Department of Mechanical Engineering at IIT Guwahati. He earned a PhD from IIT Kanpur in 2009 and served as a postdoctoral research associate at Purdue University, Indiana in 2008 and 2009.

Dr. Dalal's research focuses on computational fluid dynamics and heat transfer, finite volume methods, unstructured grid techniques, and multiphase flows. He is presently developing a versatile and robust computational fluid dynamics solver over a hybrid unstructured grid. The programming paradigm is intended to solve a wide range of problems in the areas of real-life fluid flow, heat transfer, and transport phenomena involving complex geometries.

Dr. Dalal won the Prof. K.N. Seetharamu Medal and Prize for the Best Young Researcher in Heat Transfer from the Indian Society of Heat and Mass Transfer in 2017.

Vijay K. Dhir is a distinguished professor of mechanical and aerospace engineering and former dean of the Henry Samueli School of Engineering and Applied Science at the University of California Los Angeles (UCLA). Before being named dean, Prof.

Dhir served as vice chair and chair of the Department of Mechanical and Aerospace Engineering. Dr. Dhir's efforts earned international recognition for the university's engineering school and made it a hub for interdisciplinary education and research.

He leads the Boiling Heat Transfer Laboratory at UCLA. Its pioneering work in fundamental and applied sciences involving efficient heat removal through boiling gained international attention when NASA chose one of his research teams to conduct experiments on the International Space Station. They demonstrated the effects of microgravity on boiling.

In 2006, Prof. Dhir earned one of the most prestigious honors for engineers. He was awarded membership in the National Academy of Engineering for his achievements in boiling heat transfer and nuclear reactor thermal hydraulics and safety. He received the Max Jakob Memorial Award; the American Society of Mechanical Engineers (ASME) Heat Transfer Memorial Award; the Donald Q. Kern Award of the American Institute of Chemical Engineers; and the Technical Achievement Award of the American Nuclear Society.

Prof. Dhir has been a senior technical editor of ASME's *Journal of Heat Transfer* since 2000 and received its Best Paper Award twice. He serves on advisory boards of several other journals.

His many areas of interest include two-phase heat transfer, boiling and condensation, thermal and hydrodynamic stability, thermal dynamics of nuclear reactors, and microgravity heat transfer. The Boiling Heat Transfer Laboratory is currently studying flow boiling, microgravity boiling, and thermal hydraulics of nuclear reactors.

Preface

Convective heat transfer is one of the most important areas of engineering sciences. Its principles are deployed in the designs of all types of heat exchangers and various components related to mechanical, chemical, metallurgical, and aerospace engineering devices. Electronic cooling and thermal management of data centers are accomplished based on the principles of convective heat transfer.

The first seven chapters of this book are based on material covered over several years by the authors in a convective heat transfer course at the Indian Institutes of Technology at Kanpur and Guwahati. The final three chapters are based on the lecture notes from a well-known boiling and condensation course taught at the University of California Los Angeles.

In developing this book, we decided to emphasize the systematic mathematical developments. We tried as much as possible to provide insight into the physical behaviors of these processes utilizing appropriate mathematical routes.

We welcome suggestions and critical comments from our readers.

Gautam Biswas
Amaresh Dalal
Vijay K. Dhir

Acknowledgments

This book was completed through the Global Interactive Academic Network (GIAN), a unique program of the Ministry of Human Resource Development of the Government of India. Under the sponsorship of GIAN, Prof. Vijay K. Dhir, former dean of the Henry Samueli School of Engineering and Applied Science of UCLA and one of the authors of this book, designed and taught a course titled Boiling Heat Transfer. All the authors are grateful to a number of GIAN staff members: Prof. Partha P. Chakrabarti, national coordinator and director of IIT Kharagpur; Prof. Amar Nath Samanta, national coordinator, IIT Kharagpur, of programs for IIT; and Prof. Sunil Khijwania, coordinator at IIT Guwahati.

We appreciate the help of a number of teaching assistants at IIT Guwahati who provided valuable assistance. We are grateful to Vinod Pandey, Hiranya Deka, Debendra N. Sarkar, Dhrubajyoti Das, Binita Nath, Manash P. Borthakur, Somasekhar Reddy Dantla, and Vishal Sethi. A number of faculty colleagues at several universities provided advice and encouragement. Our thanks to Dr. Shaligram Tiwari and Dr. K. Arul Prakash, IIT Madras, Dr. Gaurav Tomar, Indian Institute of Science Bangalore, Dr. Bahni Ray, Dr. Debabrata Dasgupta, and Dr. B. Premachandran, IIT Delhi, Dr. Dipankar Bandyopadhyay and Dr. Pranab Mondal, IIT Guwahati, Dr. Partha P. Mukherjee, Purdue University, Prof. Kirti C Sahu of IIT Hyderabad, and Dr. Sandipan Ghosh Moulic, IIT Kharagpur.

We also wish to thank the members of our respective families for their patience and encouragement during the process of researching and writing this book.

Special thanks are due to Mr. Kaushik Sharma and Mr. Shashi Kumar, who created the attractive typescript using LaTeX. We also thank Dr. Gagandeep Singh, Ms. Mouli Sharma, and Ms. Michele Dimont of CRC Press/Taylor & Francis, for their superb support of this project.

Gautam Biswas
Amaresh Dalal
Vijay K. Dhir

## CHAPTER 1

Preliminary Concepts and Basic Equations

1.1 REYNOLDS TRANSPORT THEOREM

The rate of change on N for a system equals the sum of change of N inside any arbitrary control volume in continuum and the rate of efflux of N across surface. This can be represented as

$$\frac{DN}{Dt} = \frac{\partial}{\partial t} \int \int \int_{\forall} \eta(\rho d\forall) + \int \int_{A} \eta(\rho \mathbf{V}.dA) \tag{1.1}$$

N ≡ extensive property; η ≡ specific property; V ≡ velocity vector. We apply Eq. (1.1) for the purpose of mass conservation. The quantity M stands for mass of the system and the corresponding specific property is 1. The Reynolds transport theorem can be written as

$$\frac{DM}{Dt} = \frac{\partial}{\partial t} \int \int \int_{\forall} \rho d\forall + \int \int_{A} \rho \mathbf{V} \cdot dA \tag{1.2}$$

$$\int \int \int_{\forall} \frac{\partial \rho}{\partial t} d\forall + \int \int \int_{\forall} \nabla \cdot (\rho \mathbf{V}) d\forall = 0 \tag{1.3}$$

$$\int \int \int_{\forall} \left[\frac{\partial \rho}{\partial t} + \nabla \cdot (\rho \mathbf{V}) \right] d\forall = 0 \tag{1.4}$$

$$\frac{\partial \rho}{\partial t} + \nabla \cdot (\rho \mathbf{V}) = 0 \tag{1.5}$$

$$\frac{\partial \rho}{\partial t} + (\mathbf{V} \cdot \nabla)\rho + \rho(\nabla \cdot \mathbf{V}) = 0$$

or

$$\frac{D\rho}{Dt} + \rho(\nabla \cdot \mathbf{V}) = 0 \quad \text{(General Form)} \tag{1.6}$$

Incompressible flows, $(D\rho/Dt) = 0$ (condition). Therefore, for incompressible flows

$$\rho(\nabla \cdot \mathbf{V}) = 0 \quad \text{(governing equation)} \tag{1.7}$$

or

$$\nabla \cdot \mathbf{V} = 0 \tag{1.8}$$

Equation (1.8) is valid, irrespective of the flow condition, steady or unsteady.

1.2   COMPRESSIBLE AND INCOMPRESSIBLE FLOWS

Let us consider change of volume of a fluid under the action of external forces.

$$E = \frac{-\Delta p}{\Delta \forall / \forall} \tag{1.9}$$

E for water is $2 \times 10^6 \; kN/m^2$, E for air $= 101 \; kN/m^2$. Air is 20,000 times more compressible than water.

The mass continuity may be written as

$$(\forall + \Delta \forall)(\rho + \Delta \rho) = \forall \rho \tag{1.10}$$

which gives

$$\frac{\Delta \forall}{\forall} = -\frac{\Delta \rho}{\rho} \tag{1.11}$$

Now, the modulus of elasticity, E is expressed as

$$E \approx \frac{-\Delta p}{-(\Delta \rho / \rho)} \approx \frac{\Delta p}{\Delta \rho / \rho} \tag{1.12}$$

or

$$\frac{\Delta \rho}{\rho} \approx \frac{\Delta p}{E} \approx \frac{1}{2} \frac{\rho V^2}{E} \quad [\, V \text{ is the velocity} \,] \tag{1.13}$$

or

$$\frac{\Delta \rho}{\rho} \sim \frac{1}{2} \frac{V^2}{a^2} \tag{1.14}$$

where, $a = \sqrt{\frac{E}{\rho}}$ = local acoustic speed
  or

$$\frac{\Delta \rho}{\rho} \sim \frac{1}{2} M^2 \tag{1.15}$$

where, M = Mach number.

Considering a maximum relative change in density of 5 percent as the criterion of an incompressible flow, the upper limit of Mach number is

$$M \approx 0.316$$

which means, flow of air up to a velocity of 110 m/s under standard condition, can be considered as incompressible flow. In the above calculation, we consider local acoustic velocity as 345 m/s.

1.3   ENERGY EQUATION USING SPECIFIC COORDINATE SYSTEM

Let us consider a differential control volume (Fig. 1.1) in a flow field and account for the energy crossing the control volume (CV) boundary.

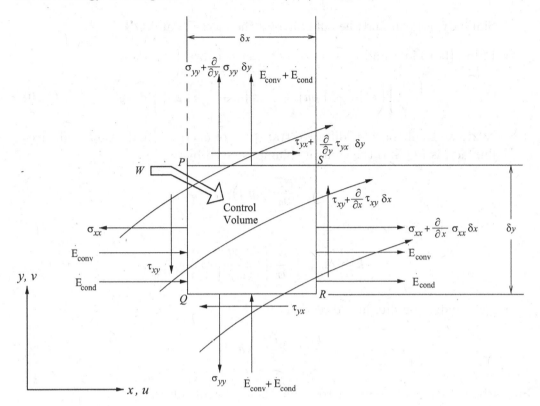

Figure 1.1   Differential control volume for energy conservation

- Thermal energy per unit mass $\equiv e$

- Kinetic energy per unit mass $= V^2/2$
  where $V^2 = u^2 + v^2$

- Rate of change of total energy within the CV

$$\frac{\partial}{\partial t}\left\{\left[\rho\left(e + \frac{V^2}{2}\right)\right]\delta x \delta y\right\} \tag{1.16}$$

rate of mass crossing $PQ = \rho u(\delta y.1)$

rate of mass crossing $RS = \rho u(\delta y.1) + \frac{\partial}{\partial x}[\rho u(\delta y.1)]\delta x$

rate of energy crossing $PQ = \rho u\, \delta y\left(e + \frac{V^2}{2}\right)$

rate of energy crossing RS

$$\left\{\left(e + \frac{V^2}{2}\right)\rho u + \frac{\partial}{\partial x}\left[\left(e + \frac{V^2}{2}\right)\rho u\right]\delta x\right\}(\delta y.1) \tag{1.17}$$

Net convective efflux of energy across this pair of faces (PQ and RS)

$$\frac{\partial}{\partial x}\left[\left(e+\frac{V^2}{2}\right)\rho u\right]\delta x \delta y \tag{1.18}$$

Similarly, we can find the efflux across the faces PS and QR.

- Efflux (both in x and y direction) of energy convected:

$$\left\{\frac{\partial}{\partial x}\left[\left(e+\frac{V^2}{2}\right)\rho u\right]+\frac{\partial}{\partial y}\left[\left(e+\frac{V^2}{2}\right)\rho v\right]\right\}\delta x \delta y \tag{1.19}$$

Next, we shall consider diffusive transport of energy. Conduction (diffusive) flux across PQ into the control volume is given by

$$-k\frac{\partial T}{\partial x}\ (\delta y.1) \tag{1.20}$$

Conduction flux across RS=

$$-k\frac{\partial T}{\partial x}\delta y+\left\{\frac{\partial}{\partial x}\left[-k\frac{\partial T}{\partial x}\right]\delta x\right\}\delta y \tag{1.21}$$

- Net conduction flux in x direction,

$$-\frac{\partial}{\partial x}\left(k\frac{\partial T}{\partial x}\right)\delta x \delta y \tag{1.22}$$

Similarly, net conduction flux across the whole control volume

$$-\left[\frac{\partial}{\partial x}\left(k\frac{\partial T}{\partial x}\right)+\frac{\partial}{\partial y}\left(k\frac{\partial T}{\partial y}\right)\right]\delta x \delta y \tag{1.23}$$

Total efflux of energy is now:

$$\left[\frac{\partial}{\partial x}\left\{\left(e+\frac{V^2}{2}\right)\rho u\right\}+\frac{\partial}{\partial y}\left\{\left(e+\frac{V^2}{2}\right)\rho v\right\}\right]\delta x \delta y$$

$$-\left[\frac{\partial}{\partial x}\left(k\frac{\partial T}{\partial x}\right)+\frac{\partial}{\partial y}\left(k\frac{\partial T}{\partial y}\right)\right]\delta x \delta y \tag{1.24}$$

In vector notation we can write that efflux of energy through the control surface (CS)

$$\left[\nabla.\left\{\rho\vec{V}\left(e+\frac{V^2}{2}\right)\right\}-\nabla.(k\nabla T)\right]\delta x \delta y \tag{1.25}$$

The Reynolds transport theorem can be stated as rate of change of energy for the system = rate of change of energy in the control volume (CV) + rate of efflux of energy through the control surface (CS) of the CV.

It can be mentioned here that the system is an open system and the system boundary and the control volume boundary are same.

According to the First Law of Thermodynamics, rate of energy production within the system + rate of work done on the system = rate of change of energy for the system.

Let us evaluate the quantity, rate of work done on the system. The rate at which body force $\vec{f}$ does work is given by force time velocity

$$\rho\left[f_x u + f_y v\right] \delta x \delta y = \rho \vec{f} . \vec{V} \delta x \delta y \tag{1.26}$$

Work done (rate) by the surface forces on the PQ plane of the system:

$$-\sigma_{xx} u \delta y - \tau_{xy} v \delta y \tag{1.27}$$

Work done (rate) by the surface forces on RS plane is:

$$\left[(\sigma_{xx} u) + \frac{\partial}{\partial x}(\sigma_{xx} u)\,\delta x + \tau_{xy} v + \frac{\partial}{\partial x}(\tau_{xy} v)\,\delta x\right]\delta y \tag{1.28}$$

Net work done (rate) by the surface forces on the pair of faces, PQ and RS

$$\left[\frac{\partial}{\partial x}(\sigma_{xx} u) + \frac{\partial}{\partial x}(\tau_{xy} v)\right]\delta x \delta y \tag{1.29}$$

Similarly, we can determine the net work done (rate) by the surface forces in y direction, and adding these two we get work done (rate) on the CV as

$$\left[\frac{\partial}{\partial x}(\sigma_{xx} u) + \frac{\partial}{\partial y}(\sigma_{yy} v) + \frac{\partial}{\partial x}(\tau_{xy} v) + \frac{\partial}{\partial y}(\tau_{yx} u)\right]\delta x \delta y \tag{1.30}$$

we assume the sources of heat such that the strength per unit volume is $\dot{S}_{th}$, the energy production within the system $= \dot{S}_{th}\delta x \delta y$. The conservation of energy requires (invoking Eqs. (1.16), (1.24), (1.26) and (1.30))

$$\frac{\partial}{\partial t}\left\{\rho\left(e + \frac{V^2}{2}\right)\right\} + \frac{\partial}{\partial x}\left\{\rho u\left(e + \frac{V^2}{2}\right)\right\}$$

$$+\frac{\partial}{\partial y}\left\{\rho v\left(e + \frac{V^2}{2}\right)\right\} - \frac{\partial}{\partial x}\left(k\frac{\partial T}{\partial x}\right) - \frac{\partial}{\partial y}\left(k\frac{\partial T}{\partial y}\right)$$

$$= \rho\left\{f_x u + f_y v\right\} + \left[\frac{\partial}{\partial x}\left\{\sigma_{xx} u + \tau_{xy} v\right\} + \frac{\partial}{\partial y}\left\{\sigma_{yy} v + \tau_{yx} u\right\}\right] + \dot{S}_{th} \tag{1.31}$$

In order to simplify these terms, let us take a recourse to the mechanical energy part through the NS equation. Multiplying the x momentum equation by u and y momentum equation by v and adding, we get

$$\rho\left[\frac{\partial}{\partial t}\left(\frac{V^2}{2}\right) + u^2\frac{\partial u}{\partial x} + uv\frac{\partial u}{\partial y} + uv\frac{\partial v}{\partial x} + v^2\frac{\partial v}{\partial y}\right]$$

$$= \rho\left\{f_x u + f_y v\right\} + u\left\{\frac{\partial}{\partial x}\sigma_{xx} + \frac{\partial}{\partial y}\tau_{xy}\right\} + v\left\{\frac{\partial}{\partial x}\tau_{yx} + \frac{\partial}{\partial y}\sigma_{yy}\right\} \tag{1.32}$$

First two convective terms of the left-hand side (LHS) of Eq. (1.31) can be rearranged as

$$\frac{\partial}{\partial x}\left\{\rho u\left(e+\frac{V^2}{2}\right)\right\}+\frac{\partial}{\partial y}\left\{\rho v\left(e+\frac{V^2}{2}\right)\right\}=\rho u\frac{\partial e}{\partial x}+e\frac{\partial}{\partial x}(\rho u)+\frac{V^2}{2}\frac{\partial}{\partial x}(\rho u)$$

$$+\rho u\frac{\partial}{\partial x}\left(\frac{V^2}{2}\right)+\rho v\frac{\partial e}{\partial y}+e\frac{\partial}{\partial y}(\rho v)+\frac{V^2}{2}\frac{\partial}{\partial y}(\rho v)+\rho v\frac{\partial}{\partial y}\left(\frac{V^2}{2}\right) \tag{1.33}$$

The temporal term of Eq. (1.31) is

$$\frac{\partial}{\partial t}\left\{\rho\left(e+\frac{V^2}{2}\right)\right\}=\rho\frac{\partial e}{\partial t}+e\frac{\partial\rho}{\partial t}+\rho\frac{\partial}{\partial t}\left(\frac{V^2}{2}\right)+\frac{V^2}{2}\left(\frac{\partial\rho}{\partial t}\right) \tag{1.34}$$

First three terms of Eq. (1.31) will yield

$$\rho\left(u\frac{\partial e}{\partial x}+v\frac{\partial e}{\partial y}+\frac{\partial e}{\partial t}\right)+e\left(\frac{\partial\rho}{\partial t}+\frac{\partial}{\partial x}(\rho u)+\frac{\partial}{\partial y}(\rho v)\right)$$

$$+\frac{V^2}{2}\left(\frac{\partial\rho}{\partial t}+\frac{\partial}{\partial x}(\rho u)+\frac{\partial}{\partial y}(\rho v)\right)+\rho\frac{\partial}{\partial t}\left(\frac{V^2}{2}\right)$$

$$+\left[\rho\left\{u\frac{\left(2u\frac{\partial u}{\partial x}+2v\frac{\partial v}{\partial x}\right)}{2}\right\}+\rho\left\{v\frac{\left(2u\frac{\partial u}{\partial y}+2v\frac{\partial v}{\partial y}\right)}{2}\right\}\right]$$

$$=\rho\left[\frac{\partial}{\partial t}\left(\frac{V^2}{2}\right)+u^2\frac{\partial u}{\partial x}+uv\frac{\partial u}{\partial y}+uv\frac{\partial v}{\partial x}+v^2\frac{\partial v}{\partial y}\right]+\rho\left[\frac{\partial e}{\partial t}+u\frac{\partial e}{\partial x}+v\frac{\partial e}{\partial y}\right] \tag{1.35}$$

Invoking this and considering homogeneous isotropic medium, the energy equation (Eq. (1.31)) can be written as

$$\rho\left[\frac{\partial e}{\partial t}+u\frac{\partial e}{\partial x}+v\frac{\partial e}{\partial y}\right]+\rho\left[\frac{\partial}{\partial t}\left(\frac{V^2}{2}\right)+u^2\frac{\partial u}{\partial x}+uv\frac{\partial u}{\partial y}+uv\frac{\partial v}{\partial x}+v^2\frac{\partial v}{\partial y}\right]$$

$$+\left(-k\frac{\partial^2 T}{\partial x^2}-k\frac{\partial^2 T}{\partial y^2}\right)=\rho\{f_x u+f_y v\}+\sigma_{xx}\frac{\partial u}{\partial x}+u\frac{\partial}{\partial x}\sigma_{xx}+\tau_{xy}\frac{\partial v}{\partial x}$$

$$+v\frac{\partial}{\partial x}\tau_{xy}+\sigma_{yy}\frac{\partial v}{\partial y}+v\frac{\partial}{\partial y}\sigma_{yy}+\tau_{yx}\frac{\partial u}{\partial y}+u\frac{\partial}{\partial y}\tau_{yx}+\dot{S}_{th} \tag{1.36}$$

Subtracting Eq. (1.32) from Eq. (1.36) and considering the stress tensor symmetric we get

$$\rho\left[\frac{\partial e}{\partial t}+u\frac{\partial e}{\partial x}+v\frac{\partial e}{\partial y}\right]-k\left[\frac{\partial^2 T}{\partial x^2}+\frac{\partial^2 T}{\partial y^2}\right]$$

$$=\left\{\sigma_{xx}\frac{\partial u}{\partial x}+\tau_{xy}\frac{\partial v}{\partial x}+\sigma_{yy}\frac{\partial v}{\partial y}+\tau_{yx}\frac{\partial u}{\partial y}\right\}+\dot{S}_{th} \tag{1.37}$$

Let us look at the terms of Eq. (1.37), within braces { }

$$\frac{\partial u}{\partial x}\sigma_{xx}=\left(-p+2\mu\frac{\partial u}{\partial x}-\frac{2}{3}\mu\left\{\frac{\partial u}{\partial x}+\frac{\partial v}{\partial y}\right\}\right)\frac{\partial u}{\partial x}$$

$$=-p\frac{\partial u}{\partial x}+2\mu\left(\frac{\partial u}{\partial x}\right)^2-\frac{2}{3}\mu\left(\frac{\partial u}{\partial x}\right)^2-\frac{2}{3}\mu\left(\frac{\partial u}{\partial x}\right)\left(\frac{\partial v}{\partial y}\right) \tag{1.38}$$

$$\frac{\partial v}{\partial y}\sigma_{yy} = \left(-p + 2\mu\frac{\partial v}{\partial y} - \frac{2}{3}\mu\left\{\frac{\partial u}{\partial x} + \frac{\partial v}{\partial y}\right\}\right)\frac{\partial v}{\partial y}$$

$$= -p\frac{\partial v}{\partial y} + 2\mu\left(\frac{\partial v}{\partial y}\right)^2 - \frac{2}{3}\mu\left(\frac{\partial v}{\partial y}\right)^2 - \frac{2}{3}\mu\left(\frac{\partial u}{\partial x}\right)\left(\frac{\partial v}{\partial y}\right) \tag{1.39}$$

$$\tau_{xy}\frac{\partial v}{\partial x} = \mu\left(\frac{\partial v}{\partial x} + \frac{\partial u}{\partial y}\right)\left(\frac{\partial v}{\partial x}\right) = \mu\left(\frac{\partial v}{\partial x}\right)^2 + \mu\left(\frac{\partial v}{\partial x}\cdot\frac{\partial u}{\partial y}\right) \tag{1.40}$$

$$\tau_{yx}\frac{\partial u}{\partial y} = \mu\left(\frac{\partial v}{\partial x} + \frac{\partial u}{\partial y}\right)\left(\frac{\partial u}{\partial y}\right) = \mu\left(\frac{\partial u}{\partial y}\right)^2 + \mu\left(\frac{\partial v}{\partial x}\cdot\frac{\partial u}{\partial y}\right) \tag{1.41}$$

Equation (1.37) may now be written as

$$\rho\left[\frac{\partial e}{\partial t} + u\frac{\partial e}{\partial x} + v\frac{\partial e}{\partial y}\right] = k\left[\frac{\partial^2 T}{\partial x^2} + \frac{\partial^2 T}{\partial y^2}\right] - p\left(\frac{\partial u}{\partial x} + \frac{\partial v}{\partial y}\right)$$

$$+\mu\underbrace{\left[2\left\{\left(\frac{\partial u}{\partial x}\right)^2 + \left(\frac{\partial v}{\partial y}\right)^2\right\} + \left(\frac{\partial u}{\partial y} + \frac{\partial v}{\partial x}\right)^2 - \frac{2}{3}\left(\frac{\partial u}{\partial x} + \frac{\partial v}{\partial y}\right)^2\right]}_{\text{Viscous dissipation}\equiv\mu\Phi} + \dot{S}_{th} \tag{1.42}$$

Equation (1.42) is the general form of energy equation, valid for both the compressible and incompressible flows.

Now, let us consider enthalpy per unit mass $= i = e + \frac{p}{\rho}$
Substituting i in Eq. (1.42), we get:

$$\rho\left[\frac{\partial i}{\partial t} + u\frac{\partial i}{\partial x} + v\frac{\partial i}{\partial y}\right] - \rho\left[\frac{\partial}{\partial t}\left(\frac{p}{\rho}\right) + u\frac{\partial}{\partial x}\left(\frac{p}{\rho}\right) + v\frac{\partial}{\partial y}\left(\frac{p}{\rho}\right)\right]$$

$$= k\nabla^2 T - p\left(\frac{\partial u}{\partial x} + \frac{\partial v}{\partial y}\right) + \mu\Phi + \dot{S}_{th} \tag{1.43}$$

or

$$\rho\left[\frac{\partial i}{\partial t} + u\frac{\partial i}{\partial x} + v\frac{\partial i}{\partial y}\right] - \left[\frac{\partial p}{\partial t} - \frac{p}{\rho}\frac{\partial\rho}{\partial t} + u\frac{\partial p}{\partial x} - \frac{up}{\rho}\frac{\partial\rho}{\partial x} + v\frac{\partial p}{\partial y} - \frac{vp}{\rho}\frac{\partial\rho}{\partial y}\right]$$

$$= k\nabla^2 T - p\left(\frac{\partial u}{\partial x} + \frac{\partial v}{\partial y}\right) + \mu\Phi + \dot{S}_{th} \tag{1.44}$$

or

$$\rho\left[\frac{\partial i}{\partial t} + u\frac{\partial i}{\partial x} + v\frac{\partial i}{\partial y}\right] = k\nabla^2 T + \underbrace{\left(\frac{\partial p}{\partial t} + u\frac{\partial p}{\partial x} + v\frac{\partial p}{\partial y}\right)}_{\frac{\partial p}{\partial t}+\vec{V}.\nabla(p)}$$

$$-\frac{p}{\rho}\left[\frac{\partial\rho}{\partial t} + u\frac{\partial\rho}{\partial x} + v\frac{\partial\rho}{\partial y} + \rho\frac{\partial u}{\partial x} + \rho\frac{\partial v}{\partial y}\right] + \mu\Phi + \dot{S}_{th} \tag{1.45}$$

or

$$\rho \left[\frac{\partial i}{\partial t} + u \frac{\partial i}{\partial x} + v \frac{\partial i}{\partial y} \right] = k\nabla^2 T + \frac{Dp}{Dt}$$

$$-\frac{p}{\rho} \left[\frac{\partial \rho}{\partial t} + \frac{\partial}{\partial x}(\rho u) + \frac{\partial}{\partial y}(\rho v) \right] + \mu\Phi + \dot{S}_{th} \tag{1.46}$$

or

$$\rho \left[\frac{\partial i}{\partial t} + \vec{V}.\nabla(i) \right] = k\nabla^2 T + \frac{Dp}{Dt} + \mu\Phi + \dot{S}_{th} \tag{1.47}$$

or

$$\rho \frac{Di}{Dt} = k\nabla^2 T + \frac{Dp}{Dt} + \mu\Phi + \dot{S}_{th} \tag{1.48}$$

Substituting, $i = c_p T$

$$\rho c_p \frac{DT}{Dt} = \frac{Dp}{Dt} + k\nabla^2 T + \mu\Phi + \dot{S}_{th} \tag{1.49}$$

This is the general form of the energy equation.

Special Case: for liquids, we can directly write from Eq. (1.42)

$$\rho c \frac{DT}{Dt} = k \left[\frac{\partial^2 T}{\partial x^2} + \frac{\partial^2 T}{\partial y^2} \right] + \mu\Phi + \dot{S}_{th} \tag{1.50}$$

For liquids, $c_p = c_v = c$ and $p \left(\frac{\partial u}{\partial x} + \frac{\partial v}{\partial y} \right) = 0$. In the case of liquids, the viscous dissipation term will have to be modified by invoking the incompressibility condition.

1.4 GENERALIZED APPROACH FOR DERIVATION OF ENERGY EQUATION

Let us apply Reynolds transport theorem to an arbitrary control volume (Fig. 1.2) for the conservation of energy. We may mention here that the system boundary and the control volume (CV) boundary are same. We can write rate of change of energy for the system, (**A**) = rate of change of energy in the CV, (**B**) + rate of efflux of energy through the control surface of the CV, (**C**). Applying first law of thermodynamics, we can also write rate of change of energy for the system, (**A**) = rate of work done on the system (**D**) + rate of energy production within the system, (**E**).

It is ubiquitous that the system is open and the system boundary and the control volume boundary are the same.

We evaluate the quantities **B** and **C** in the following manner

$$(\mathbf{B}) + (\mathbf{C}) = \int \int \int \left[\frac{\partial}{\partial t}(\rho E) + \nabla \cdot (\rho \mathbf{V} E) - \nabla \cdot (k\nabla T) \right] d\forall$$

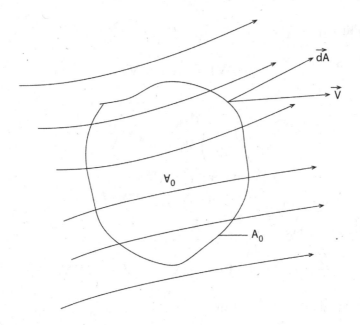

Figure 1.2 Arbitrary control volume fixed in space

Here E is defined as $e + V^2/2$. Now, **(B)** + **(C)** can be equated with the rate of work done on the system **(D)** + rate of energy production within the system **(E)**:

$$\int\int\int \left[\frac{\partial}{\partial t}(\rho E) + \nabla.\{\rho \mathbf{V}E\} - \nabla.(k\nabla T)\right] d\forall$$
$$= \int\int\int \left[\rho\, \mathbf{f}.\mathbf{V} + \nabla.(\sigma.\mathbf{V}) + \dot{S}_{th}\right] d\forall \tag{1.51}$$

In the above equation S_{th} is the volumetric heat source.

For the homogeneous and isotropic medium,

$$\underbrace{\frac{\partial}{\partial t}(\rho E) + \nabla.\{\rho \mathbf{V}E\}}_{I}\underbrace{-k\nabla^2 T}_{II} = \underbrace{\rho\mathbf{f}.\mathbf{V}}_{III} + \underbrace{\nabla.(\sigma.\mathbf{V})}_{IV} + \underbrace{\dot{S}_{th}}_{V} \tag{1.52}$$

We obtain the following from term I

$$\frac{\partial}{\partial t}(\rho E) + \nabla\cdot\{\rho \mathbf{V}E\} = E\frac{\partial\rho}{\partial t} + \rho\frac{\partial E}{\partial t} + \rho\mathbf{V}.\nabla E + E(\nabla\cdot\rho\mathbf{V})$$
$$= \rho\left[\frac{\partial E}{\partial t} + (\mathbf{V}\cdot\nabla)E\right] + E\left[\frac{\partial\rho}{\partial t} + \nabla.(\rho\mathbf{V})\right]$$
$$= \rho\left[\frac{\partial E}{\partial t} + (\mathbf{V}\cdot\nabla)E\right] = \rho\frac{DE}{Dt} \tag{1.53}$$

Term IV produces,

$$\nabla \cdot (\sigma \cdot \mathbf{V}) = \frac{\partial}{\partial x_i} (\sigma \cdot \mathbf{V})_i$$

$$= \frac{\partial}{\partial x_i} (\sigma_{ij} \cdot \mathbf{V_j})$$

$$= \left(\frac{\partial}{\partial x_i} \sigma_{ij} \right) \mathbf{V_j} + \sigma_{ij} \frac{\partial \mathbf{V_j}}{\partial x_i}$$

$$= (\nabla \cdot \sigma)_j \mathbf{V_j} + \sigma_{ij} (\nabla \mathbf{V})_{ij}$$

$$= (\nabla \cdot \sigma) \cdot \mathbf{V} + \sigma : \nabla \mathbf{V} \tag{1.54}$$

Energy equation becomes

$$\rho \frac{DE}{Dt} = k\nabla^2 T + \rho \, \mathbf{f} \cdot \mathbf{V} + (\nabla \cdot \sigma) \cdot \mathbf{V} + \sigma : \nabla \mathbf{V} + \dot{S}_{th} \tag{1.55}$$

Navier-Stokes equations are given by

$$\rho \frac{D\mathbf{V}}{Dt} = \rho \, \mathbf{f} + \nabla \cdot \sigma \tag{1.56}$$

or taking dot product of both side with $\mathbf{V}$

$$\rho \left(\mathbf{V} \cdot \frac{D\mathbf{V}}{Dt} \right) = \rho \, \mathbf{f} \cdot \mathbf{V} + (\nabla \cdot \sigma) \cdot \mathbf{V} \tag{1.57}$$

or

$$\rho \frac{D}{Dt} \left(\frac{\mathbf{V^2}}{2} \right) = \rho \, \mathbf{f} \cdot \mathbf{V} + (\nabla \cdot \sigma) \cdot \mathbf{V} \tag{1.58}$$

and subtracting Eq. (1.58) from Eq. (1.55) one gets

$$\rho \frac{De}{Dt} = k\nabla^2 T + \sigma : \nabla \mathbf{V} + \dot{S}_{th} \tag{1.59}$$

Here,

$$\sigma : \nabla \mathbf{V} = \left[-\left\{ p + \frac{2}{3}\mu \nabla \cdot \mathbf{V} \right\} I + \tau \right] : \nabla \mathbf{V} \tag{1.60}$$

Again,

$$\tau = 2\mu \text{ Def } \mathbf{V} \tag{1.61}$$

and

$$\text{Def } \mathbf{V} = \frac{1}{2} \left[\frac{\partial V_j}{\partial x_i} + \frac{\partial V_i}{\partial x_j} \right] \tag{1.62}$$

So,

$$\sigma : \nabla \mathbf{V} = -pI : \nabla \mathbf{V} - \frac{2}{3}\mu(\nabla \cdot \mathbf{V})I : \nabla \mathbf{V} + \tau : \nabla \mathbf{V} \tag{1.63}$$

$$= -p\left(\nabla \cdot \mathbf{V}\right) - \frac{2}{3}\mu(\nabla \cdot \mathbf{V})^2 + \tau : \nabla \mathbf{V}$$

$$= -p(\nabla \cdot \mathbf{V}) + \mu\phi \tag{1.64}$$

where

$$\Phi = \left[\left\{\frac{\partial V_i}{\partial x_j} + \frac{\partial V_j}{\partial x_i}\right\}\left(\frac{\partial V_i}{\partial x_j}\right) - \frac{2}{3}\left(\frac{\partial V_i}{\partial x_i}\right)^2\right] \tag{1.65}$$

From continuity,

$$\frac{\partial \rho}{\partial t} + \nabla.\left(\rho \mathbf{V}\right) = 0$$

$$\frac{\partial \rho}{\partial t} + \left(\mathbf{V}.\nabla\right)\rho + \rho\left(\nabla.\mathbf{V}\right) = 0 \tag{1.66}$$

or

$$-\nabla.\mathbf{V} = \frac{1}{\rho}\left[\frac{\partial \rho}{\partial t} + \left(\mathbf{V}.\nabla\right)\rho\right] = \frac{1}{\rho}\frac{D\rho}{Dt}$$

$$-p\left(\nabla.\mathbf{V}\right) = \frac{p}{\rho}\left[\frac{\partial \rho}{\partial t} + \left(\mathbf{V}.\nabla\right)\rho\right] = \frac{p}{\rho}\frac{D\rho}{Dt} \tag{1.67}$$

The energy equation Eq. (1.59) can be written as

$$\rho\frac{De}{Dt} = k\nabla^2 T + \frac{p}{\rho}\frac{D\rho}{Dt} + \mu\Phi + \dot{S}_{th} \tag{1.68}$$

$$i = e + \frac{p}{\rho} \tag{1.69}$$

$$\rho\frac{De}{Dt} = \rho\left[\frac{Di}{Dt} - \frac{D}{Dt}\left(\frac{p}{\rho}\right)\right] = \rho\left[\frac{Di}{Dt} + \frac{p}{\rho^2}\frac{D\rho}{Dt} - \frac{1}{\rho}\frac{Dp}{Dt}\right] \tag{1.70}$$

Now, the energy equation becomes

$$\rho\frac{Di}{Dt} = -\frac{p}{\rho}\frac{D\rho}{Dt} + \frac{Dp}{Dt} + k\nabla^2 T + \frac{p}{\rho}\frac{D\rho}{Dt} + \mu\Phi + \dot{S}_{th} \tag{1.71}$$

or

$$\rho\frac{Di}{Dt} = \frac{Dp}{Dt} + k\nabla^2 T + \mu\Phi + \dot{S}_{th} \quad \text{(General)} \tag{1.72}$$

For liquids, one can write (from Eqs. (1.59) and (1.64)),

$$\rho\frac{De}{Dt} = k\nabla^2 T + \mu\Phi + \dot{S}_{th} \tag{1.73}$$

where the general viscous dissipation function is

$$\Phi = \left[\left\{ \frac{\partial V_i}{\partial x_j} + \frac{\partial V_j}{\partial x_i} \right\} \left(\frac{\partial V_i}{\partial x_j} \right) - \frac{2}{3} \left(\frac{\partial V_i}{\partial x_i} \right)^2 \right] = \left[2\frac{\partial u}{\partial x}\frac{\partial u}{\partial x} + \left(\frac{\partial u}{\partial y} + \frac{\partial v}{\partial x} \right)\frac{\partial u}{\partial y} \right.$$

$$+ \left(\frac{\partial u}{\partial z} + \frac{\partial w}{\partial x} \right)\frac{\partial u}{\partial z} + \left(\frac{\partial v}{\partial x} + \frac{\partial u}{\partial y} \right)\frac{\partial v}{\partial x} + 2\frac{\partial v}{\partial y}\frac{\partial v}{\partial y} + \left(\frac{\partial v}{\partial z} + \frac{\partial w}{\partial y} \right)\frac{\partial v}{\partial z}$$

$$\left. + \left(\frac{\partial w}{\partial x} + \frac{\partial u}{\partial z} \right)\frac{\partial w}{\partial x} + \left(\frac{\partial w}{\partial y} + \frac{\partial v}{\partial z} \right)\frac{\partial w}{\partial y} + 2\frac{\partial w}{\partial z}\frac{\partial w}{\partial z} - \frac{2}{3}\left(\frac{\partial u}{\partial x} + \frac{\partial v}{\partial y} + \frac{\partial w}{\partial z} \right)^2 \right]$$

Having rearranged, one gets

$$\Phi = \left[2\left(\frac{\partial u}{\partial x} \right)^2 + 2\left(\frac{\partial v}{\partial y} \right)^2 + 2\left(\frac{\partial w}{\partial z} \right)^2 + \left(\frac{\partial v}{\partial x} + \frac{\partial u}{\partial y} \right)^2 + \left(\frac{\partial w}{\partial y} + \frac{\partial v}{\partial z} \right)^2 \right.$$

$$\left. + \left(\frac{\partial u}{\partial z} + \frac{\partial w}{\partial x} \right)^2 - \frac{2}{3}\left(\frac{\partial u}{\partial x} + \frac{\partial v}{\partial y^2} + \frac{\partial w}{\partial z} \right)^2 \right] \tag{1.74}$$

However, in absence of heat sources ($\dot{S}_{th} = 0$) and negligible viscous dissipation ($\mu\Phi = 0$, which is often true for liquids), the energy equation for the liquids can be written as (from Eq. (1.73))

$$\rho c \left(\frac{\partial T}{\partial t} + u\frac{\partial T}{\partial x} + v\frac{\partial T}{\partial y} + w\frac{\partial T}{\partial z} \right) = k \left(\frac{\partial^2 T}{\partial x^2} + \frac{\partial^2 T}{\partial y^2} + \frac{\partial^2 T}{\partial z^2} \right) \tag{1.75}$$

For gases, in general the energy equation can be written as (from Eq. (1.72)).

$$\rho c_p \frac{DT}{Dt} = \frac{Dp}{Dt} + k\nabla^2 T + \mu\Phi + \dot{S}_{th}$$

1.5   IMPORTANT DIMENSIONLESS NUMBERS

In order to nondimensionalize, we define

$$u^* = \frac{u}{U_\infty}, \qquad v^* = \frac{v}{U_\infty}, \qquad w^* = \frac{w}{U_\infty}, \qquad \theta = \frac{T - T_\infty}{T_w - T_\infty},$$

$$x^* = \frac{x}{L}, \qquad y* = \frac{y}{L}, \qquad z^* = \frac{z}{L}, \qquad p^* = \frac{p}{\rho U_\infty^2}, \qquad t^* = t\frac{U_\infty}{L}$$

Assumption of $\dot{S}_{th}=0$ produces,

$$\rho c_p \left(\frac{\partial T}{\partial t} + u\frac{\partial T}{\partial x} + v\frac{\partial T}{\partial y} + w\frac{\partial T}{\partial z} \right) = \left(\frac{\partial p}{\partial t} + u\frac{\partial p}{\partial x} + v\frac{\partial p}{\partial y} + w\frac{\partial p}{\partial z} \right)$$

$$+ k \left(\frac{\partial^2 T}{\partial x^2} + \frac{\partial^2 T}{\partial y^2} + \frac{\partial^2 T}{\partial z^2} \right) + \mu\Phi \tag{1.76}$$

The time derivatives will vanish for the steady state situation. In any case,

$$\frac{\partial \theta}{\partial t^*} + u^*\frac{\partial \theta}{\partial x^*} + v^*\frac{\partial \theta}{\partial y^*} + w^*\frac{\partial \theta}{\partial z^*} = \frac{U_\infty^2}{c_p(T_w - T_\infty)}\left(\frac{\partial p^*}{\partial t^*} + u^*\frac{\partial p^*}{\partial x^*} + v^*\frac{\partial p^*}{\partial y^*} + w^*\frac{\partial p^*}{\partial z^*} \right)$$

$$+\frac{k}{\rho c_p U_\infty L}\left(\frac{\partial^2\theta}{\partial x^{*2}}+\frac{\partial^2\theta}{\partial y^{*2}}+\frac{\partial^2\theta}{\partial z^{*2}}\right)+\frac{\mu U_\infty^2}{L}\frac{1}{\rho c_p(T_w-T_\infty)U_\infty}\bar{\bar{\Phi}} \qquad (1.77)$$

$$\frac{\partial\theta}{\partial t^*}+u^*\frac{\partial\theta}{\partial x^*}+v^*\frac{\partial\theta}{\partial y^*}+w^*\frac{\partial\theta}{\partial z^*}=Ec\left(\frac{\partial p^*}{\partial t^*}+u^*\frac{\partial p^*}{\partial x^*}+v^*\frac{\partial p^*}{\partial y^*}+w^*\frac{\partial p^*}{\partial z^*}\right)$$

$$+\frac{1}{Re.Pr}\left(\frac{\partial^2\theta}{\partial x^{*2}}+\frac{\partial^2\theta}{\partial y^{*2}}+\frac{\partial^2\theta}{\partial z^{*2}}\right)+\frac{Ec}{Re}\bar{\bar{\Phi}} \qquad (1.78)$$

$$Ec=\frac{U_\infty^2}{c_p(T_w-T_\infty)},\quad Pr=\frac{\mu c_p}{k},\quad Re=\rho\frac{U_\infty L}{\mu} \qquad (1.79)$$

Let us look at the term Ec (Eckert number)

$$Ec=\frac{U_\infty^2}{c_p(T_w-T_\infty)}=\frac{\text{kinetic energy of the flow}}{\text{boundary layer enthalpy difference}} \qquad (1.80)$$

$$\frac{1}{Ec}=\frac{c_p(T_w-T_\infty)}{U_\infty^2}=\frac{c_pT_\infty}{U_\infty^2}\left(\frac{T_w}{T_\infty}-1\right)$$

$$=\frac{c_p}{\gamma(c_p-c_v)}\frac{a^2}{U_\infty^2}\left(\frac{T_w}{T_\infty}-1\right)=\frac{1}{(\gamma-1)}\frac{1}{M^2}\left(\frac{T_w}{T_\infty}-1\right) \qquad (1.81)$$

So,

$$Ec=\frac{(\gamma-1)M^2}{[(T_w/T_\infty)-1)]} \qquad (1.82)$$

where M = (fluid velocity) / (local sound speed) and a is the local sound speed, and M is known as Mach Number.

For low Mach number situations, the energy equation in nondimensional form is

$$\frac{\partial\theta}{\partial t^*}+u^*\frac{\partial\theta}{\partial x^*}+v^*\frac{\partial\theta}{\partial y^*}+w^*\frac{\partial\theta}{\partial z^*}=\frac{1}{Re.Pr}\left[\frac{\partial^2\theta}{\partial x^{*2}}+\frac{\partial^2\theta}{\partial y^{*2}}+\frac{\partial^2\theta}{\partial z^{*2}}\right] \qquad (1.83)$$

Dimensional form will be

$$\rho c_p\left(u\frac{\partial T}{\partial x}+v\frac{\partial T}{\partial y}+w\frac{\partial T}{\partial z}\right)=k\left[\frac{\partial^2 T}{\partial x^2}+\frac{\partial^2 T}{\partial y^2}+\frac{\partial^2 T}{\partial z^2}\right] \qquad (1.84)$$

For forced convection, the variables can be presented as

$$\theta=\theta(u,\ v,\ T,\ x,\ y,\ Re,\ Pr)$$

We can write the expression for conduction through the fluid layer adjacent to the wall as

$$q_w''=-k_f\frac{\partial T}{\partial n}=h(T_w-T_\infty) \qquad (1.85)$$

where n is the dimension in the direction normal to the surface. The thermal conductivity of the fluid is k_f. Historically $q_w'' = -k_f(\partial T/\partial n)$ was introduced by Fourier [1]. He also introduced the concept of heat transfer coefficient as $h(T_w - T_\infty) = q_w''$ or

$$h = \frac{-k_f \frac{\partial T}{\partial n}}{(T_w - T_\infty)} = \frac{-k_f \frac{(T_w - T_\infty)}{L} \frac{\partial \theta}{\partial \bar{n}}}{T_w - T_\infty} \tag{1.86}$$

where $\bar{n} = n/L$ or

$$\left(\frac{hL}{k}\right) = Nu = -\left(\frac{\partial \theta}{\partial \bar{n}}\right) \tag{1.87}$$

The Nusselt number, Nu is the nondimensional temperature gradient at the surface on which heat transfer takes place.

1.6 BOUNDARY LAYERS

1.6.1 Velocity Boundary Layer

Transition of zero velocity at the surface to the freestream velocity U_∞ takes place through a very thin layer δ. With the increase in y from the surface, the x velocity component, u, must increase until it approaches U_∞. The quantity δ is the boundary layer thickness (Fig. 1.3) and it is formally defined as the value of y for which $u = 0.99U_\infty$.

The flow field has two regions:
(a) The region where $(\partial u/\partial y)$ and consequently shear stress is significant.
(b) The region where $(\partial u/\partial y)$ and shear stress is negligible.

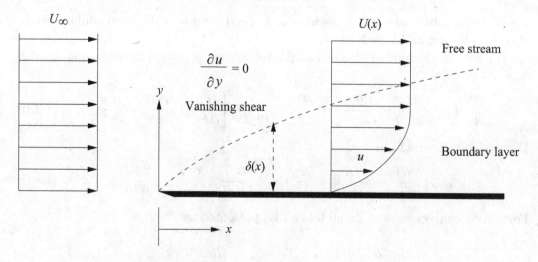

Figure 1.3 Velocity boundary layer on a flat plate

Here $\partial u/\partial y$ determines the local friction coefficient, and $\delta_x = 5.0x/\sqrt{Re_x}$

$$C_f = \tau_w / \frac{1}{2}\rho U_\infty^2 \tag{1.88}$$

$$\tau_w = \mu \frac{\partial u}{\partial y}\Big|_{y=0} \tag{1.89}$$

The slope (du/dy) changes with x as shown in Fig. 1.4.

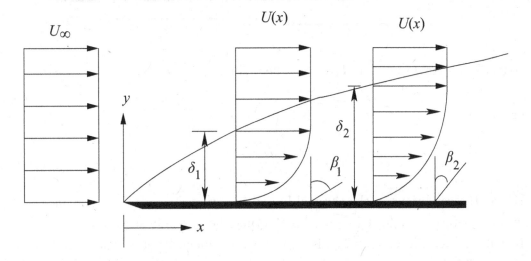

Figure 1.4 Change of slope in the downstream

1.6.2 Thermal Boundary Layer

As with the velocity boundary layer, a thermal boundary layer should develop if the temperatures at the fluid free stream and the solid-surface differ. Fluid particles that come into contact with the plate achieve thermal equilibrium of the plate's surface temperature and the temperature gradient develops in the fluid. The region of the fluid in which these temperature gradients exist, is thermal boundary layer (Fig. 1.5). We can write

$\delta_T \equiv$ value of y for which, $(T_w - T)/(T_w - T_\infty) = 0.99$

At any distance from leading edge, $q''_x = -k_f \frac{\partial T}{\partial y}\Big|_{y=0}$, because at the surface, energy transfer is through conduction. The expression is exact because at the surface, there is no fluid motion. So, we can write

$$h(T_w - T_\infty) = -k_f \frac{\partial T}{\partial y}\Big|_{y=0} \tag{1.90}$$

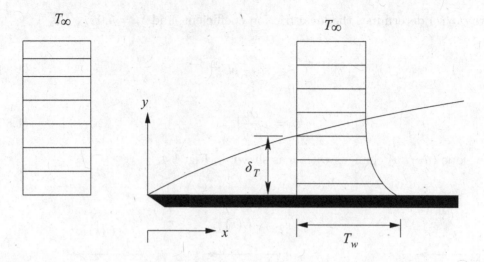

Figure 1.5 Thermal boundary layer on a heated plate

or

$$h = \frac{-k_f \left(\frac{\partial T}{\partial y} \right) |_{y=0}}{(T_w - T_\infty)} \qquad (1.91)$$

Hence, the conditions in the thermal boundary layer which strongly influence the wall temperature gradient $\left(\frac{\partial T}{\partial y} |_{y=0} \right)$, determine the rate of heat transfer across the boundary layer. Since $(T_w - T_\infty)$ is a constant (independent of x), and δ_T increases with increasing x, temperature gradient in the boundary layer must decrease with increasing x (Fig. 1.6). This means $\frac{\partial T}{\partial y} |_{y=0}$ decreases with increase in x, whereby q_x'' and h also decrease with increase in x.

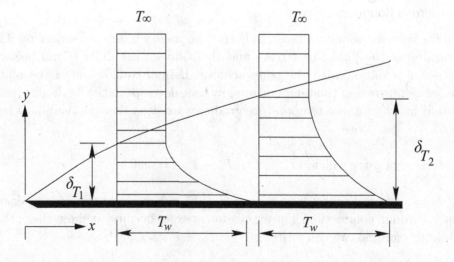

Figure 1.6 Change of temperature gradient within the thermal boundary layer

1.6.3 More about Velocity Boundary Layer and Thermal Boundary Layer

The Navier-Stokes equations along with the equation of continuity is:

$$\frac{\partial u}{\partial t} + u\frac{\partial u}{\partial x} + v\frac{\partial u}{\partial y} = -\frac{1}{\rho}\frac{\partial p}{\partial x} + \nu\left[\frac{\partial^2 u}{\partial x^2} + \frac{\partial^2 u}{\partial y^2}\right] \tag{1.92}$$

$$\frac{\partial v}{\partial t} + u\frac{\partial v}{\partial x} + v\frac{\partial v}{\partial y} = -\frac{1}{\rho}\frac{\partial p}{\partial y} + \nu\left[\frac{\partial^2 v}{\partial x^2} + \frac{\partial^2 v}{\partial y^2}\right] \tag{1.93}$$

$$\frac{\partial u}{\partial x} + \frac{\partial v}{\partial y} = 0 \tag{1.94}$$

- In the region, very near to the wall, fluid motion is retarded until it adheres to the surface. The transition of main-stream velocity from zero at the surface to full magnitude takes place through the boundary layer. Its thickness is δ which is a function of x.

- In order to find out velocity distribution in the field and velocity gradients at the wall, full Navier-Stokes equations should be solved. This is almost impossible analytically and can be solved only by numerical techniques.

- Easier approach would be to divide the flow field in two regions, - inviscid and viscous (boundary layer zone). The inviscid flow is irrotational in this case. Here we solve different governing equations in the inviscid zone and the boundary layer zone.

- To solve the governing equation in the boundary layer zone, boundary layer equation should be derived.

1.6.4 Steady Flow over Flat Plate

The boundary layer theory was first conceptualized by Prandtl in 1904 [2]. In this section, we will derive boundary layer equations using order of magnitude analysis. We define the thickness of the boundary layer δ and note that as $Re \to \infty, \delta$ tends to zero. We must understand that proper scale for viscous effects is not L (the plate length is L). The natural scale in y-direction is δ, within which the viscous effect is present.

Let the dimensionless variables be defined as the following

$$u^* = u/U_\infty, v^* = v/U_\infty, p^* = p/\rho U_\infty^2, x^* = x/L, y^* = y/L$$

The Eqs. (1.92) through (1.94) become

$$u^* \frac{\partial u^*}{\partial x^*} + v^* \frac{\partial u^*}{\partial y^*} = -\frac{\partial p^*}{\partial x^*} + \frac{1}{Re}\left[\frac{\partial^2 u^*}{\partial x^{*2}} + \frac{\partial^2 u^*}{\partial y^{*2}}\right] \tag{1.95}$$

$$u^* \frac{\partial v^*}{\partial x^*} + v^* \frac{\partial v^*}{\partial y^*} = -\frac{\partial p^*}{\partial y^*} + \frac{1}{Re}\left[\frac{\partial^2 v^*}{\partial x^{*2}} + \frac{\partial^2 v^*}{\partial y^{*2}}\right] \tag{1.96}$$

$$\frac{\partial u^*}{\partial x^*} + \frac{\partial v^*}{\partial y^*} = 0 \tag{1.97}$$

Following are the scales for the boundary layer variables:

Variable		Dimensional scale		Nondimensional scale
u	:	U_∞	:	1
x	:	L	:	1
y	:	δ	:	$\epsilon = (\delta/L); \epsilon \ll 1$

Consider, the continuity equation: $\partial u^*/\partial x^* \sim O(1) \Rightarrow \partial v^*/\partial y^*$ should be $O(1)$. We are not allowed to drop any term from the continuity equation, we do not allow accumulation or annihilation of mass. Now, v^* has to be of order ϵ because y^* at its maximum is ϵ $(\delta/L \ll 1)$.

Let us return to the Navier-Stokes equations:

x momentum

$$u^* \frac{\partial u^*}{\partial x^*} + v^* \frac{\partial u^*}{\partial y^*} \quad = \quad -\frac{\partial p^*}{\partial x^*} + \frac{1}{Re}\left[\frac{\partial^2 u^*}{\partial x^{*2}} + \frac{\partial^2 u^*}{\partial y^{*2}}\right] \tag{1.98}$$

$$(1)\frac{(1)}{(1)} \quad (\epsilon)\frac{(1)}{(\epsilon)} \quad = \quad (1) \quad (\epsilon^2)\left[\frac{(1)}{(1)} \quad \frac{1}{(\epsilon^2)}\right]$$

y momentum

$$u^* \frac{\partial v^*}{\partial x^*} + v^* \frac{\partial v^*}{\partial y^*} \quad = \quad -\frac{\partial p^*}{\partial y^*} + \frac{1}{Re}\left[\frac{\partial^2 v^*}{\partial x^{*2}} + \frac{\partial^2 v^*}{\partial y^{*2}}\right] \tag{1.99}$$

$$(1)\frac{(\epsilon)}{(1)} \quad (\epsilon)\frac{(\epsilon)}{(\epsilon)} \quad = \quad (?) \quad (\epsilon^2)\left[\frac{(\epsilon)}{(1)} \quad \frac{\epsilon}{(\epsilon^2)}\right]$$

Continuity

$$\frac{\partial u^*}{\partial x^*} + \frac{\partial v^*}{\partial y^*} \quad = \quad 0 \tag{1.100}$$

$$\frac{(1)}{(1)} \quad \frac{(\epsilon)}{(\epsilon)}$$

Within the boundary layer $I_f \sim V_f$ (inertia force $\approx$ viscous force)

$$\rho \frac{U_\infty U_\infty}{L} \approx \mu \frac{U_\infty}{\delta^2} \tag{1.101}$$

or

$$\frac{\rho U_\infty L}{\mu} \approx \frac{L^2}{\delta^2} \quad \Rightarrow \quad Re \approx \frac{1}{\epsilon^2} \tag{1.102}$$

Two outcomes: x-momentum equation can be re-written based on order of magnitude approximation (retaining the terms of order 1).

$$u^* \frac{\partial u^*}{\partial x^*} + v^* \frac{\partial u^*}{\partial y^*} = -\frac{\partial p^*}{\partial x^*} + \frac{1}{Re} \left[\frac{\partial^2 u^*}{\partial y^{*2}} \right] \tag{1.103}$$

and the y-momentum equation becomes

$$\frac{\partial p^*}{\partial y^*} = O(\epsilon) \tag{1.104}$$

The meaning of Eq. (1.104) is the following. There is no variation in pressure in y direction within the boundary layer. Pressure is impressed on the boundary layer by the outer flow. The pressure, p is only a function of x within the boundary layer.

At the outer edge of the boundary layer if we substitute x-momentum equation, we shall obtain

$$u^* \frac{du^*}{dx^*} = -\frac{dp^*}{dx^*} \tag{1.105}$$

In dimensional form

$$u \frac{du}{dx} = -\frac{1}{\rho} \frac{dp}{dx} \tag{1.106}$$

or

$$p + \frac{1}{2} \rho u^2 = \text{constant} \tag{1.107}$$

Boundary conditions:

$$\text{at} \quad y = 0, \ u = 0 = v$$

or

$$\text{at } y^* = 0, \ u^* = 0 = v^*$$

$$\text{at} \quad y = \delta, \ u = U_\infty(x)$$

or

$$\text{at} \quad y^* = \epsilon, \ u^* = 1$$

1.6.5 Derivation of the Equation for Thermal Boundary Layer

$$Pr = \frac{\mu c_p}{k} = \frac{\mu}{\rho} \frac{\rho c_p}{k} = \frac{\nu}{\alpha} = \frac{\text{kinematic viscosity}}{\text{thermal diffusivity}} \sim \left(\frac{\delta}{\delta_T}\right) \quad (1.108)$$

Having defined non-dimensional temperature as $\theta = (T - T_\infty)/(T_w - T_\infty)$, the energy equation, for the incompressible flows can be written as

$$u^* \frac{\partial \theta}{\partial x^*} + v^* \frac{\partial \theta}{\partial y^*} = \frac{1}{Re\ Pr} \left[\frac{\partial^2 \theta}{\partial x^{*2}} + \frac{\partial^2 \theta}{\partial y^{*2}} \right]$$

$$(1)\frac{(1)}{(1)} \quad (\epsilon)\frac{(1)}{(\epsilon)} = \frac{(\epsilon^2)}{1} \left[\frac{(1)}{(1)} \quad \frac{1}{(\epsilon^2)} \right] \quad (1.109)$$

where,

$$\theta = \frac{T - T_\infty}{T_w - T_\infty} \quad (1.110)$$

It can be said the minimum temperature is T_{min} and maximum temperature is T_{max}

$$T_{min} = T_\infty, \quad T_{max} = T_w \text{ and } 0 < \theta < 1$$

The Prandtl number varies in a wide range from value of order of 0.01 for liquid metals to a value of order of 1000 for viscous oils. Simplifications are possible for very small or very large Prandtl numbers. We shall avoid such simplifications in order to keep the boundary layer equations general. However, Re is always large in our consideration. We can finally write the thermal boundary layer equation as

$$u^* \frac{\partial \theta}{\partial x^*} + v^* \frac{\partial \theta}{\partial y^*} = \frac{1}{Re\ Pr} \left[\frac{\partial^2 \theta}{\partial y^{*2}} \right] \quad \text{(nondimensional)} \quad (1.111)$$

or

$$\rho c_p \left(u \frac{\partial T}{\partial x} + v \frac{\partial T}{\partial y} \right) = k \frac{\partial^2 T}{\partial y^2} \quad \text{(dimensional)} \quad (1.112)$$

Physical significance of $\frac{\partial^2 T}{\partial y^2} >> \frac{\partial^2 T}{\partial x^2}$ is that the axial conduction in fluids much less than that of the transport rate in the normal direction (diffusion). Please note that $\frac{\partial T}{\partial x}$ cannot be zero as long as heat transfer is taking place. However, $\frac{\partial T}{\partial y} >> \frac{\partial T}{\partial x}$.

$Nu = (hL/k) =$ Nusselt number = Dimensionless temperature gradient at the surface.

$Nu = Nu(u^*, v^*, \theta, x^*, y^*, Re, Pr)$ for forced convection in the absence of dissipation and volumetric heat generation.

We can also show that in the presence of viscous dissipation, the non-dimensional form of the thermal boundary layer equation becomes

$$u^* \frac{\partial \theta}{\partial x^*} + v^* \frac{\partial \theta}{\partial y^*} = \frac{1}{RePr} \left[\frac{\partial^2 \theta}{\partial y^{*2}} \right] + \frac{Ec}{Re} \left[\left(\frac{\partial u^*}{\partial y^*} \right)^2 \right] \tag{1.113}$$

Similarly in the presence of viscous dissipation the subsequent dimensional form can be written as

$$\rho c_p \left(u \frac{\partial T}{\partial x} + v \frac{\partial T}{\partial y} \right) = k \left[\frac{\partial^2 T}{\partial y^2} \right] + \mu \left[\left(\frac{\partial u}{\partial y} \right)^2 \right] \tag{1.114}$$

In the above analysis, $\left(\frac{\partial u}{\partial y} \right)^2$ has been considered as the leading representative term of the viscous dissipation function. Obviously, for such flows,

$$Nu = Nu(u^*, v^*, \theta, x^*, y^*, Re, Pr, Ec)$$

1.7   IMPORTANT DEFINITIONS

- Boundary layer thickness $\equiv \delta = 5.0x/\sqrt{Re_x}$ (for laminar flow past a flat plate)

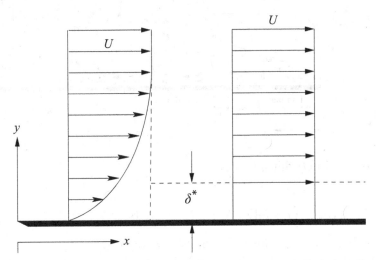

Figure 1.7   Displacement thickness

- Displacement thickness $\equiv \delta^* =$ distance by which the external potential flow is displaced outwards as a consequence of decrease in velocity in the boundary layer (Fig. 1.7).

$$U\delta^* = \int_0^\infty (U - u)dy \tag{1.115}$$

$$\delta^* = \int_0^\infty \left(1 - \frac{u}{U} \right) dy \tag{1.116}$$

- Momentum thickness, δ^{**}, is defined as the loss of momentum in the boundary layer as compared with that of potential flow

$$\rho U^2 \delta^{**} = \int_0^\infty \rho u(U - u)dy \tag{1.117}$$

$$\delta^{**} = \int_0^\infty \left(1 - \frac{u}{U}\right)\frac{u}{U}dy \tag{1.118}$$

- Laminar flow: Fluid motion is well ordered. Random fluctuations of the velocity components are absent and the fluid layers slide over each other.

- Turbulent flow: Fluid motion is highly irregular and characterized by velocity fluctuations (Fig. 1.8). The critical Reynolds number is defined as $Re_c = \rho U x_c / \mu \approx 5 \times 10^5$ (for external flows)

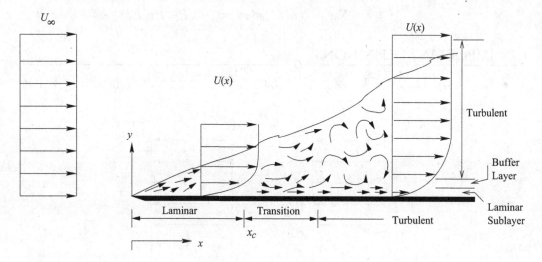

Figure 1.8 Transition in a flat plate flow

1.8 PRANDTL NUMBER AND RATIO OF BOUNDARY LAYERS

We have already seen in Section 1.6.5 that $P_r \sim (\delta/\delta_T)$.

Let us consider a case for $P_r < 1$ or $\delta_T > \delta$ (see Fig. 1.9). From the thermal boundary layer consideration, a scale analysis within the boundary layer gives

$$u\frac{\partial T}{\partial x} \sim \alpha\frac{\partial^2 T}{\partial y^2}$$

$$\text{or,} \ U_\infty\frac{\Delta T}{L} \sim \alpha\frac{\Delta T}{(\delta_T)^2}$$

$$\text{or,} \ \left(\frac{\delta_T}{L}\right)^2 \sim \frac{\alpha}{U_\infty L}$$

$$\text{or,} \ \left(\frac{\delta_T}{L}\right)^2 \sim \frac{\nu}{U_\infty L} \cdot \frac{\alpha}{\nu} \tag{1.119}$$

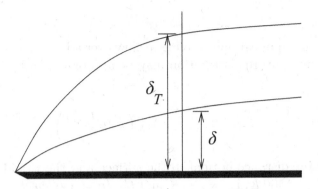

Figure 1.9 The situation $\delta_T > \delta(P_r < 1)$

From the velocity boundary layer consideration, a scale analysis within the boundary layer gives

$$u\frac{\partial u}{\partial x} \quad \sim \quad \nu\frac{\partial^2 u}{\partial y^2}$$

$$\text{or, } U_\infty\frac{U_\infty}{L} \quad \sim \quad \nu\frac{U_\infty}{\delta^2}$$

$$\text{or, } \left(\frac{\delta}{L}\right)^2 \quad \sim \quad \frac{\nu}{U_\infty L} \tag{1.120}$$

From Eqs. (1.119) and (1.120) we get

$$\left(\frac{\delta_T}{L}\right)^2 \quad \sim \quad \left(\frac{\delta}{L}\right)^2 \cdot \frac{\alpha}{\nu}$$

$$\text{or, } \left(\frac{\delta}{\delta_T}\right)^2 \quad \sim \quad Pr$$

$$\text{or, } (Pr)^{\frac{1}{2}} \quad \sim \quad \left(\frac{\delta}{\delta_T}\right) \tag{1.121}$$

Through a similar analysis, it can be shown that for $P_r > 1$.

$$(Pr)^{\frac{1}{3}} \quad \sim \quad \left(\frac{\delta}{\delta_T}\right) \quad \text{for} \quad \delta > \delta_T \tag{1.122}$$

REFERENCES

1. J. Fourier, Analytical Theory of Heat, translated by A. Freeman, G. E. Stechert and Co., New York, 1978.

2. L. Prandtl, Uber Flussigkeitsbewegung bei sehr Kleiner Reibung, Proceedings of 3rd International Mathematics Congress, Heidelberg, pp. 484-491, 1904, also NACA TM 452, 1928.

EXERCISES

1. In a particular application involving airflow over a heated surface, the boundary layer temperature distribution may be approximated by

$$\frac{T - T_w}{T_\infty - T_w} = 1 - exp\left(-Pr\frac{U_\infty y}{\nu}\right)$$

where y is the distance normal to the surface and the Prandtl number (Pr) is 0.7. If $T_\infty = 400\,K, T_w = 300\,K$ and $U_\infty/\nu = 5000\,m^{-1}$, what is the heat flux? Which direction is the heat flux directed to? (k for air is $0.0263\,W/m.K$)

2. When is heat transfer brought about by conduction in a fluid body and when is it caused by convection? For which case is the rate of heat transfer higher? How does the convection heat transfer coefficient differ from the thermal conductivity of a fluid? What should be non-dimensional group if in the expression for Nusselt number, the k is replaced by the thermal conductivity of solid?

3. Experiments have shown that, for airflow at $T_\infty = 35°C$ and $V_1 = 100\ m/s$, the rate of heat transfer q_1 from a turbine blade of characteristic length $L_1 = 0.15\ m$ and surface temperature $T_{W1} = 300°C$ is $1500\ W$. What would be the heat transfer rate from a second turbine blade of characteristic length $L_2 = 0.3\ m$ operating at $T_{W2} = 400°C$ in air flow of $T_\infty = 35°C$ and $V_2 = 50\ m/s$? The surface area of the blade may be assumed to be proportional to characteristic length.

4. The boundary layer theory assumes high Reynolds number flow. High Reynolds number signifies inertia force is much higher than the viscous force. On the other hand, the viscous force and the inertia force are of the same order of magnitude within the boundary layer. Start with the idea that the inertia force and the viscous force are comparable within the boundary layer and show that this reconciles the fact that the Reynolds number is reasonably high in the flow field where the boundary layer exists.

5. In boundary layer theory, a boundary layer can be characterized by any of the following quantities (i) boundary layer thickness (ii) displacement thickness (iii) momentum thickness. How do these quantities differ in their physical as well as mathematical definitions? For the flow over a flat plate, which of these is expected to have the highest value at a given location on the wall, and which is expected to have the lowest?

6. Consider flow over a heated flat plate at a constant wall temperature, T_w. The free stream temperature is T_∞. The y-coordinate is in the normal direction of flow. Show that the Nusselt number is the non-dimensional temperature gradient at the wall.

7. If the thermal boundary layer thickness is smaller than the velocity boundary layer, show that the following relationship holds

$$Pr^{\frac{1}{3}} \approx \left(\frac{\delta}{\delta_T}\right)$$

8. The upper surface of a 50 cm thick solid plate ($k = 215\,W/mK$) is being cooled by water with temperature of 20°C. The upper and lower surfaces of the solid plate maintained at constant temperatures of 50°C and 100°C, respectively. Determine the convection heat transfer coefficient and the water temperature gradient at the upper plate surface (k for water is 0.6 W/mK).

9. The thermal energy equation may be expressed as

$$\rho\frac{De}{Dt} = k\nabla^2 T + \sigma : \nabla\mathbf{V} + S_{th}$$

All the symbols have their usual meaning. The symbol σ is the stress tensor and given by

$$\sigma = -\left(p + \frac{2}{3}\mu(\nabla.\mathbf{V})\right)I + \tau$$

and τ, the shear stress, is given by

$$\tau = 2\mu\left(\frac{1}{2}\left[\frac{\partial V_j}{\partial x_i} + \frac{\partial V_i}{\partial x_j}\right]\right)$$

Show that the final form of the energy equation is given by

$$\rho\frac{De}{Dt} = k\nabla^2 T + \frac{p}{\rho}\frac{D\rho}{Dt} + \mu\phi + S_{th}$$

The symbol ϕ is the viscous dissipation function. Write down a complete form for the viscous dissipation function for compressible flows.

10. Prove that

$$I : (\nabla\mathbf{V}) = \nabla \cdot \mathbf{V}$$

## CHAPTER 2

External Flows

2.1 BLASIUS SOLUTION

Blasius [1] presented a similarity solution of boundary layer equation. This was primarily based on a series expansion.

The simplest example of boundary layer flow is flow over a flat plate as shown in Fig. 2.1.

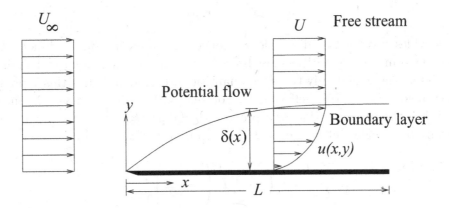

Figure 2.1 Flat plate boundary layer flow

The governing equations are:

$$u\frac{\partial u}{\partial x} + v\frac{\partial u}{\partial y} = \nu\frac{\partial^2 u}{\partial y^2} \tag{2.1}$$

$$\frac{\partial u}{\partial x} + \frac{\partial v}{\partial y} = 0 \tag{2.2}$$

The boundary conditions are : (a) at $y = 0, u = v = 0$,   (b) at $y = \infty, u = U_\infty$.

Substitution of $-\frac{1}{\rho}\frac{dp}{dx}$ in the boundary layer momentum equation in terms of free stream velocity produces $U_\infty\left(dU_\infty/dx\right)$ which is equal to zero. Hence the governing equation does not contain any pressure gradient term. The characteristics parameters of this problem are u, U_∞, ν, x, y. Before we proceed further, let us discuss laws of similarity.

- The u component of velocity has the property that two velocity profiles of $u(x, y)$ at different x locations differ only by a scale factor.

- The velocity profiles $u(x, y)$ at all values of x can be made same if they are plotted in coordinates which have been made dimensionless with reference to the scale factors.

- The local free stream velocity $U(x)$ at x is an obvious scale factor for u, because the dimensionless $u(x)$ varies with y between zero and unity at all sections. The scale factor for y, denoted by $g(x)$, is proportional to local boundary layer thickness so that y itself varies between zero and unity . Finally

$$\frac{u[x_1, (y/g(x_1))]}{U(x_1)} = \frac{u[x_2, (y/g(x_2))]}{U(x_2)} \tag{2.3}$$

Again, let us consider the statement of the problem

$$u = u(U_\infty, \nu, x, y) \tag{2.4}$$

Five variables involve two dimensions. Hence it is reducible for a dimensionless relation in terms of 3 quantities. For boundary layer equations a special similarity variable is available and only two such quantities are needed. When this is possible, the flow field is said to be self similar. For self similar flows the x-component of velocity has the property that two profiles of $u(x, y)/U_\infty$ at different x locations differ only by a scale factor that is at best a function of x.

$$\frac{u}{U_\infty} = F\left(\frac{y}{\delta}\right) = F\left(\frac{y}{\sqrt{\frac{\nu x}{U_\infty}}}\right) \tag{2.5}$$

For the Blasius problem the similarity law is

$$\frac{u}{U_\infty} = F\left[\frac{y}{\sqrt{\frac{\nu x}{U_\infty}}}\right] = F(\eta) \tag{2.6}$$

where

$$\eta = \frac{y}{\delta}, \quad \delta \sim \sqrt{\frac{\nu x}{U_\infty}} \tag{2.7}$$

or

$$\eta = \frac{y}{\sqrt{\frac{\nu x}{U_\infty}}} \tag{2.8}$$

$$\psi = \int u \, dy = \int U_\infty F(\eta) \sqrt{\frac{\nu x}{U_\infty}} d\eta = \sqrt{U_\infty \nu x} \int F(\eta) d\eta \qquad (2.9)$$

or

$$\psi = \sqrt{U_\infty \nu x} \, f(\eta) + C(x) \qquad (2.10)$$

where $f(\eta) = \int F(\eta) d\eta$, and $C(x) = 0$ if the stream function at the solid surface is set equal to 0.

$$u = \frac{\partial \psi}{\partial y} = \frac{\partial \psi}{\partial \eta} \frac{\partial \eta}{\partial y} = U_\infty f'(\eta) \qquad (2.11)$$

$$v = -\left(\frac{\partial \psi}{\partial x} + \frac{\partial \psi}{\partial \eta} \frac{\partial \eta}{\partial x} \right) = \frac{1}{2} \sqrt{\frac{\nu U_\infty}{x}} \left[\eta f'(\eta) - f(\eta) \right] \qquad (2.12)$$

Similarly,

$$\frac{\partial u}{\partial x} = -\frac{U_\infty}{2} \frac{\eta}{x} f''(\eta) \qquad (2.13)$$

$$\frac{\partial u}{\partial y} = U_\infty \sqrt{\frac{U_\infty}{\nu x}} f''(\eta) \qquad (2.14)$$

$$\frac{\partial^2 u}{\partial y^2} = \frac{U_\infty^2}{\nu x} f'''(\eta) \qquad (2.15)$$

Substituting these terms in Eq. (2.1) and simplifying we get

$$2f'''(\eta) + f(\eta) f''(\eta) = 0 \qquad (2.16)$$

The boundary conditions are : at $y = 0$, $u = 0$ and $v = 0$. As a consequence, we can write

$$at \; \eta = 0, f'(\eta) = 0 \;\; \text{and} \;\; f(\eta) = 0$$

Similarly at $y = \infty$, $u = U_\infty$ results in

$$at \; \eta = \infty, \; f'(\eta) = 1$$

Equation (2.16) is a third order nonlinear differential equation. Blasius obtained this solution in the form of a series expanded around $\eta = 0$.

Let us assume a power series expansion (for small values of η) of $f(\eta)$

$$f(\eta) = A_0 + A_1\eta + \frac{A_2}{2!}\eta^2 + \frac{A_3}{3!}\eta^3 + \frac{A_4}{4!}\eta^4 + \ldots\ldots$$

$$f'(\eta) = A_1 + A_2\eta + \frac{A_3}{2!}\eta^2 + \frac{A_4}{3!}\eta^3 + \ldots\ldots$$

$$f''(\eta) = A_2 + A_3\eta + \frac{A_4}{2!}\eta^2 + \frac{A_5}{3!}\eta^3 + \ldots\ldots$$

$$f'''(\eta) = A_3 + A_4\eta + \frac{A_5}{2!}\eta^2 + \frac{A_6}{3!}\eta^3 + \ldots\ldots \tag{2.17}$$

Boundary conditions at $\eta = 0$, $f'(\eta) = 0$ and at $\eta = 0$, $f(\eta) = 0$ applied to above will produce $A_0 = 0$; $A_1 = 0$.

We derive another boundary condition from the physics of the problem: $y = 0$, $(\partial^2 u/\partial y^2) = 0$ which leads to $\eta = 0 : f'''(\eta) = 0$; invoking this into above we get $A_3 = 0$. Finally Eq. (2.16) is substituted for f, f'', f''' into the Blasius equation, we find

$$2\left[A_4\eta + \frac{A_5}{2!}\eta^2 + \frac{A_6}{3!}\eta^3 + \ldots\right] + \left[\frac{A_2}{2!}\eta^2 + \frac{A_4}{4!}\eta^4 + \frac{A_5}{5!}\eta^5 + \ldots\ldots\right]$$

$$\times \left[A_2 + \frac{A_4}{2!}\eta^2 + \frac{A_5}{3!}\eta^3\right] = 0$$

Collecting different powers of η and equating the corresponding coefficients equal to zero, we obtain

$$2A_4 = 0, \quad 2\frac{A_5}{2!} + \frac{A_2^2}{2!} = 0, \quad 2\frac{A_6}{3!} = 0$$

$$2\frac{A_7}{4!} + \frac{A_2 A_4}{4!} + \frac{A_2 A_4}{2!\,2!} = 0$$

$$2\frac{A_8}{5!} + \frac{A_2 A_5}{5!} + \frac{A_2 A_5}{2!\,3!} = 0$$

This will finally yield : $A_4 = A_6 = A_7 = 0$

$$A_5 = -\frac{A_2^2}{2}, \quad A_8 = \frac{11}{4}A_2^3, \quad A_{11} = -\frac{375}{8}A_2^4$$

Now substituting the series for $f(\eta)$ in terms of η and A_2

$$f(\eta) = \frac{A_2}{2!}\eta^2 - \frac{1}{2}\frac{A_2^2}{5!}\eta^5 + \frac{11}{4}\frac{A_2^3}{8!}\eta^8 - \frac{1}{8}\frac{375}{11!}A_2^4\eta^{11}$$

$$f(\eta) = A_2^{1/3}\left[\frac{(A_2^{1/3}\eta)^2}{2!} - \frac{1}{2}\frac{(A_2^{1/3}\eta)^5}{5!} + \frac{11}{4}\frac{(A_2^{1/3}\eta)^8}{8!} - \frac{1}{8}\frac{375(A_2^{1/3}\eta)^{11}}{11!} + \ldots\right] \tag{2.18}$$

or

$$f(\eta) = A_2^{1/3} \, F(A_2^{1/3}\eta)$$

Equation (2.18) satisfies boundary conditions at $\eta = 0$. Applying boundary conditions at $\eta = \infty$, we have

$$\lim_{\eta \to \infty} \left[A_2^{2/3} F'\left(A_2^{1/3}\eta\right) \right] = f'(\infty) = 1$$

or

$$A_2 = \left[\frac{1}{\lim_{\eta \to \infty} F'(\eta)} \right]^{3/2} \tag{2.19}$$

The value of A_2 can be determined numerically to a good degree of accuracy. Howarth [2] found $A_2 = 0.33206$.

Numerical Approach

Let us rewrite Eq. (2.16)

$$2f'''(\eta) + f(\eta)\, f''(\eta) = 0$$

as three first order differential equations in the following way:

$$f' = G \tag{2.20}$$

$$G' = H \tag{2.21}$$

$$H' = -\frac{1}{2}fH \tag{2.22}$$

The condition $f(0) = 0$ [or $\eta = 0, f(\eta) = 0$] remains useful. $f'(0) = 0$ [or $\eta = 0, f'(\eta) = 0$] , means $G(0) = 0$. Finally $\eta = \infty$, $f'(\eta) = 1$ gives $\Rightarrow G(\infty) = 1$.

Note that the equations for f and G have initial values. The value for $H(0)$ is not known. This is an unusual initial value problem. We handle this as an initial value problem by choosing values of $H(0)$ and solving by numerical methods $f(\eta), G(\eta)$ and $H(\eta)$. In general $G(\infty) = 1$ will not be satisfied for the function G arising from the numerical solution. We then choose other initial values of H so that we find an $H(0)$ which results in $G(\infty) = 1$. This method is called the shooting technique. In Eqs. (2.20) through (2.22) the primes refer to differentiation with respect to η. The integration steps following a Runge-Kutta method are given below

$$f_{n+1} = f_n + \frac{1}{6}\left(k_1 + 2k_2 + 2k_3 + k_4\right) \tag{2.23}$$

$$G_{n+1} = G_n + \frac{1}{6}\left(l_1 + 2l_2 + 2l_3 + l_4\right) \tag{2.24}$$

$$H_{n+1} = H_n + \frac{1}{6}\left(m_1 + 2m_2 + 2m_3 + m_4\right) \tag{2.25}$$

as one moves from η_n to $\eta_{n+1} = \eta_n + h$. The values of k, l and m are as follows

$$
\frac{df}{d\eta} = f' = F_1\left(f, G, H, \eta\right)
$$
$$
\frac{dG}{d\eta} = G' = F_2\left(f, G, H, \eta\right)
$$
$$
\frac{dH}{d\eta} = H' = F_3\left(f, G, H, \eta\right) \tag{2.26}
$$

$$
k_1 = h\, F_1\left(f_n, G_n, H_n, \eta_n\right)
$$
$$
l_1 = h\, F_2\left(f_n, G_n, H_n, \eta_n\right)
$$
$$
m_1 = h\, F_3\left(f_n, G_n, H_n, \eta_n\right) \tag{2.27}
$$

$$
k_2 = h\, F_1\left(f_n + \frac{1}{2}k_1,\ G_n + \frac{1}{2}l_1,\ H_n + \frac{1}{2}m_1,\ \eta_n + \frac{h}{2}\right)
$$
$$
l_2 = h\, F_2\left(f_n + \frac{1}{2}k_1,\ G_n + \frac{1}{2}l_1,\ H_n + \frac{1}{2}m_1,\ \eta_n + \frac{h}{2}\right)
$$
$$
m_2 = h\, F_3\left(f_n + \frac{1}{2}k_1,\ G_n + \frac{1}{2}l_1,\ H_n + \frac{1}{2}m_1,\ \eta_n + \frac{h}{2}\right) \tag{2.28}
$$

In a similar way k_3, l_3, m_3 and k_4, l_4, m_4 are calculated following standard formulae for Runge-Kutta integration.

The functions F_1, F_2 and F_3 are G, H and $-fH/2$. Then at a distance $\Delta\eta$ from the wall, we have

$$
f\left(\Delta\eta\right) = f(0) + G(0)\Delta\eta
$$
$$
G\left(\Delta\eta\right) = G(0) + H(0)\Delta\eta
$$
$$
H\left(\Delta\eta\right) = H(0) + H'(0)\Delta\eta
$$
$$
H'\left(\Delta\eta\right) = -\frac{1}{2}\,f\left(\Delta\eta\right) H\left(\Delta\eta\right) \tag{2.29}
$$

As mentioned, $f''(0) = H(0)$ is unknown. $H(0) = S$ must be determined such that the condition $f'(\infty) = G(\infty) = 1$ is satisfied. The condition at infinity is usually approximated at a finite value of η (around $\eta = 10$).

The value of $H(0)$ now can be calculated by finding $\tilde{H}(0)$ at which $G(\infty)$ crosses unity (Fig. 2.2). Refer to Fig. 2.2(b)

$$
\frac{\tilde{H}(0) - H(0)_1}{1 - G(\infty)_1} = \frac{H(0)_2 - H(0)_1}{G(\infty)_2 - G(\infty)_1}
$$

and repeat the process by using $\tilde{H}(0)$ and better of two initial values $H(0)$. Thus the correct initial value will be determined.

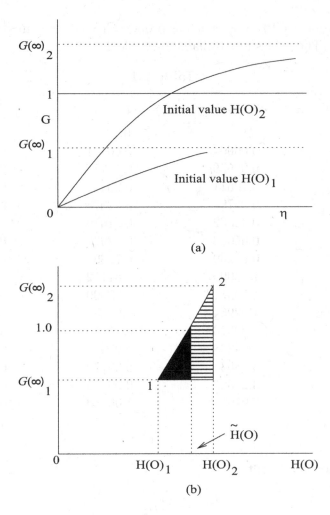

Figure 2.2   Correction of the initial guess for $H(0)$

Shear Stress

$$\tau_{wall} = \mu \frac{\partial u}{\partial y}\Big|_{y=0}$$

or

$$\tau_{wall} = \mu U_\infty \frac{\partial}{\partial \eta} f'(n) \frac{\partial \eta}{\partial y}\Big|_{\eta=0}$$

or

$$\tau_{wall} = \mu U_\infty [0.33206] \frac{1}{\sqrt{\nu x / U_\infty}}$$

or

$$\tau_{wall} = \frac{0.332 \rho U_\infty^2}{\sqrt{Re_x}}$$

Each time examine $G(\eta)$ versus η for proper $G(\infty)$. Compare the values with Schlichting [3]. The values are available in Table 2.1

Table 2.1

η	f	$f' = G$	$f'' = H$
0	0	0	0.33206
0.2	0.00664	0.06641	0.33199
0.4	0.02656	0.13277	0.33147
0.8	0.10611	0.26471	0.32739
1.2	0.23795	0.39378	0.31659
1.6	0.42032	0.51676	0.29667
2.0	0.65003	0.62977	0.26675
2.4	0.92230	0.72899	0.22809
2.8	1.23099	0.81152	0.18401
3.2	1.56911	0.87609	0.13913
3.6	1.92954	0.92333	0.09809
4.0	2.30576	0.95552	0.06424
4.4	2.69238	0.97587	0.03897
4.8	3.08534	0.98779	0.02187
5.0	3.28329	0.99155	0.01591
8.8	7.07923	1.00000	0.00000

Local skin friction coefficient

$$C_{fx} = \frac{\tau_{wall}}{\frac{1}{2}\rho U_\infty^2}$$

$$C_{fx} = \frac{0.664}{\sqrt{Re_x}}$$

Total friction force per unit width

$$F = \int_0^L \tau_{wall}\, dx = \int_0^L \frac{0.332\,\rho\,U_\infty^2}{\sqrt{\frac{U_\infty}{\nu}}} \frac{dx}{\sqrt{x}} = \left[\frac{0.332\,\rho\,U_\infty^2 x^{1/2}}{\sqrt{\frac{U_\infty}{\nu}}\,\frac{1}{2}} \right]_0^L$$

$$F = 0.664\,\rho\,U_\infty^2 \sqrt{\frac{\nu L}{U_\infty}}$$

Average skin friction coefficient

$$\overline{C}_f = \frac{F}{1/2\rho U_\infty^2 L} = \frac{1.328}{\sqrt{Re_L}}$$

From the table, it is seen that $f' = G = 0.99$ for $\eta = 5$. So, u/U_∞ reaches 0.99 at $\eta = 5$. It is possible to write

$$\frac{\delta}{\sqrt{\frac{\nu x}{U_\infty}}} = 5; \quad \text{or} \quad \frac{\delta}{x} = \frac{5.0}{\sqrt{Re_x}}$$

2.2 TEMPERATURE DISTRIBUTION OVER FLAT PLATE BOUNDARY LAYER

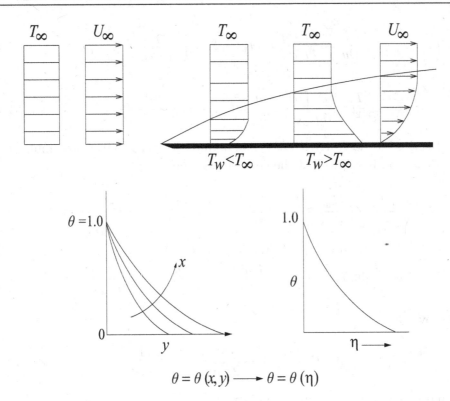

Figure 2.3 The growth of thermal boundary layer

Thermal boundary layer equation

$$u\frac{\partial T}{\partial x} + v\frac{\partial T}{\partial y} = \frac{k}{\rho c_p}\frac{\partial^2 T}{\partial y^2} \tag{2.30}$$

$$u = U_\infty f'(\eta), v = -\frac{1}{2}\sqrt{\frac{\nu U_\infty}{x}}\left\{f(\eta) - \eta f'(\eta)\right\} \tag{2.31}$$

$$\theta = \frac{T - T_\infty}{T_w - T_\infty} \tag{2.32}$$

Figure 2.3 describes the growth of thermal boundary layer on a heated flat plate.

The boundary condition for the situation described above is $\theta(0) = 1$, $\theta(\infty) = 0$. Therefore, $T = T_\infty + \theta(T_w - T_\infty)$.

Let us assume $T_w = T_w(x)$ (a general case). On substituting $\eta = (y/\sqrt{\frac{\nu x}{U_\infty}})$, we can write different derivatives as

$$\frac{\partial T}{\partial x} = \theta \frac{dT_w}{dx} + (T_w - T_\infty)\left(-\frac{y}{2x}\right)\sqrt{\frac{U_\infty}{\nu x}}\frac{d\theta}{d\eta} \tag{2.33}$$

$$\frac{\partial T}{\partial y} = \frac{\partial T}{\partial \eta}\frac{\partial \eta}{\partial y} = (T_w - T_\infty)\frac{\partial \theta}{\partial \eta}\sqrt{\frac{U_\infty}{\nu x}} \tag{2.34}$$

$$\frac{\partial^2 T}{\partial y^2} = \frac{\partial}{\partial \eta}\left[\frac{\partial T}{\partial y}\right]\frac{\partial \eta}{\partial y} = (T_w - T_\infty)\frac{\partial^2 \theta}{\partial \eta^2}\frac{U_\infty}{\nu x} \tag{2.35}$$

On substituting the derivatives in thermal boundary layer equation and multiplying by $\frac{x}{U_\infty(T_w - T_\infty)}$ on both sides of the equation, we get

$$\frac{f'\theta x}{(T_w - T_\infty)}\frac{dT_w}{dx} - \frac{f'y}{2}\sqrt{\frac{U_\infty}{\nu x}}\theta' - \frac{(f - \eta f')}{2}\theta' = \frac{k}{\rho c_p}\frac{\theta''}{\nu} \tag{2.36}$$

$$f'\theta x \frac{d}{dx}[\ln(T_w - T_\infty)] - \frac{f'\eta\theta'}{2} - \frac{(f - \eta f')\theta'}{2} = \frac{k}{\mu c_p}\theta'' \tag{2.37}$$

Dividing by $f'\theta$

$$x\frac{d}{dx}[\ln(T_w - T_\infty)] - \frac{f\theta'}{2f'\theta} = \frac{k}{\mu c_p}\theta''\frac{1}{f'\theta} \tag{2.38}$$

This equation could be written in the form

$$\frac{x(dT_w/dx)}{T_w(x) - T_\infty} = \frac{k\theta''(\eta)}{\mu c_p f'(\eta)\theta(\eta)} + \frac{f(\eta)\theta'(\eta)}{2f'(\eta)\theta(\eta)} = \lambda \tag{2.39}$$

This equation can be solved by using the method of separation of variables. This leads to

$$\frac{x(dT_w/dx)}{T_w(x) - T_\infty} = \lambda \tag{2.40}$$

Wall temperature has to follow this relation

$$d\{\ln(T_w - T_\infty)\} = \frac{dx}{x}\lambda \tag{2.41}$$

Integrating we get,

$$\ln(T_w - T_\infty) = \lambda \ln x + C \tag{2.42}$$

or

$$(T_w - T_\infty) = Cx^\lambda$$

when

$$\lambda = 0, \; T_w = T_\infty + C \qquad \text{(constant wall temperature)}$$

Again,

$$\frac{k\theta''(\eta)}{\mu c_p f'(\eta)\theta(\eta)} + \frac{f(\eta)\theta'(\eta)}{2f'(\eta)\theta(\eta)} = \lambda \tag{2.43}$$

or

$$\frac{\theta''(\eta)}{Pr} + \frac{1}{2}f\theta' = \lambda f'\theta \tag{2.44}$$

or

$$\theta'' + \frac{Pr}{2}f\theta' = \lambda Pr f'\theta \tag{2.45}$$

$$\theta'' + \frac{Pr}{2}f\theta' - \lambda Pr f'\theta = 0 \tag{2.46}$$

In this equation f is known from the Blasius solution. The boundary conditions are:

$$y = 0, T = T_w \to \eta = 0, \theta = 1$$

$$y = \infty, T = T_\infty \to \eta = \infty, \theta = 0$$

$$T = T_\infty + \theta(T_w - T_\infty)$$

The final forms of the boundary conditions are $\theta(0) = 1$ and $\theta(\infty) = 0$.
Let $Y_1 = \theta$ and $Y_2 = \theta'$. Equation (2.46) can be written as

$$Y_2' = -\frac{Pr}{2}fY_2 + \lambda Pr f'Y_1 \tag{2.47}$$

with $Y_1(0) = 1$ and $Y_1(\infty) = 0$.
    The numerical solution using Runge-Kutta method can be obtained via following steps

1. Estimate a value of $Y_2(0)$ or $\theta'(0)$.

2. Solve for Y_1, Y_2.

3. Check if $Y_1(\infty) = 0$, i.e., $\theta(\infty) = 0$?

4. If yes, stop. The calculated $\theta(\eta)$'s are correct solution.

5. If no, correct $\theta'(0)$ or $Y_2(0)$ and repeat the calculation.

The Equation (2.16) and Equation (2.47)can be solved together numerically for obtaining the similarity profiles for velocity and temperature. As already mentioned, basic Runge-Kutta procedure is deployed for obtaining the solutions. A computer code, developed for this purpose, has been provided in the appendix at the end of this chapter.

2.3   ANALYTICAL SOLUTION FOR $\lambda = 0$

$$\theta'' + \frac{Pr}{2}f\theta' = 0 \tag{2.48}$$

or

$$\frac{\theta''}{\theta'} = -\frac{Pr}{2}f \tag{2.49}$$

or

$$\frac{d}{d\eta}(\ln \theta') = -\frac{Pr}{2}f$$

or

$$\ln \theta' = -\frac{Pr}{2}\int f(\eta)d\eta + \ln C$$

or

$$\theta' = Ce^{-\frac{Pr}{2}\int f(\eta)d\eta}$$

or

$$\theta = C\int e^{-\frac{Pr}{2}\int f(\eta)d\eta}d\eta + D$$

or

$$\theta(\eta) - \theta(0) = C\int_0^\eta e^{-\frac{Pr}{2}\int_0^\eta f(\eta)d\eta}d\eta$$

or

$$\theta(\eta) = 1 + C\int_0^\beta (e^{-\frac{Pr}{2}\int_0^\beta f(r)dr})d\beta \tag{2.50}$$

Now $\theta(\infty) = 0$ will result in

$$-1 = C\int_0^\infty (e^{-\frac{Pr}{2}\int_0^\beta f(r)dr})d\beta \tag{2.51}$$

or

$$C = -\frac{1}{\int_0^\infty (e^{-\frac{Pr}{2}\int_0^\beta f(r)dr})d\beta} \tag{2.52}$$

$$\theta(\eta) = 1 - \frac{\int_0^\eta (e^{-\frac{Pr}{2}\int_0^\beta f(r)dr})d\beta}{\int_0^\infty (e^{-\frac{Pr}{2}\int_0^\beta f(r)dr})d\beta}$$

$$\theta(\eta) = 1 - \frac{\int_0^\eta \left(e^{-\frac{Pr}{2}\int_0^\eta f(\eta)d\eta}\right)d\eta}{\int_0^\infty \left(e^{-\frac{Pr}{2}\int_0^\eta f(\eta)d\eta}\right)d\eta} \tag{2.53}$$

Nu_x = Nusselt number = $\frac{hx}{k}$

$$h = \frac{-k(dT/dy)_{y=0}}{(T_w - T_\infty)} = \frac{-k\left[\frac{\partial T}{\partial \eta}\frac{\partial \eta}{\partial y}\right]_{\eta=0}}{(T_w - T_\infty)} \tag{2.54}$$

We know,

$$T = T_\infty + \theta\,(T_w - T_\infty) \text{ we can write, } \frac{\partial T}{\partial \eta} = \frac{\partial \theta}{\partial \eta}\,(T_w - T_\infty)$$

$$h = -k\,\frac{\partial \theta}{\partial \eta}\frac{\partial \eta}{\partial y}\,|_{\eta=0} = -k\theta'(0)\sqrt{\frac{U_\infty}{\nu x}} \tag{2.55}$$

$$Nu_x = \frac{hx}{k} = -\theta'(0)\sqrt{\frac{U_\infty x}{\nu}} = -\theta'(0)(Re_x)^{1/2} \tag{2.56}$$

$$\frac{Nu_x}{(Re_x)^{1/2}} = -\theta'(0) \simeq 0.332\,Pr^{1/3} \quad (0.6 \le Pr \le 10) \tag{2.57}$$

The solution was obtained by Pohlhausen [4].

2.4   MORE DISCUSSIONS ON SOLUTION VIA NUMERICAL ROUTE

The heat transfer rate at the wall is given by

$$q_w'' = -k\frac{\partial T}{\partial y}|_{y=0} \tag{2.58}$$

Hence,

$$\frac{q_w''}{k(T_w - T_\infty)} = -\frac{\partial \theta}{\partial y}|_{y=0} = -\frac{\partial \theta}{\partial \eta}\frac{\partial \eta}{\partial y}|_{y=0} \tag{2.59}$$

i.e.,

$$\frac{q_w''x}{k(T_w - T_\infty)} = -\theta'|_{\eta=0}\sqrt{Re_x} \tag{2.60}$$

i.e.,

$$Nu_x = -\theta'|_{\eta=0}\sqrt{Re_x} \tag{2.61}$$

Nu_x and Re_x, are, the local Nusselt and Reynolds numbers based on x. Because θ depends only on η for a given Pr, $\theta'|_{\eta=0}$ depends only on Pr and its values can be obtained from the solution for the variation θ with η for any value of Pr. It is convenient to define.

$$A(Pr) = -\theta'|_{\eta=0} \tag{2.62}$$

In terms of this function A, Eq. (2.61) can be written as

$$Nu_x = A\sqrt{Re_x} \tag{2.63}$$

Values of A for various values of Pr are shown in Table 2.2.

Table 2.2

Values of A for various values of Pr

Pr	$A = -\theta'|_{\eta=0}$	$0.332Pr^{1/3}$
0.6	0.276	0.280
0.7	0.293	0.295
0.8	0.307	0.308
0.9	0.320	0.321
1.0	0.332	0.332
1.1	0.344	0.343
7.0	0.645	0.635

Over the range of Prandtl numbers covered in the table, it can be found that A varies very nearly as $Pr^{1/3}$ and, as will be seen from the results given in the table, is quite closely represented by the approximate relation:

$$A = 0.332Pr^{1/3} \tag{2.64}$$

We calculate the total heat transfer rate from a plate of length, L. Unit width of the plate is considered because the flow is, by assumption, two-dimensional. The total heat transfer rate per unit width Q_w will, of course, be related to the local heat transfer rate, q_w'' by:

$$Q_w = \int_0^L q_w'' dx \tag{2.65}$$

But Eq. (2.60) gives the local heat transfer rate as

$$q_w'' = Ak(T_w - T_\infty)\sqrt{\frac{U_\infty}{x\nu}} \tag{2.66}$$

Substituting this result Eq. (2.65) then continues the integration:

$$Q_w = 2Ak(T_w - T_\infty)\sqrt{\frac{UL}{\nu}} \tag{2.67}$$

If a mean heat transfer coefficient for the whole plate is defined such that

$$\bar{h} = \frac{Q_w}{L(T_w - T_\infty)} \tag{2.68}$$

Then, since unit width of the plate is being considered, Eq. (2.67) gives

$$\bar{h} = \frac{2Ak}{L}\sqrt{\frac{U_\infty L}{\nu}}$$

It is possible to calculate average Nusselt number as

$$\overline{Nu} = \frac{\bar{h}L}{k} = 2A\sqrt{Re_L} \tag{2.69}$$

Let us revisit the thermal conditions for the wall again. We consider the case where:

$$T_w = T_\infty + Cx^\lambda$$

i.e.,

$$T_w - T_\infty = Cx^\lambda \tag{2.70}$$

C and λ being constants. The heat transfer rate at the wall is as before given by

$$q_w'' = -k\frac{\partial T}{\partial y}|_{y=0} \tag{2.71}$$

$$q_w'' = -k(T_w - T_\infty)\frac{\partial \theta}{\partial \eta}|_{\eta=0}\frac{\partial \eta}{\partial y}$$

$$q_w'' = -kC\, x^\lambda\sqrt{\frac{U_\infty}{\nu x}}\frac{\partial \theta}{\partial \eta}|_{\eta=0}$$

$$q_w'' = \frac{-kC\sqrt{U_\infty}}{\sqrt{\nu}}\,\theta'(0)x^{\lambda-1/2} \tag{2.72}$$

For the constant heat flux along the wall, the expression should be independent of x or in other words, we can say

$$\lambda = \frac{1}{2} \tag{2.73}$$

2.5   APPROXIMATE METHODS FOR FLAT PLATE BOUNDARY LAYER

Flow over a heated flat plate for different Prandtl numbers $\left(\frac{\delta}{\delta_T} \sim (Pr)^{1/3} \text{ for } Pr > 1 \right.$ or $\frac{\delta}{\delta_T} \sim (Pr)^{1/2} \text{ for } Pr < 1 \right)$ has been illustrated in Fig. 2.4. Governing equations are

$$\frac{\partial u}{\partial x} + \frac{\partial v}{\partial y} = 0 \tag{2.74}$$

$$u\frac{\partial u}{\partial x} + v\frac{\partial v}{\partial y} = -\frac{1}{\rho}\frac{dp}{dx} + \nu\frac{\partial^2 u}{\partial y^2} \tag{2.75}$$

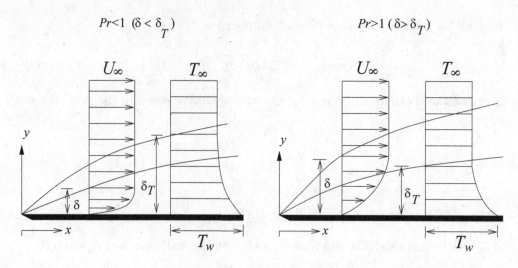

Figure 2.4 Velocity and thermal boundary layers (a) for $\delta < \delta_T$ (b) for $\delta > \delta_T$

$$u\frac{\partial T}{\partial x} + v\frac{\partial T}{\partial y} = \frac{k}{\rho c_p}\frac{\partial^2 T}{\partial y^2} \tag{2.76}$$

The boundary conditions are:

$$at\ y = 0, u = 0 = v, T = T_w$$
$$at\ y = \delta,\ u = U_\infty,\ at\ y = \delta_T, T = T_\infty$$

Integral method due to von Karman and Pohlhausen can be used to integrate Eq. (2.75)

$$\underbrace{\int_0^\delta u\frac{\partial u}{\partial x}dy}_{\text{I}} + \underbrace{\int_0^\delta v\frac{\partial u}{\partial y}dy}_{\text{II}} = \underbrace{\int_0^\delta -\frac{1}{\rho}\frac{dp}{dx}dy}_{\text{III}} + \underbrace{\int_0^\delta \nu\frac{\partial^2 u}{\partial y^2}dy}_{\text{IV}} \tag{2.77}$$

Term III is zero for flow over flat plate (because $dp/dx = 0$). Term II may be written as

$$[vu]_0^\delta - \int_0^\delta \frac{\partial v}{\partial y}udy = \left[vu|_\delta - vu|_0\right] + \int_0^\delta u\frac{\partial u}{\partial x}dy \quad \text{(from continuity)}$$

From continuity equation, one can write

$$\frac{\partial u}{\partial x} = -\frac{\partial v}{\partial y}$$

or

$$\int_0^\delta \frac{\partial v}{\partial y}dy = -\int_0^\delta \frac{\partial u}{\partial x}dy$$

or

$$v|_{y=\delta} = -\int_0^\delta \frac{\partial u}{\partial x}dy$$

Term II finally becomes

$$-U_\infty \int_0^\delta \frac{\partial u}{\partial x} dy + \int_0^\delta u \frac{\partial u}{\partial x} dy$$

Equation (2.77) can be written as

$$2 \int_0^\delta u \frac{\partial u}{\partial x} dy - U_\infty \int_0^\delta \frac{\partial u}{\partial x} dy = -\nu \frac{\partial u}{\partial y}\bigg|_{y=0}$$

or

$$\frac{d}{dx} \int_0^\delta u^2 dy - U_\infty \frac{d}{dx} \int_0^\delta u \, dy = -\nu \frac{\partial u}{\partial y}\bigg|_{y=0} \qquad (2.78)$$

Equation (2.78) is the momentum integral equation for flow over a flat plate. The terms on the left-hand side are to be noted. In these terms, differentiation with respect to x, and integration with respect to y is interchanged as the upper limit of integration is independent of x. Momentum integral equation for a non-zero pressure gradient surface will be discussed later.

Let us assume the velocity profile as

$$u = C_0 + C_1 y + C_2 y^2 + C_3 y^3$$
$$\frac{u}{U_\infty} = C_0 + C_1 \eta + C_2 \eta^2 + C_3 \eta^3, \quad \text{where } \eta = y/\delta$$

We already know that η is called similarity parameter; despite the growth of boundary layer in x direction, u/U_∞ remains similar for same $\frac{y}{\delta}$ at any x.

Boundary conditions:
Application of the condition,

$$at \ y = 0, u = 0, \ \ or, \ at \ \eta = 0, \frac{u}{U_\infty} = 0, \ \ \text{makes} \ C_0 = 0$$

Application of the condition,

$$at \ y = 0, \frac{\partial^2 u}{\partial y^2} = 0 \ \left(\text{comes from} \ u \frac{\partial u}{\partial x} + v \frac{\partial u}{\partial y} = \nu \frac{\partial^2 u}{\partial y^2}\right),$$
$$or, \ at \ \eta = 0, \frac{\partial^2 (u/U_\infty)}{\partial \eta^2} = 0,$$
$$or, \ [2C_2 + 6C_3 \eta]_{at \ \eta=0} = 0 \ \ \text{yields} \ C_2 = 0$$

Application of the condition,

$$at \ y = \delta, u = U_\infty, \ \ or, \ at \ \eta = 1, \frac{u}{U_\infty} = 1$$
$$or, \ \left[C_1 \eta + C_3 \eta^3\right]_{\eta=1} = 1, \ \ \text{makes,} \ C_1 + C_3 = 1$$

$$at \ y = \delta, \frac{\partial u}{\partial y} = 0 \ \ or, \ at \ \eta = 1, \frac{\partial}{\partial \eta} \left[\frac{u}{U_\infty} \right] = 0$$

$$or, \ [C_1 + 3C_3\eta^2]_{\eta=1} = 0, \ \ yields \ C_1 + 3C_3 = 0$$

Finally we get

$C_1 = \dfrac{3}{2}, C_3 = -\dfrac{1}{2}$. The velocity profile can be written as, $\quad \dfrac{u}{U_\infty} = \dfrac{3}{2}\eta - \dfrac{1}{2}\eta^3$.

By making use of the velocity profile, the first term in Eq. (2.78) can be evaluated as

$$\int_0^\delta u^2 dy = \frac{U_\infty^2}{4} \int_0^1 [9\eta^2 + \eta^6 - 6\eta^4] \, \delta d\eta$$

$$= \frac{\delta U_\infty^2}{4} \left[\frac{9}{3} + \frac{1}{7} - \frac{6}{5} \right] = \frac{68}{35} \frac{\delta U_\infty^2}{4}$$

Evaluation of the second term of Eq. (2.78) produces

$$\int_0^\delta u dy = \frac{U_\infty}{2} \int_0^1 [3\eta - \eta^3] \, \delta d\eta = \frac{5}{4} \frac{U_\infty \delta}{2}$$

The third term of equation (2.78) $-\nu\frac{\partial u}{\partial y}\big|_{y=0}$ becomes

$$= -\frac{\nu}{\delta} \frac{\partial}{\partial \eta} \left[U_\infty \left(\frac{3}{2}\eta - \frac{1}{2}\eta^3 \right) \right] \big|_{\eta=0}$$

$$= -\frac{3}{2} \frac{\nu U_\infty}{\delta}$$

Finally Eq. (2.78) can be written as

$$\frac{U_\infty^2}{4} \frac{d\delta}{dx} \left[\frac{68}{35} - \frac{5}{2} \right] = -\frac{3}{2} \frac{U_\infty \nu}{\delta}$$

which reduces to

$$\delta \frac{d\delta}{dx} = \frac{140}{13} \frac{\nu}{U_\infty}$$

On integration this gives

$$\frac{1}{2} \frac{13}{140} \delta^2 = \frac{\nu x}{U_\infty} + C \tag{2.79}$$

The initial condition: $at \ x = 0, \delta = 0$ gives $C = 0$. Finally,

$$\delta^2 = \frac{280}{13} \frac{\nu x}{U_\infty}$$

which gives

$$\delta = 4.64 \sqrt{\frac{\nu x}{U_\infty}}$$

On substituting $Re_x = \frac{U_\infty x}{\nu}$ we get

$$\delta = 4.64 \frac{x}{\sqrt{Re_x}} \tag{2.80}$$

The integral form of thermal boundary layer equation may be written as

$$\int_0^{\delta_T} u \frac{\partial T}{\partial x} dy + \int_0^{\delta_T} v \frac{\partial T}{\partial y} dy = \int_0^{\delta_T} \frac{k}{\rho c_p} \frac{\partial^2 T}{\partial y^2} dy \tag{2.81}$$

Integrating the second term by parts and applying continuity, we get

$$\int_0^{\delta_T} u \frac{\partial T}{\partial x} dy + [vT]_0^{\delta_T} - \int_0^{\delta_T} T \frac{\partial v}{\partial y} dy = \left[\frac{k}{\rho c_p} \frac{\partial T}{\partial y} \right]_0^{\delta_T}$$

or

$$\int_0^{\delta_T} u \frac{\partial T}{\partial x} dy + \int_0^{\delta_T} T \frac{\partial u}{\partial x} dy + T_\infty \, v|_{\delta_T} = -\frac{k}{\rho c_p} \frac{\partial T}{\partial y} \Big|_{y=0}$$

where $v|_{\delta_T} = -\int_0^{\delta_T} \frac{\partial u}{\partial x} dy$ or

$$\int_0^{\delta_T} \left(u \frac{\partial T}{\partial x} + T \frac{\partial u}{\partial x} \right) dy - T_\infty \int_o^{\delta_T} \frac{\partial u}{\partial x} dy = -\frac{k}{\rho c_p} \frac{\partial T}{\partial y} \Big|_{y=0}$$

or

$$\int_0^{\delta_T} \frac{\partial}{\partial x} [u(T - T_\infty)] dy = -\frac{k}{\rho c_p} \frac{\partial T}{\partial y} \Big|_{y=0}$$

or

$$\frac{d}{dx} \int_0^{\delta_T} [u(T - T_\infty)] dy = -\frac{k}{\rho c_p} \frac{\partial T}{\partial y} \Big|_{y=0} \tag{2.82}$$

This equation is called the energy integral equation. In order to solve this, let us assume the temperature profile, $\theta = \frac{T - T_w}{T_\infty - T_w} = C_0 + C_1 \zeta + C_2 \zeta^2 + C_3 \zeta^2$ where $\zeta = \frac{y}{\delta_T}$ is a similarity parameter.

Applying thermal boundary conditions;

at $\quad y = 0; T = T_w$, or $\quad$ at $\quad \zeta = 0, \ \theta = 0$ we get $C_0 = 0$

at $\quad y = 0; \frac{\partial^2 T}{\partial y^2} = 0$, or $\quad$ at $\quad \zeta = 0, \frac{d^2\theta}{d\zeta^2} = 0$ we get $\ C_2 = 0$

at $\quad y = \delta_T; T = T_\infty$, or $\quad$ at $\quad \zeta = 1, \ \theta = 1$ we get $\ C_1 + C_3 = 1$

at $\quad y = \delta_T; \frac{\partial T}{\partial y} = 0$, or $\quad$ at $\quad \zeta = 1, \frac{d\theta}{d\zeta} = 0$ we get $C_1 + 3C_3 = 0$

Solving for C_1 and C_3 we get $C_1 = \frac{3}{2}$, and $C_3 = -\frac{1}{2}$. Therefore,

$$\frac{T - T_w}{T_\infty - T_w} = \frac{3}{2}\zeta - \frac{1}{2}\zeta^3 \tag{2.83}$$

We can also write

$$\theta - 1 = \frac{T - T_w - T_\infty + T_w}{T_\infty - T_w} = \frac{T - T_\infty}{T_\infty - T_w}$$

or

$$(T - T_\infty) = (\theta - 1)(T_\infty - T_w)$$

First term of the energy integral (Eq. (2.82)) can be reduced to

$$(T_\infty - T_w)\frac{d}{dx}\int_0^{\delta_T} u(\theta - 1)dy$$

$$U_\infty(T_\infty - T_w)\frac{d}{dx}\int_0^{\delta_T}\left[\frac{3}{2}\eta - \frac{1}{2}\eta^3\right]\left[\frac{3}{2}\zeta - \frac{1}{2}\zeta^3 - 1\right]dy$$

$$U_\infty(T_\infty - T_w)\frac{d}{dx}\int_0^{\delta_T}\left[\frac{3}{2}\left(\frac{y}{\delta}\right) - \frac{1}{2}\left(\frac{y}{\delta}\right)^3\right]\left[\frac{3}{2}\left(\frac{y}{\delta_T}\right) - \frac{1}{2}\left(\frac{y}{\delta_T}\right)^3 - 1\right]dy$$

Substituting $\xi = \frac{\delta_T}{\delta}$, rearranging and integrating, we get

$$U_\infty(T_\infty - T_w)\frac{d}{dx}\left[\delta\left\{\frac{3}{4}\xi^2 - \frac{3}{20}\xi^4 - \frac{3}{20}\xi^2 + \frac{1}{28}\xi^4 - \frac{3}{4}\xi^2 + \frac{1}{8}\xi^4\right\}\right]$$

$$= U_\infty(T_\infty - T_w)\frac{d}{dx}\left[\delta\left\{-\frac{3}{20}\xi^2 + \frac{3}{280}\xi^4\right\}\right]$$

The RHS of energy integral equation (Eq. (2.82)) is evaluated in the following way:

$$(T - T_w) = (T_\infty - T_w)\left[\frac{3}{2}\left(\frac{y}{\delta_T}\right) - \frac{1}{2}\left(\frac{y}{\delta_T}\right)^3\right]$$

$$\frac{\partial T}{\partial y} = (T_\infty - T_w)\left[\frac{3}{2}\frac{1}{\delta_T} - \frac{3}{2}\frac{y^2}{\delta_T^3}\right]$$

or

$$-\frac{k}{\rho c_p}\frac{\partial T}{\partial y}\Big|_{y=0} = -\alpha\frac{3}{2}\frac{(T_\infty - T_w)}{\delta_T}$$

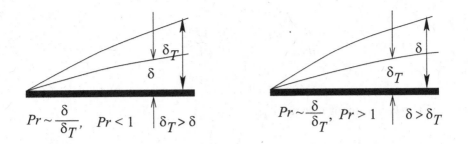

$$Pr \sim \frac{\delta}{\delta_T}, \quad Pr < 1 \quad \Big| \, \delta_T > \delta \qquad\qquad Pr \sim \frac{\delta}{\delta_T}, \quad Pr > 1 \quad \Big| \, \delta > \delta_T$$

Figure 2.5   Different situations with $Pr < 1$ and $Pr > 1$

Now Eq. (2.82) becomes

$$(T_\infty - T_w)U_\infty \frac{d}{dx}\left[\delta\left\{-\frac{3}{20}\xi^2 + \frac{3}{280}\xi^4\right\}\right] = -\frac{3}{2}\frac{\alpha(T_\infty - T_w)}{\delta_T} \tag{2.84}$$

Figure 2.5 explains different situations pertaining to the growth of boundary layers for various Prandtl numbers.

For $Pr > 1$, $\delta > \delta_T$, $\xi = \frac{\delta_T}{\delta}$ becomes small, and so $\xi^4 << \xi^2$. Therefore, for $Pr > 1$ (Fig. 2.5), Eq. (2.84) becomes

$$\delta U_\infty \frac{d}{dx}\left[\delta\frac{\xi^2}{20}\right] = \frac{\alpha}{2\xi}$$

or

$$\frac{10\alpha}{U_\infty \xi} = \delta\frac{d\delta}{dx}\xi^2 + \delta^2\frac{d}{dx}\xi^2$$

where

$$\delta\frac{d\delta}{dx} = \frac{140}{13}\frac{\nu}{U_\infty}, \quad and \quad \delta^2 = \frac{280}{13}\frac{\nu x}{U_\infty} \quad \text{(from Eq.(2.79))}$$

Finally, on substitution of the above two relations, Eq. (2.84) becomes

$$\xi^3 + 2x\xi\frac{d}{dx}(\xi^2) = \frac{13}{14}\frac{1}{Pr} \tag{2.85}$$

Put $\chi = \xi^3$; $\frac{d\chi}{dx} = 3\xi^2\frac{d\xi}{dx}$; substituting in Eq. (2.85)

$$\chi + \frac{4x}{3}\frac{d\chi}{dx} = \frac{13}{14}\frac{1}{Pr}$$

or

$$\frac{d\chi}{dx} + \frac{3}{4x}\chi = \frac{13}{14}\frac{1}{Pr}\frac{3}{4x}$$

This linear differential equation for which the integrating factor is evaluated as $e^{\int\frac{3}{4}\frac{dx}{x}} = e^{(3/4)ln\,x} = x^{3/4}$

$$x^{3/4}\frac{d\chi}{dx} + \frac{3}{4}\frac{\chi}{x}x^{3/4} = \frac{13\times3}{14\times4}\frac{1}{Pr}x^{-1/4} \tag{2.86}$$

or

$$\frac{d}{dx}(x^{3/4}\chi) = \frac{39}{56}\frac{1}{Pr}x^{-1/4} \tag{2.87}$$

$$\chi = \frac{13}{14}\frac{1}{Pr} + bx^{-3/4} \tag{2.88}$$

where b is the constant of integration.

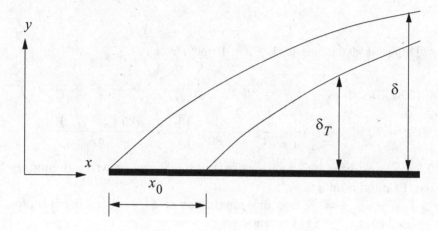

Figure 2.6 Flat plate situation with unheated initial length

Analyzing the boundary conditions as described in Fig. 2.6, we consider a short unheated length x_0 so that at $x = x_0$, $\delta_T = 0$, $\xi = 0$; and so, at $x = x_0$, $\chi = 0$. Substituting in Eq. (2.88), we get

$$b = -\frac{13}{14}\frac{1}{Pr}x_0^{3/4} \tag{2.89}$$

After substituting in Eq. (2.88) for b, we get

$$\xi^3 = \chi = \frac{13}{14}\frac{1}{Pr} - \frac{13}{14}\frac{1}{Pr}\left(\frac{x_0}{x}\right)^{3/4}$$

or

$$\frac{\delta_T}{\delta} = \xi = \frac{1}{1.026}(Pr)^{-1/3}\left[1 - \left(\frac{x_0}{x}\right)^{3/4}\right]^{1/3}$$

For $x_0 \to 0$, we finally have

$$\delta_T = \frac{\delta}{1.026}(Pr)^{-1/3} \tag{2.90}$$

Again we know that

$$h(T_w - T_\infty) = -k\frac{\partial T}{\partial y}\Big|_{y=0}$$

On substituting for $\frac{\partial T}{\partial y}\big|_{y=0}$, we get

$$h = \frac{-k\frac{(T_\infty - T_w)}{\delta_T}\frac{3}{2}}{(T_w - T_\infty)}$$

or

$$h = k\frac{3}{2}\frac{1.026}{4.64x}(Pr)^{1/3}(Re_x)^{1/2}$$

(after invoking Eqs. (2.80) and (2.90)). The above expression leads to

$$\frac{hx}{k} = 0.332(Re_x)^{1/2}(Pr)^{1/3}$$

The local Nusselt number can be expressed as

$$Nu_x = 0.332(Re_x)^{1/2}(Pr)^{1/3} \tag{2.91}$$

Now the average heat transfer coefficient $\bar{h} = \int_0^L h_x dx / \int_0^L dx$ and so the average Nusselt number

$$\overline{Nu_L} = \frac{\bar{h}L}{k} = 0.664(Re_L)^{1/2}(Pr)^{1/3} \tag{2.92}$$

where $Re_L - \frac{\rho U_\infty L}{\mu}$

2.6 VISCOUS DISSIPATION EFFECTS ON LAMINAR BOUNDARY LAYER FLOW OVER FLAT PLATE

For the high speed flows, the effects due to viscous dissipation have to be considered. Now, $u = U\,f'(\eta)$ and $\eta = y/\sqrt{(\nu x)/U}$, and so the the viscous dissipation term is

$$\mu\left[\left(\frac{\partial u}{\partial y}\right)^2 + \ldots\ldots\right]$$

$$= \mu\left[\left(\frac{\partial u}{\partial \eta}\frac{\partial \eta}{\partial y}\right)^2 + \ldots\ldots\right]$$

$$= \mu\left[\left(U\,f''\sqrt{\frac{U}{\nu x}}\right)^2 + \ldots\ldots\right]$$

$$= \mu U^2(f'')^2\,\frac{U}{\nu x}$$

The general energy equation is simplified as

$$u\frac{\partial T}{\partial x} + v\frac{\partial T}{\partial y} = \frac{k}{\rho c_p}\left[\frac{\partial^2 T}{\partial y^2}\right] + \frac{\mu}{\rho c_p}\left[\left(\frac{\partial u}{\partial y}\right)^2 + \ldots\ldots\right]$$

(2.93)

or

$$U\,f'\left[\theta\frac{dT_w}{dx} + (T_w - T_\infty)\left(\frac{\partial\theta}{\partial\eta}\right)\left(-\frac{y}{2x}\right)\sqrt{\frac{U}{\nu x}}\right]$$

$$-\frac{1}{2}\sqrt{\frac{\nu\,U}{x}}(f - \eta f')\left\{(T_w - T_\infty)\frac{\partial\theta}{\partial\eta}\sqrt{\frac{U}{\nu x}}\right\}$$

$$= \frac{k}{\rho c_p}\frac{\partial^2\theta}{\partial\eta^2}(T_w - T_\infty)\frac{U}{\nu x} + \frac{\mu}{\rho c_p}U^2(f'')^2\frac{U}{\nu x}$$

Multiply by $\frac{x}{U(T_w-T_\infty)}$ to obtain

$$\frac{f'\theta x}{(T_w - T_\infty)}\frac{dT_w}{dx} - \frac{f\theta'}{2} = \frac{\theta''}{Pr} + Ec(f'')^2$$

(2.94)

If we assume $(dT_w/dx) = 0$, the general equation becomes

$$\frac{\theta''}{Pr} + f\frac{\theta'}{2} + Ec(f'')^2 = 0$$

(2.95)

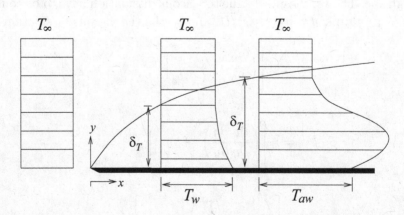

Figure 2.7 Temperature profile within the boundary layer (a) for low speed flows (b) for high speed flows

As defined earlier, $\theta = (T - T_\infty)/(T_w - T_\infty)$ and the boundary conditions are $\theta(0) = 1$, $\theta(\infty) = 0$ with T_w = wall temperature. *If the heat flux at the wall is assumed to be zero, i.e., $q_w'' = 0$, the wall temperature becomes adiabatic wall temperature.*

In other words, T_{aw} = adiabatic wall temperature, if the surface temperature is the temperature of a perfectly insulated surface. As a consequence of the dissipation in the boundary layer, even if there is no heat transfer to a body in a flow, a thermal boundary layer forms at the body. If the surface of the body is impermeable to heat, i.e., adiabatic, the dissipation dictates the distribution of the wall temperature in such a way that it is above the surrounding temperature (Fig 2.7).

For the temperature field at an adiabatic body, if $\phi \sim \frac{\theta}{Ec/2}$, the general equation reduces to the form

$$\frac{Ec}{2}\frac{1}{Pr}\phi'' + \frac{Ec}{2}\frac{f}{2}\phi' + Ec(f'')^2 = 0$$

or

$$\frac{\phi''}{Pr} + \frac{f\phi'}{2} + 2(f'')^2 = 0 \qquad (2.96)$$

with the adiabatic boundary conditions $\phi'(0) = 0, and \ \phi(\infty) = 0$. As already defined,

$$\phi \sim \frac{\theta}{Ec/2} = \frac{T - T_\infty}{T_w - T_\infty}\frac{2c_p(T_w - T_\infty)}{U^2} = \frac{T - T_\infty}{U^2/2c_p}$$

Thus we get

$$T = T_\infty + \frac{U^2}{2c_p}\phi(\eta) \qquad (2.97)$$

which becomes,

$$T_{aw} = T_\infty + \frac{U^2}{2c_p}\phi(0) \qquad (2.98)$$

T_{aw} can be determined if $\phi(0)$ is known, where $\phi(0)$ is also called recovery factor, r and

$$T_{aw} = T_\infty + r\frac{U^2}{2c_p}$$

It is found from experiments,

$$\phi(0) = \text{fn}(Pr), \quad \phi(0) = \sqrt{Pr} \quad \text{(see Fig. 2.8)}$$

For the low speed flows, $\phi(0) = r = 0$ as no kinetic energy is recovered as thermal energy.

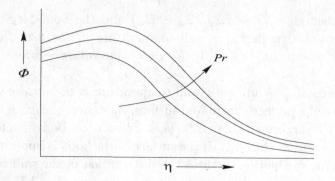

Figure 2.8 ϕ as a function of Pr and η

For ideal gases: i.e., $i = e + pv = e + RT$, we can write

$$\frac{di}{dT} = \frac{de}{dT} + R$$

or

$$c_p = c_v + R$$

which gives on simplification,

$$c_p = \frac{\gamma R}{\gamma - 1}$$

Now,

$$T_{aw} = T_\infty \left[1 + \frac{rU^2(\gamma - 1)}{2\gamma RT_\infty} \right]$$

i.e.,

$$T_{aw} = T_\infty \left[1 + \frac{r(\gamma - 1)}{2} M_\infty^2 \right] \tag{2.99}$$

For air, $Pr = 0.72$, $\phi(0) = \sqrt{0.72} = 0.85 = r$. Putting $\gamma = 1.4$, we get

$$T_{aw} = T_\infty [1 + 0.17 M_\infty^2]$$

Now for $T_\infty = 400K$, and $M_\infty = 0.1$, we obtain $T_{aw} = 400.68K$. Similarly for $M_\infty = 0.8$, $T_{aw} = 443.52K$; for $M_\infty = 3.0$, $T_{aw} = 1012K$.

Consider the general solution for

$$\frac{\theta''}{Pr} + \frac{1}{2} f\theta' + Ec(f'')^2 = 0$$

with $\eta = 0, \theta = 1$ and $\eta = \infty, \theta = 0$.

Now, let us propose

$$\theta(\eta) \sim \frac{E_c}{2}\phi(\eta)$$

at $\eta = 0$, we get

$$\theta(0) \sim \frac{E_c}{2}\phi(0)$$

We can write

$$[\theta(\eta) - \theta(0)] \sim \left[\frac{E_c}{2}\phi(\eta) - \frac{E_c}{2}\phi(0)\right]$$

$$\theta(\eta) \sim \left[1 - \frac{E_c}{2}\phi(0)\right] + \frac{E_c}{2}\phi(\eta) \tag{2.100}$$

If the viscous dissipation is neglected, the dimensionless temperature can be defined as $\tilde{\theta}$.

Now the proportionality (2.100) can be changed to equality

$$\theta = \left[1 - \frac{Ec}{2}\phi(0)\right]\tilde{\theta} + \frac{1}{2}Ec\,\phi(\eta) \tag{2.101}$$

The derivatives terms can be written as

$$\theta'' = \left[1 - \frac{Ec}{2}\phi(0)\right]\tilde{\theta}'' + \frac{1}{2}Ec\,\phi''(\eta) \tag{2.102}$$

$$\theta' = \left[1 - \frac{Ec}{2}\phi(0)\right]\tilde{\theta}' + \frac{1}{2}Ec\,\phi'(\eta) \tag{2.103}$$

On substitution in the general equation (2.95), we get

$$\left[1 - \frac{Ec}{2}\phi(0)\right]\frac{\tilde{\theta}''}{Pr} + \frac{1}{2}f\left[1 - \frac{Ec}{2}\phi(0)\right]\tilde{\theta}' + \frac{1}{2}Ec\left[\frac{\phi''(\eta)}{Pr} + \frac{f}{2}\phi'(\eta)\right] + Ec(f'')^2 = 0$$

Known conditions are $\theta(0) = 1$ and $\theta(\infty) = 0$. Let us revisit the relationship between θ and $\tilde{\theta}$, using Eq. (2.101).

$$\theta(\eta) = \left[1 - \frac{Ec}{2}\phi(0)\right]\tilde{\theta}(\eta) + \frac{Ec}{2}\phi(\eta)$$

at $\eta = 0, \theta(\eta) = 1$ is the known boundary condition. Applying this we get

$$\left[1 - \frac{Ec}{2}\phi(0)\right]\tilde{\theta}(0) + \frac{Ec}{2}\phi(0) = 1$$

$$\tilde{\theta}(0) = \left[1 - \frac{Ec}{2}\phi(0)\right] / \left[1 - \frac{Ec}{2}\phi(0)\right] = 1$$

In a similar manner we can use the boundary condition at $\eta = \infty, \theta(\infty) = 0$ in Eq. (2.101)

$$\left[1 - \frac{Ec}{2}\phi(0)\right]\tilde{\theta}(\infty) + \frac{Ec}{2}\phi(\infty) = 0$$

$$\tilde{\theta}(\infty) = \frac{-\frac{Ec}{2}\phi(\infty)}{\left[1 - \frac{Ec}{2}\phi(0)\right]} = \frac{-\frac{Ec}{2}\frac{\theta(\infty)}{(Ec/2)}}{\left[1 - \frac{Ec}{2}\phi(0)\right]} = 0$$

The general equation now becomes

$$\left[1 - \frac{Ec}{2}\phi(0)\right]\left[\frac{\tilde{\theta}''}{Pr} + \frac{1}{2}f\tilde{\theta}'\right] + \frac{Ec}{2}\left[2(f'')^2 + \frac{\phi''}{Pr} + \frac{f}{2}\phi'\right] = 0 \qquad (2.104)$$

we know

$$\frac{\phi''}{Pr} + \frac{f}{2}\phi' + 2(f'')^2 = 0 \qquad (2.96, \quad \text{rewritten})$$

The available conditions are: adiabatic condition $\phi'(0) = 0$ and $\phi(\infty) = 0$. The Eq. (2.96) with the boundary conditions can be solved by the shooting technique.

After simplification and reduction of Eq. (2.104) we get

$$\frac{\tilde{\theta}''}{Pr} + \frac{1}{2}f\tilde{\theta}' = 0$$

Now we try for the general solution of the homogeneous part.

We have seen, $\theta(0) = 1$ means $\tilde{\theta}(0) = 1$; and $\theta(\infty) = 0$ means $\tilde{\theta}(\infty) = 0$ [since $\phi(\infty) = 0$].

The solution of this equation that results finally is of the form

$$\tilde{\theta}(\eta) = 1 - \frac{\int_0^\eta exp\left(-\frac{Pr}{2}\int_0^\eta f d\eta\right) d\eta}{\int_0^\infty exp\left(-\frac{Pr}{2}\int_0^\eta f(\eta)d\eta\right) d\eta} \qquad (2.105)$$

Now the heat transfer for the wall is

$$q_w'' = -k\frac{\partial T}{\partial y} = -k(T_w - T_\infty)\frac{\partial \theta}{\partial y}\bigg|_{y=0} = -k(T_w - T_\infty)\frac{\partial \theta}{\partial \eta}\frac{\partial \eta}{\partial y}\bigg|_{\eta=0}$$

or

$$q_w'' = -k(T_w - T_\infty)\sqrt{\frac{U}{\nu x}}\theta'(0) \qquad (2.106)$$

Following the adiabatic condition, $\phi'(0) = 0$, we get $\theta'(0) = \left[1 - \frac{Ec}{2}\phi(0)\right]\tilde{\theta}'(0)$. Substituting $\theta'(0)$ and Ec, we get

$$q_w'' = -k(T_w - T_\infty)\sqrt{\frac{U}{\nu x}}\left[1 - \frac{U^2}{2c_p(T_w - T_\infty)}\phi(0)\right]\tilde{\theta}'(0)$$

After rearranging and replacing appropriately using value of T_{aw} from before, we get

$$q_w'' = -k(T_w - T_\infty)\sqrt{\frac{U}{\nu x}}\left(\frac{T_w - T_{aw}}{T_w - T_\infty}\right)\tilde{\theta}'(0)$$

or

$$\frac{q_w''}{(T_w - T_{aw})}\frac{x}{k} = -\tilde{\theta}'(0)\sqrt{Re_x} = Nu_x \qquad (2.107)$$

This expression is the same as for the case without viscous dissipation, but the only difference is that here we have $(T_w - T_{aw})$ in place of $(T_w - T_\infty)$. As a result, there will be occurrence of heat transfer reversal. Further insight into the problem of interest is available in Eckert [5].

2.7 MORE ABOUT SIMILARITY SOLUTION (VELOCITY BOUNDARY LAYER)

A systematic study of flow past a wedge shaped body (Fig. 2.9) suggests

$$\eta = Ayx^a \quad \text{and} \quad \frac{u}{U} = f'(\eta) \qquad (2.108)$$

ψ is obtained by integration as $\psi = \frac{1}{A}x^{-a}[Uf(\eta)]$. The momentum equation for a flat plate boundary layer reduces to

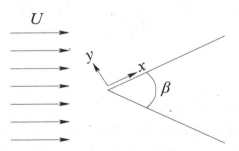

Figure 2.9 Flow past a wedge

$$\frac{\nu A^2}{U}x^{2a}f''' - \frac{a}{x}ff'' = 0 \qquad (2.109)$$

Choosing $a = -1/2$ cancels x throughout the equation and choosing $A = \sqrt{U/\nu}$ keeps the equation free of the flow and fluid parameters.

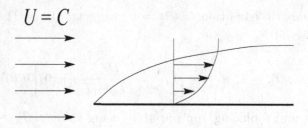

$$U = C$$

Figure 2.10 Flow past a flat plate for $m = 0$

The similarity solution can be obtained for a wider class of problems where $U = Cx^m$. This form of U represents flow past a wedge-shaped surface as shown earlier (Fig. 2.9). The relationship between m and β is $m = \frac{\beta/\pi}{2-\beta/\pi}$. For $\beta = 0$ and $m = 0$ the flat plate problem is recovered (Fig. 2.10). For $\beta = \pi, m = 1$ and this is stagnation point flow (Fig. 2.11). The pressure field is $p(x) = \beta - \frac{1}{2}\rho U^2(x)$, hence

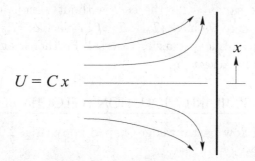

$$U = Cx$$

Figure 2.11 Stagnation point flow for $m = 1$

the pressure gradient is

$$\frac{dp}{dx} = -\rho U \frac{dU}{dx} = -\rho U C m x^{m-1} = -\frac{\rho U^2 m}{x} \tag{2.110}$$

and the x momentum equation becomes

$$u\frac{\partial u}{\partial x} + v\frac{\partial u}{\partial y} = \frac{mU^2}{x} + \nu\frac{\partial^2 u}{\partial y^2}, \quad U = Cx^m$$

Note that x and y no longer refer to Cartesian coordinates. Instead they are boundary layer coordinates, parallel and perpendicular to the solid surface. The similarly variable $\eta = y\sqrt{U/\nu x}$ and the stream function $\psi = \sqrt{\nu x U}f(\eta)$ reduce the momentum equation to the two point boundary value problem

$$f''' + \frac{(m+1)}{2}ff'' + m[1 - f'^2] = 0 \tag{2.111}$$

with $f(0) = f'(0) = 0$, $f'(\infty) = 1$.

The equation for f with m as a wedge parameter is called the Falkner-Skan equation. Similarity solutions can be obtained for a wide class of problems where

$U = Cx^m$. This form of U represents flow past wedge shaped surfaces. The relationship between β and m is given by

$$m = \frac{\beta}{2\pi - \beta} = \text{ wedge parameter}$$

$$m(2\pi - \beta) = \beta$$

where $m < 0$ signifies adverse pressure gradient. Angle β is negative and U is not a constant for wedge flow (Fig. 2.12).

2.7.1   Derivation of Falkner-Skan Equation

We have $U = C\ x^m$, $\frac{dU}{dx} = C\ m\ x^{m-1}$. Therefore $\frac{x}{U}\frac{dU}{dx} = \frac{x}{U}\ C\ m\ x^{m-1} = \frac{x}{C\ x^m}C\ m\ x^{m-1} = m$.

Let us consider a wedge flow with

$$\eta = yh(x), \quad \psi(x,\eta) = g(x)f(\eta) \tag{2.112}$$

$$u\frac{\partial u}{\partial x} + v\frac{\partial u}{\partial y} = -\frac{1}{\rho}\frac{\partial p}{\partial x} + \nu\frac{\partial^2 u}{\partial y^2}$$

Here $-\frac{1}{\rho}\frac{\partial p}{\partial x}$ becomes $U\frac{dU}{dx}$ for wedge flows. We know that

$$u = \left.\frac{\partial \psi}{\partial y}\right|_x, \quad v = -\left.\frac{\partial \psi}{\partial x}\right|_y$$

Again, if $(x, y) = \text{fn } (\chi, \eta)$, the transformation is

$$\frac{\partial}{\partial x} = \frac{\partial}{\partial \chi} \cdot \frac{\partial \chi}{\partial x} + \frac{\partial}{\partial \eta} \cdot \frac{\partial \eta}{\partial x}$$

$$\frac{\partial}{\partial y} = \frac{\partial}{\partial \chi} \cdot \frac{\partial \chi}{\partial y} + \frac{\partial}{\partial \eta} \cdot \frac{\partial \eta}{\partial y}$$

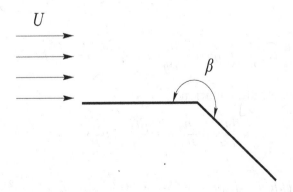

Figure 2.12   Flow over a wedge that may cause flow separation

If $(x, y) = \text{fn}\,(x, \eta)$, the von Mises transformation is

$$\left.\frac{\partial}{\partial x}\right|_y = \left.\frac{\partial}{\partial x}\right|_\eta \left(\frac{\partial x}{\partial x}\right)_y + \left.\frac{\partial}{\partial \eta}\right|_x \left(\frac{\partial \eta}{\partial x}\right)_y = \left.\frac{\partial}{\partial x}\right|_\eta + \left.\frac{\partial}{\partial \eta}\right|_x \left(\frac{\partial \eta}{\partial x}\right)_y \qquad (2.113)$$

and,

$$\left.\frac{\partial}{\partial y}\right|_x = \left.\frac{\partial}{\partial \eta}\right|_x \left(\frac{\partial \eta}{\partial y}\right)_x + \left.\frac{\partial}{\partial x}\right|_\eta \left.\frac{\partial x}{\partial y}\right|_x = \left.\frac{\partial}{\partial \eta}\right|_x \left.\frac{\partial \eta}{\partial y}\right|_x \qquad (2.114)$$

Substituting, we get

$$u = \left.\frac{\partial \psi}{\partial y}\right|_x = \left.\frac{\partial}{\partial \eta}\,[g(x)\,f(\eta)]\right|_x \frac{\partial [y\,h(x)]}{\partial y} = f'\,g\,h$$

$$v = -\left.\frac{\partial \psi}{\partial x}\right|_y = -\left[\left.\frac{\partial}{\partial \eta}\,[g(x)\,f(\eta)]\right|_x \left.\frac{\partial \eta}{\partial x}\right|_y + \left.\frac{\partial (g\,f)}{\partial x}\right|_\eta\right] = -[g\,f'y\,h' + f\,g']$$

Choose $g(x)h(x) = U(x)$

$$\frac{\partial u}{\partial x} = \left.\frac{\partial u}{\partial x}\right|_\eta + \left.\frac{\partial u}{\partial \eta}\right|_x \left.\frac{\partial \eta}{\partial x}\right|_y$$

$$= \left.\frac{\partial}{\partial x}[f'\,gh]\right|_\eta + \left.\frac{\partial}{\partial \eta}[f'\,gh]\right|_x\,yh'$$

$$= f'\,(gh' + hg') + f''\,ghyh'$$

$$\frac{\partial u}{\partial y} = \left.\frac{\partial u}{\partial \eta}\right|_x \left.\frac{\partial \eta}{\partial y}\right|_x$$

$$= \left.\frac{\partial}{\partial \eta}(f'\,gh)\right|_x h = f''\,gh^2$$

$$\frac{\partial^2 u}{\partial y^2} = \left.\frac{\partial}{\partial \eta}\,[f''\,gh^2]\right|_x \left.\frac{\partial \eta}{\partial y}\right|_x = h\,f'''\,gh^2 = gf'''\,h^3$$

Consider the x-momentum equation

$$u\frac{\partial u}{\partial x} + v\frac{\partial u}{\partial y} = U\frac{dU}{dx} + \nu\frac{\partial^2 u}{\partial y^2}$$

$$f'\,gh(f''\,ghh'y + f'\,gh' + hf'\,g') - [fg' + gf'yh'](f''\,gh^2) =$$
$$gh(g'h + gh') + \nu gf'''\,h^3 \qquad (2.115)$$

After simplification, we get

$$f''' + \frac{[g'h + gh']}{\nu h^2}(1 - f'^2) + \frac{g'}{\nu h}ff'' = 0 \tag{2.116}$$

As f is a function of η alone, both coefficients should be constants

$$\frac{gh' + hg'}{\nu h^2} = C_1 \quad \text{and} \quad \frac{g'}{\nu h} = C_2$$

$$\frac{1}{\nu h^2}\frac{dU}{dx} = C_1$$

or

$$\frac{U}{x}\frac{1}{\nu h^2}\left(\frac{x}{U}\frac{dU}{dx}\right) = C_1$$

or

$$\frac{U}{x\nu h^2}m = C_1$$

or

$$h = \sqrt{\frac{U}{\nu x}\frac{m}{C_1}}$$

Now choosing $C_1 = m$, the expression for h reduces to that of flat plate, that is

$$h = \sqrt{\frac{U}{\nu x}}$$

Again,

$$\frac{g'}{\nu h} = C_2$$

or

$$\frac{dg}{dx} = C_2\,\nu\sqrt{\frac{U}{\nu x}} = C_2\sqrt{\frac{\nu U}{x}} = C_2\sqrt{\nu\,C}\,x^{m-1}$$

or

$$\frac{dg}{dx} = \sqrt{C_2^2\,C}\,\sqrt{\nu}\,(x)^{\frac{m-1}{2}}$$

After integrating, we get

$$g = \frac{C_2}{(m+1)/2}\sqrt{\nu U x}$$

Choosing $C_2 = \frac{m+1}{2}$, we obtain $g = \sqrt{\nu U x}$. The result is compatible, because $\psi = g(x)f(\eta) = \sqrt{\nu U x} f(\eta)$. The final resulting equation becomes

$$f''' + C_1(1 - f'^2) + C_2 f f'' = 0$$

or

$$f''' + m(1 - f'^2) + \frac{m+1}{2} f f'' = 0 \tag{2.117}$$

This is called the Falkner-Skan equation and it can be solved as three initial value problems for which the boundary conditions are as follows:

$$u(0) = 0 \quad \text{or} \quad f'(0) = 0$$
$$u(\infty) = U \quad \text{or} \quad f'(\infty) = 1$$
$$v(0) = 0 \quad \text{or} \quad f(0) = 0 \tag{2.118}$$

Equation (2.117) can be solved using the shooting technique. The solutions for different values of m have been shown in Fig. 2.13.

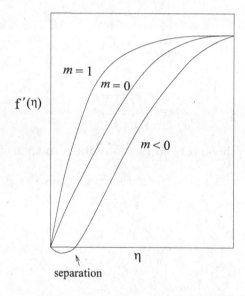

Figure 2.13 Variation of velocity profile for different values of m

2.8 MORE ABOUT SIMILARITY SOLUTION OF ENERGY EQUATION

The two-dimensional energy equation is given as

$$u \frac{\partial T}{\partial x} + v \frac{\partial T}{\partial y} = \frac{k}{\rho c_p} \frac{\partial^2 T}{\partial y^2} + \frac{\mu}{\rho c_p} \left(\frac{\partial u}{\partial y} \right)^2 \tag{2.119}$$

For similarity solution of this equation:

$$\theta = \frac{T - T_\infty}{T_w - T_\infty} = \theta(\eta)$$

We know that $\quad \eta = y\sqrt{\frac{U}{\nu x}}; \quad \psi = \sqrt{\nu U x}f(\eta); \quad \frac{u}{U} = f'(\eta),$ or, $u = f'gh.$
Also,

$$
\begin{aligned}
v &= -\left.\frac{\partial \psi}{\partial x}\right|_y = -\left[\left.\frac{\partial \psi}{\partial x}\right|_\eta + \frac{\partial \psi}{\partial \eta}\frac{\partial \eta}{\partial x}\right] \\
&= -\left[\sqrt{\nu x}\,\frac{1}{2}U^{-1/2}\frac{dU}{dx}f + \sqrt{\nu U}\,\frac{1}{2}x^{-1/2}f + \sqrt{\nu U x}\,f'\frac{\partial \eta}{\partial x}\right] \\
&= -\left[\sqrt{\frac{\nu U}{x}}\frac{f}{2} + \frac{f}{2}\sqrt{\frac{\nu x}{U}}\frac{dU}{dx} + \sqrt{\nu U x}\,f'\frac{d\eta}{dx}\right] \\
&= -\frac{1}{2}\sqrt{\frac{\nu U}{x}}(f - \eta f') - \frac{f}{2}\sqrt{\frac{\nu x}{U}}\frac{dU}{dx}
\end{aligned}
$$

The energy equation becomes

$$
\begin{aligned}
Uf'&\left[\theta\frac{\partial T_w}{\partial x} + (T_w - T_\infty)\frac{\partial \theta}{\partial \eta}\left(-\frac{y}{2x}\right)\sqrt{\frac{U}{\nu x}}\right] \\
&+ \left\{-\frac{1}{2}\sqrt{\frac{\nu U}{x}}(f - \eta f') - \frac{f}{2}\sqrt{\frac{\nu x}{U}}\frac{dU}{dx}\right\}\left\{(T_w - T_\infty)\frac{\partial \theta}{\partial \eta}\bigg/\sqrt{\frac{U}{\nu x}}\right\} \\
&= \frac{k}{\rho c_p}(T_w - T_\infty)\frac{U}{\nu x}\theta'' + \frac{\mu}{\rho c_p}\frac{U^3}{\nu x}(f'')^2
\end{aligned}
$$

or

$$
\frac{k}{\rho c_p}(T_w - T_\infty)\frac{U}{\nu x}\theta'' + \theta'\frac{f}{2}(T_w - T_\infty)\left[\frac{U}{x} + \frac{dU}{dx}\right] - Uf'\frac{dT_w}{dx}\theta + \frac{U^3}{c_p x}(f'')^2 = 0
$$

Finally it reduces to the form

$$
\frac{\theta''}{Pr} + \frac{f}{2}(1 + m)\theta' - xf'\theta\frac{(dT_w/dx)}{(T_w - T_\infty)} + Ec\,(f'')^2 = 0 \tag{2.120}
$$

It can be solved by the method of separation of variables, and so

$$
\frac{1}{f'\theta}\left\{\frac{\theta''}{Pr} + \frac{f}{2}(1 + m)\theta' + Ec(f'')^2\right\} = \frac{x(dT_w/dx)}{(T_w - T_\infty)} = \lambda \tag{2.121}
$$

the equation

$$
\frac{x(dT_w/dx)}{(T_w - T_\infty)} = \lambda \tag{2.122}
$$

defines the wall temperature condition for different values of λ.

The equation

$$\frac{\theta''}{Pr} + \frac{f}{2}(1+m)\,\theta' + Ec\,(f'')^2 - \lambda f'\theta = 0 \tag{2.123}$$

is required to be solved using the shooting technique, for different values of Pr, Ec, m and λ.

For $m = 0$, we have flat plate problem; $\lambda = 0$ leads to constant wall temperature case and $Ec = 0$ leads to the case without viscous dissipation. For $m = 0, \lambda = 0$ and $Ec = 0$, the simplified version of equation becomes

$$\frac{\theta''}{Pr} + \frac{1}{2}f\theta' = 0 \tag{2.124}$$

with $\theta(0) = 1$ and $\theta(\infty) = 0$. For this special case, $\lambda = 0$ and $T_w = $ constant, Eq 2.124 can be solved by the shooting technique.

For $Pr = 1$ the system becomes $\theta'' + \frac{1}{2}f\theta' = 0$ and $f''' + \frac{1}{2}ff'' = 0$. The boundary conditions are $f'(0) = 0$, $f'(\infty) = 1$ and $f(0) = 0$. We can solve the energy equation using $f''' + \frac{1}{2}ff'' = 0$. Let us say $\tilde{B} = f'$, then $\tilde{B}'' + \frac{1}{2}f\tilde{B}' = 0$ with the boundary conditions, $\tilde{B}(0) = 0$ and $\tilde{B}(\infty) = 1$. Now the energy equation is

$$\theta'' + \frac{1}{2}f\theta' = 0 \quad \text{with} \quad \theta(0) = 1, \theta(\infty) = 0 \tag{2.125}$$

Substituting $B = (1 - \theta)$, the equation becomes

$$-B'' - \frac{1}{2}fB' = 0 \quad \text{with} \quad B(0) = 0, B(\infty) = 1$$

which gives $\tilde{B} = B$ or, $\theta = (1 - B)$. The solution is plotted in Fig. 2.14.

2.9 APPROXIMATE METHOD FOR BOUNDARY LAYER FLOWS OVER NON-ZERO PRESSURE GRADIENT SURFACES

Consider the case of a steady, two-dimensional and incompressible flow over a curved surface. Upon integrating the dimensional form of boundary layer equation from $y = 0$ (wall) to $y = \delta$ (where δ signifies the interface of the free stream and the boundary layer), we obtain

$$\int_0^\delta \left(u\frac{\partial u}{\partial x} + v\frac{\partial u}{\partial y} \right) dy = \int_0^\delta \left(-\frac{1}{\rho}\frac{dp}{dx} + \nu\frac{\partial^2 u}{\partial y^2} \right) dy \tag{2.126}$$

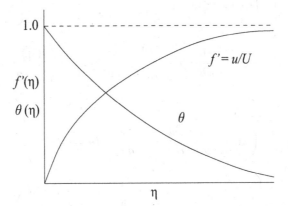

Figure 2.14 Variation of velocity profile and temperature profile with η

or

$$\int_0^\delta u\frac{\partial u}{\partial x}dy + \int_0^\delta v\frac{\partial u}{\partial y}dy = \int_0^\delta -\frac{1}{\rho}\frac{dp}{dx}dy + \int_0^\delta \nu\frac{\partial^2 u}{\partial y^2}dy \qquad (2.127)$$

The second term of the left-hand side can be expanded as

$$\int_0^\delta v\frac{\partial u}{\partial y}dy = [vu]_0^\delta - \int_0^\delta u\frac{\partial v}{\partial y}dy \qquad (2.128)$$

We discuss, in this section, a method that can be applied to a boundary layer flow with pressure gradient. The method was first introduced by Pohlhausen [6]. Further development was done by Holstein and Bohlen [7].

$$\int_0^\delta v\frac{\partial u}{\partial y}dy = U\,v_\delta + \int_0^\delta u\frac{\partial u}{\partial x}dy \quad \left(\text{since}\frac{\partial u}{\partial x} = -\frac{\partial v}{\partial y}\right) \qquad (2.129)$$

or

$$\int_0^\delta v\frac{\partial u}{\partial y}dy = -U\int_0^\delta \frac{\partial u}{\partial x}dy + \int_0^\delta u\frac{\partial u}{\partial x}dy \qquad (2.130)$$

Thus, the equation can be re-constituted as

$$\int_0^\delta 2u\frac{\partial u}{\partial x}dy - U\int_0^\delta \frac{\partial u}{\partial x}dy = -\int_0^\delta \frac{1}{\rho}\frac{dp}{dx}dy - \nu\frac{\partial u}{\partial y}\Big|_{y=0} \qquad (2.131)$$

Substituting the relation between dp/dx and the free stream velocity U for the inviscid zone in Eq. (2.131) we get

$$\int_0^\delta 2u\frac{\partial u}{\partial x}dy - U\int_0^\delta \frac{\partial u}{\partial x}dy - \int_0^\delta U\frac{dU}{dx}dy = -\left(\frac{\mu\frac{\partial u}{\partial y}|_{y=0}}{\rho}\right) \tag{2.132}$$

or

$$\int_0^\delta \left(2u\frac{\partial u}{\partial x} - U\frac{\partial u}{\partial x} - U\frac{dU}{dx}\right)dy = -\frac{\tau_w}{\rho} \tag{2.133}$$

which is reduced to

$$\int_0^\delta \frac{\partial}{\partial x}[u(U-u)]dy + \frac{dU}{dx}\int_0^\delta (U-u)dy = \frac{\tau_w}{\rho} \tag{2.134}$$

Since the integrals vanish outside the boundary layer, we are allowed to put $\delta = \infty$.

$$\int_0^\infty \frac{\partial}{\partial x}[u(U-u)]dy + \frac{dU}{dx}\int_0^\infty (U-u)dy = \frac{\tau_w}{\rho} \tag{2.135}$$

or

$$\frac{d}{dx}\int_0^\infty [u(U-u)]dy + \frac{dU}{dx}\int_0^\infty (U-u)dy = \frac{\tau_w}{\rho} \tag{2.136}$$

Substituting the definitions of displacement thickness and momentum thickness, we obtain

$$\frac{d}{dx}\left(U^2\delta^{**}\right) + \delta^*U\frac{dU}{dx} = \frac{\tau_w}{\rho} \tag{2.137}$$

Equation (2.137) is known as the momentum integral equation for two-dimensional incompressible laminar boundary layer over a non-zero pressure gradient surface. The same remains valid for turbulent boundary layers as well. Needless to say, the wall shear stress (τ_w) will be different for laminar and turbulent flows. The term $\left(U\frac{dU}{dx}\right)$ signifies spacewise acceleration of the free stream. It indicates the presence of means of free stream pressure gradient in the flow direction.

Equation (2.137) can be further arranged as

$$U^2\frac{d\delta^{**}}{dx} + (2\delta^{**} + \delta^*)U\frac{dU}{dx} = \frac{\tau_w}{\rho} \tag{2.138}$$

The velocity profile of the boundary layer is considered to be a fourth-order polynomial in terms of the dimensionless distance $\eta = y/\delta$, and is expressed as

$$u/U = \alpha\eta + b\eta^2 + c\eta^3 + d\eta^4 \tag{2.139}$$

The boundary conditions are

at $\eta = 0$: $u = 0, v = 0$ and $\frac{\nu}{\delta^2}\frac{\partial^2 u}{\partial\eta^2} = \frac{1}{\rho}\frac{dp}{dx} = -U\frac{dU}{dx}$

at $\eta = 1$: $u = U,$ and $\frac{\partial u}{\partial\eta} = 0, \frac{\partial^2 u}{\partial\eta^2} = 0$

A dimensionless quantity, known as shape factor, is introduced as

$$\Lambda = \frac{\delta^2}{\nu}\frac{dU}{dx} \qquad (2.140)$$

The following relations are obtained

$$a = 2 + \frac{\Lambda}{6}, \quad b = -\frac{\Lambda}{2}, \quad c = -2 + \frac{\Lambda}{2}, \quad d = 1 - \frac{\Lambda}{6} \qquad (2.141)$$

Now, the velocity profile can be expressed as

$$u/U = F(\eta) + \Lambda G(\eta), \qquad (2.142)$$

where

$$F(\eta) = 2\eta - 2\eta^3 + \eta^4, \quad G(\eta) = \frac{1}{6}\eta(1-\eta)^3 \qquad (2.143)$$

The shear stress $\tau_w = \mu(\partial u/\partial y)_{y=0}$ is given by

$$\frac{\tau_w \delta}{\mu U} = 2 + \frac{\Lambda}{6} \qquad (2.144)$$

We use the following dimensionless parameters, namely

$$\text{shear correlation,} \qquad L = \frac{\tau_w \delta^{**}}{\mu U} \qquad (2.145)$$

$$\text{pressure gradient correlation,} \quad K = \frac{(\delta^{**})^2}{\nu}\frac{dU}{dx} \qquad (2.146)$$

and

$$\text{shape factor correlation,} \quad H = \delta^*/\delta^{**} \qquad (2.147)$$

The integrated momentum Eq. (2.137) reduces to

$$U^2\frac{d\delta^{**}}{dx} + (2\delta^{**} + \delta^*)U\frac{dU}{dx} = \frac{\tau_w}{\rho}$$

or

$$U\frac{d\delta^{**}}{dx} + \delta^{**}\left(2 + \frac{\delta^*}{\delta^{**}}\right)\frac{dU}{dx} = \frac{\tau_w}{\rho U}$$

The term $\tau_w/\rho U$ can be written as $\tau_w/\rho U = [\tau_w\delta^{**}/\mu U][\nu/\delta^{**}] = \nu L/\delta^{**}$.
From the above two expressions, one can write

$$U\frac{d\delta^{**}}{dx} + \delta^{**}(2 + H)\frac{dU}{dx} = \frac{\nu L}{\delta^{**}} \qquad (2.148)$$

or

$$\delta^{**} \frac{U}{\nu} \frac{d\delta^{**}}{dx} + \frac{(\delta^{**})^2}{\nu} \frac{dU}{dx}(H + 2) = L$$

or

$$2\delta^{**} \frac{U}{\nu} \frac{d\delta^{**}}{dx} = 2[L - K(H + 2)]$$

or

$$U \frac{d}{dx} \left[\frac{(\delta^{**})^2}{\nu} \right] = 2[L - K(H + 2)] \tag{2.149}$$

The parameter L is related to the skin friction and K is linked to the pressure gradient. If we take K as the independent variable, L and H can be shown to be the functions of K since

$$U\delta^* = \int_0^\delta (U - u)dy$$

or

$$U\delta^* = \delta \int_0^1 (U - u)d\eta$$

or

$$\frac{\delta^*}{\delta} = \int_0^1 \left(1 - \frac{u}{U}\right) d\eta$$

$$\frac{\delta^*}{\delta} = \int_0^1 [1 - F(\eta) - \Lambda G(\eta)] \; d\eta = \frac{3}{10} - \frac{\Lambda}{120} \tag{2.150}$$

$$\frac{\delta^{**}}{\delta} = \int_0^1 (F(\eta) + \Lambda G(\eta))(1 - F(\eta) - \Lambda G(\eta)) \; d\eta \tag{2.151}$$

or

$$\frac{\delta^{**}}{\delta} = \frac{37}{315} - \frac{\Lambda}{945} - \frac{\Lambda^2}{9072} \tag{2.152}$$

$$K = \frac{[\delta^{**}]^2}{\delta^2} \Lambda = \Lambda \left(\frac{37}{315} - \frac{\Lambda}{945} - \frac{\Lambda^2}{9072} \right)^2 \tag{2.153}$$

Therefore,

$$L = \left(2 + \frac{\Lambda}{6}\right) \frac{\delta^{**}}{\delta} = \left(2 + \frac{\Lambda}{6}\right) \left(\frac{37}{315} - \frac{\Lambda}{945} - \frac{\Lambda^2}{9072} \right) = f_1(K) \tag{2.154}$$

$$H = \frac{\delta^*}{\delta^{**}} = \frac{(3/10) - (\Lambda/120)}{(37/315) - (\Lambda/945) - (\Lambda^2/9072)} = f_2(K) \qquad (2.155)$$

The right-hand side of Eq. (2.149) is thus a function of K alone. Walz [8] pointed out that this function can be approximated with a good degree of accuracy by a linear function of K so that

$$2[L - K(H + 2)] = a - bK$$

Equation (2.149) can now be written as

$$\frac{d}{dx}\left[\frac{U(\delta^{**})^2}{\nu}\right] = a - \frac{bU(\delta^{**})^2}{\nu}\frac{1}{U}\frac{dU}{dx} + \frac{U(\delta^{**})^2}{\nu}\frac{1}{U}\frac{dU}{dx} \qquad (2.156)$$

or

$$\frac{d}{dx}\left(\frac{U(\delta^{**})^2}{\nu}\right) = a - (b-1)\frac{U[\delta^{**}]^2}{\nu}\frac{1}{U}\frac{dU}{dx} \qquad (2.157)$$

Solution of this differential equation for the dependent variable $(U[\delta^{**}]^2/\nu)$ subject to the boundary condition $U = 0$ when $x = 0$, gives

$$\frac{U(\delta^{**})^2}{\nu} = \frac{a}{U^{b-1}}\int_0^x U^{b-1}\, dx \qquad (2.158)$$

With $a = 0.47$ and $b = 6$, the approximation is particularly close between the stagnation point and the point of maximum velocity. Finally, the value of the dependent variable is

$$(\delta^{**})^2 = \frac{0.47\nu}{U^6}\int_0^x U^5\, dx \qquad (2.159)$$

at any x. If U is a complicated function of x or known only pointwise, the integral can be evaluated numerically, say by Simpson's rule. Once δ^{**} is known, other quantities can be evaluated. Alternatively by taking the limit of Eq. (2.159), it can be shown that

$$(\delta^{**})^2|_{x=0} = 0.47\nu/6U'(0) \qquad (2.160)$$

This corresponds to $K = 0.0783$. It may be mentioned that $[\delta^{**}]$ is not equal to zero at the stagnation point. If $([\delta^{**}]^2/\nu)$ is determined from Eq. (2.159), $K(x)$ can be obtained from Eq. (2.146). Table 2.3 gives the necessary parameters for obtaining results, such as velocity profile and shear stress τ_w. The approximate method can be applied successfully to a wide range of boundary layer flows over a non-zero pressure gradient surface.

As mentioned earlier, K and Λ are related to the pressure gradient and the shape factor. Introduction of K and Λ in the integral analysis enables extension of the Karman and Pohlhausen method for solving flows over curved geometry. However, the analysis is not valid for the geometries, where $\Lambda < -12$ and $\Lambda > +12$.

Table 2.3
Auxiliary functions after Holstein and Bohlen [7]

Λ	K	$f_1(K)$	$f_2(K)$
12	0.0948	2.250	0.356
10	0.0919	2.260	0.351
8	0.0831	2.289	0.340
7.6	0.0807	2.297	0.337
7.2	0.0781	2.305	0.333
7.0	0.0767	2.309	0.331
6.6	0.0737	2.318	0.328
6.2	0.0706	2.328	0.324
5.0	0.0599	2.361	0.310
3.0	0.0385	2.427	0.283
1.0	0.0135	2.508	0.252
0	0	2.554	0.235
-1	-0.0140	2.604	0.217
-3	-0.0429	2.716	0.179
-5	-0.0720	2.847	0.140
-7	-0.0999	2.999	0.100
-9	-0.1254	3.176	0.059
-11	-0.1474	3.383	0.019
-12	-0.1567	3.500	0

Thwaites [9] modified and improved the method due to Holstein and Bohlen [7]. He rewrote the momentum integral equation in terms of a parameter, λ, defined as

$$\lambda = \frac{(\delta^{**})^2 U'}{\nu} = \left(\frac{\delta^{**}}{\delta}\right)^2 \Lambda$$

The method due to Thwaites is not discussed in this text. The advanced readers may like to refer to the original work of Thwaites [9].

2.10 EFFECT OF PRESSURE GRADIENT ON EXTERNAL FLOWS

For the case of a boundary layer on a flat plate, the pressure gradient of the external stream is zero. Let us consider a body with curved surface (Fig. 2.15). Upstream of the highest point the streamlines of the outer flow converge, resulting in an increase of the free stream velocity $U(x)$ and a consequent fall of pressure with x. Downstream of the highest point the streamlines diverge, resulting in a decrease of $U(x)$ and a rise in pressure. In this section we shall investigate the effect of such a pressure gradient on the shape of the boundary layer profile $u(x,y)$. The boundary layer equation is

$$u\frac{\partial u}{\partial x} + v\frac{\partial u}{\partial y} = -\frac{1}{\rho}\frac{\partial p}{\partial x} + \nu\frac{\partial^2 u}{\partial y^2}$$

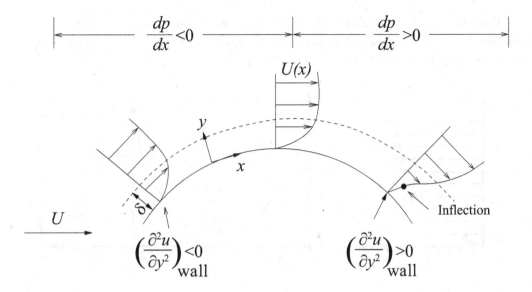

Figure 2.15 Velocity profile associated with separation in a cross flow over a circular cylinder

where the pressure gradient is found from the external velocity field as $dp/dx = -\rho U(dU/dx)$, with x taken along the surface of the body. At the wall, the boundary layer equation becomes

$$\mu\left(\frac{\partial^2 u}{\partial y^2}\right)_w = \frac{\partial p}{\partial x} \tag{2.161}$$

In an accelerating stream $dp/dx < 0$ (see Fig. 2.15),

$$\left(\frac{\partial^2 u}{\partial y^2}\right)_{wall} < 0 \qquad \text{(accelerating)} \tag{2.162}$$

As the velocity profile has to merge smoothly with the external flow, the slope $\partial u/\partial y$ slightly below the edge of the boundary layer decreases with y from a positive value to zero. Therefore, $\partial^2 u/\partial y^2$ slightly below the boundary layer edge is negative. Eq. (2.162) then shows that $\partial^2 u/\partial y^2$ has the same sign at both the wall and the boundary layer edge, and presumably throughout the boundary layer (Fig. 2.16). In contrast, for a decelerating external stream, the curvature of the velocity profile at the wall is

$$\left(\frac{\partial^2 u}{\partial y^2}\right)_{wall} > 0 \qquad \text{(decelerating)} \tag{2.163}$$

so that the curvature changes sign somewhere within the boundary layer. In other words, the boundary layer profile in a decelerating flow ($dp/dx > 0$) has a point of inflection where $\partial^2 u/\partial y^2 = 0$ (Fig. 2.17).

The shape of the velocity profiles in the figures suggests that *an adverse pressure gradient tends to increase the thickness of the boundary layer*. This can also be seen from the continuity equation.

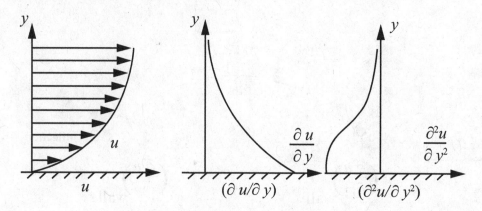

Figure 2.16   Profiles for $dp/dx < 0$ (favorable pressure gradient)

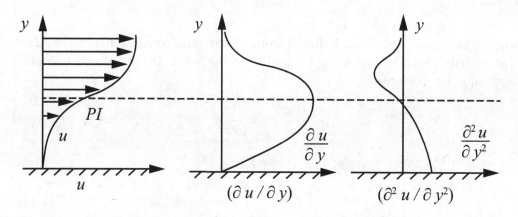

Figure 2.17   Profiles for $dp/dx > 0$ (adverse pressure gradient)

$$v(y) = -\int_0^y \frac{\partial u}{\partial x}\, dy \qquad (2.164)$$

Compared to a flat plate, a decelerating external stream causes a larger $-\partial u/\partial x$ within the boundary layer because the deceleration of the outer flow adds to the viscous deceleration within the boundary layer. From the above equation we observe that the v-field, directed away from the surface, is larger for a decelerating flow. The boundary layer therefore thickens not only by viscous diffusion but also by advection away from the surface, resulting in a rapid increase in the boundary layer thickness with x. If pressure falls along the direction of flow, $dp/dx < 0$ and we say that the pressure gradient is favorable. If, on the other hand, the pressure rises along the directions of flow, $dp/dx > 0$ and we say that the pressure gradient is adverse. The rapid growth of the boundary layer thickness in a decelerating stream, and the associated large v-field, causes the important phenomena of separation, in which the external stream ceases to flow nearly parallel to the boundary surface.

2.11   DESCRIPTION OF FLOW PAST CIRCULAR CYLINDER

Let us start with a consideration of the creeping flow around a circular cylinder, characterized by $Re < 1$. (Here we shall define $Re = U_\infty d/\nu$, based on the upstream velocity and the cylinder diameter.) Vorticity is generated close to the surface because of the no-slip boundary condition. In the Stokes approximation this vorticity is simply diffused, not advected, which results in a fore and aft symmetry.

As Re is increased beyond 1, the vorticity is increasingly confined behind the cylinder because of advection. For $Re > 4$ two small attached or standing eddies appear behind the cylinder. The wake is completely laminar and the vortices act like rollers over which the main stream flows (Fig. 2.18). The eddies get larger as Re in increased. A very interesting sequence of events begins to develop when

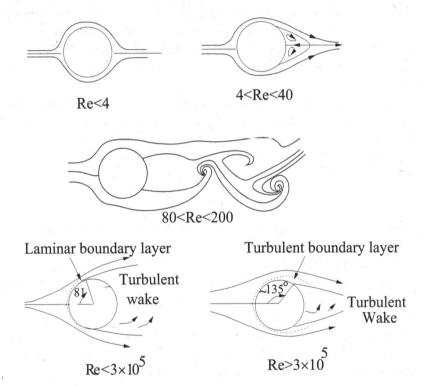

Figure 2.18   Influence of Reynolds number on the wake of a circular cylinder

the Reynolds number is increased beyond 40, at which point the wake behind the cylinder becomes unstable. The eddies start to oscillate in time and asymmetry is brought about in the wake structure. The wake develops a slow oscillation in which the velocity is periodic in time, with the amplitude of the oscillation increasing downstream. The oscillating wake starts shedding two staggered rows of vortices with opposite sense of rotation, in the stream. Karman investigated the phenomena and concluded that a nonstaggered row of vortices is unstable, and a staggered row is stable only if the ratio of lateral distance between the vortices to their longitudinal distance is 0.28. The staggered row of vortices behind a blunt body is called a von Karman vortex street. The structure is amazingly stable and it is possible to see a large number of vortices downstream of the body. The vortices move downstream at a speed smaller than the upstream velocity U. At this stage periodicity is induced in the entire flow field due to the vortex shedding phenomena. During the vortex shedding process, when an eddy on one side is shed, another forms on the other side, resulting in an unsteady flow near the cylinder. As vortices of opposite circulations are shed off alternately from the two sides, the circulation around the cylinder changes sign, resulting in an oscillating lift or lateral force. If the frequency of vortex shedding is close to the natural frequency of some mode of vibration of the cylinder body, then an appreciable lateral vibration takes place. You may recall the collapse of the Tacoma Narrows Bridge in Washington State in the U.S. A torsional mode of oscillation in the structure synchronized with the vortex shedding frequency to build up the oscillations to destructive magnitude. Engineering structures such as chimney stacks, suspension bridges and oil drilling platforms are designed to break up a coherent shedding of vortices from the cylindrical structures. This is done by using helical strakes on the cylinder surface, which break up the spanwise coherence of shedding of the structure.

The passage of regular vortices causes velocity in the wake to have a dominant periodicity. The frequency f is expressed as a nondimensional parameter known as the Strouhal number, named after V. Strouhal, a German engineer. The Strouhal number is defined as

$$S \equiv \frac{fd}{U}$$

For a circular cylinder the value of S remains close to 0.21 for a large range of Reynolds numbers. For small values of cylinder diameter and moderate values of U, the resulting frequencies of the vortex shedding and oscillating lift lie in the acoustic range. For example, at $U = 10$ m/s and a wire diameter of 2 mm, the frequency corresponding to a Strouhal number of 0.21 is $f = 1050$ cycles per second and the corresponding Reynolds number is 1200.

Strouhal made his first measurements in 1878. Roshko, in 1954, measured the frequencies using a hot-wire probe. For low Reynolds number laminar regions Roshko consolidated his results to an equation of the form

$$S = 0.212(1 - 21.2/Re) \qquad (2.165)$$

At about Reynolds number of 500, multiple frequencies start to appear and the wake tends to become turbulent.

2.12   EXPERIMENTAL RESULTS FOR CIRCULAR CYLINDER FLOW

In this section, we analyze the heat transfer results for the case of flow over a heated circular cylinder (Fig. 2.19). In order to understand heat transfer at a moderate

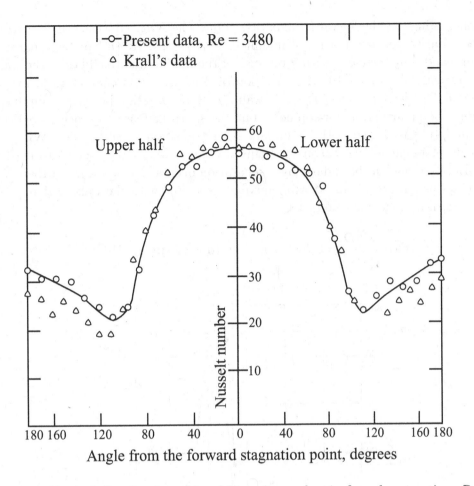

Figure 2.19 Typical distribution of local Nusselt number in forced convection, $Re = 3480$ (courtesy of Sarma and Sukhatme [10])

Reynolds number ($Re = 3480$), we take a recourse to the investigation of Sarma and Sukhatme [10]. Starting with a high value at the stagnation point, Nu_θ decreases with increasing θ as a result of laminar boundary layer development. The minimum is reached at the point of separation. The point of separation is symmetrical about the forward stagnation point and occurs at about 110°. After the separation point, Nu_θ increases with θ because of the mixing associated with the vortex formation in the wake. It may be mentioned that the values of Nu_θ are the time-averaged quantities. Sarma and Sukhatme [10] also found the correlation between the Nusselt number at the forward stagnation point and Reynolds number as

$$Nu_{fsp} = 0.91(Re)^{0.5} \tag{2.166}$$

Since working fluid was air only, the variation of Prandtl number was not considered. Next we take up the heat transfer for higher Reynolds numbers [11]. Consider the results for $Re_D < 10^5$ (Figure 2.20). Following the earlier trend, the Nusselt number is high at the forward stagnation point. The Nu_θ reduces with the growth of boundary layer and reaches a minimum at $\theta = 80°$. At this point, separation occurs, and Nu_θ increases with θ because of the mixing due to eddies in the wake. In contrast, for $Re_D > 10^5$ the variation of Nu_θ with θ is characterized by two minima. The decline in Nu_θ from the value at the stagnation point is again due to laminar boundary layer development, but the sharp increase that occurs between 80° and 100° is owing to the boundary layer transition to turbulence. With further development of the turbulent boundary layer, Nu_θ again begins to decline. Eventually turbulent boundary layer separation occurs at $\theta \approx 140°$ and then Nu_θ increases as a result of considerable mixing associated with the wake region.

Overall average condition is:

$$\overline{Nu_D} = \frac{\bar{h}D}{k} = CRe_D^m\ Pr^{1/3} \qquad \text{(due to Hilpert [12])} \qquad (2.167)$$

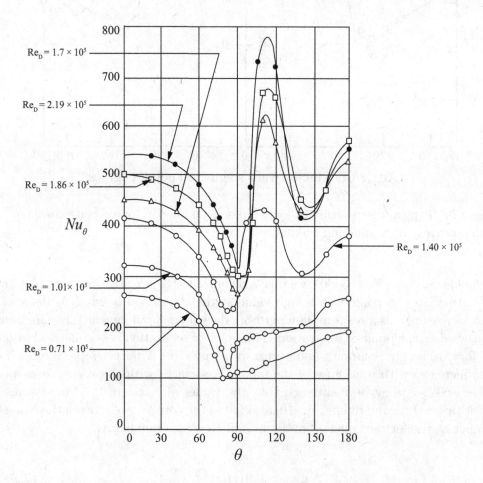

Figure 2.20 Local Nusselt number distribution for flow over a circular cylinder for different Reynolds numbers (courtesy of Giedt [11])

A more recent general correlation is:

$$\overline{Nu}_D = C Re_D^m \; Pr^n \; \left(\frac{Pr_\infty}{Pr_w}\right)^{1/4} \quad \text{(by Zhukauskas [13])} \tag{2.168}$$

The values of C and m are given in Table 2.4.

Table 2.4 [12,14]

Re_D	C	m
4 - 40	0.911	0.385
40 - 4000	0.683	0.466
4000 - 40000	0.193	0.618
40000 - 400000	0.027	0.805

- All properties are evaluated at T_∞, except Pr_w, which is evaluated at T_w. For $Pr \leq 10$, $n = 0.37$ and for $Pr > 10$, $n = 0.36$.

2.13   OTHER IMPORTANT CORRELATIONS

For small Prandtl number fluids (liquid metals) the local heat transfer coefficient on a flat plate is given by

$$Nu_x = 0.565 \, Pe_x^{1/2}, \quad \text{for} \quad Pe_x \leq 0.01 \tag{2.169}$$

From experiments (see Schlichting [3]) the local friction coefficients for turbulent flow over a flat plate are given by

$$C_{f\,x} = 0.0592 \, Re_x^{-1/5} \quad 5 \times 10^5 < Re_x < 10^7 \tag{2.170}$$

To a reasonable approximation, the velocity boundary layer thickness for turbulent flow over a flat plate may be expressed as

$$\frac{\delta}{x} = \frac{0.37}{(Re_x)^{-1/5}} \tag{2.171}$$

(Observe that δ varies as $x^{4/5}$ in contrast to $x^{1/2}$ for laminar flow.) The local Nusselt number for turbulent flow over a flat plate is:

$$Nu_x = 0.0296 \, Re_x^{4/5} \, Pr^{1/3} \tag{2.172}$$

In mixed boundary layer situations, the average heat transfer coefficient for the

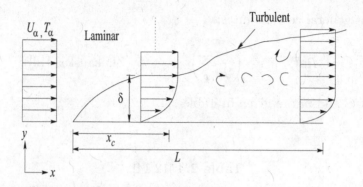

Figure 2.21 Transitional flow over a flat plate

flow over a flat plate (Fig. 2.21) is given by

$$\overline{h}_L = \frac{1}{L} \left(\int_0^{x_c} h_{lam} \ dx + \int_{x_c}^{L} h_{turb} \ dx \right) \tag{2.173}$$

or

$$\overline{h}_L = \frac{k}{L} \left[0.332 \left(\frac{U_\infty}{\nu} \right)^{1/2} \int_0^{x_c} \frac{dx}{x^{1/2}} + 0.0296 \left(\frac{U_\infty}{\nu} \right)^{4/5} \int_{x_c}^{L} \frac{dx}{x^{1/5}} \right] Pr^{1/3}$$

or

$$\overline{Nu}_L = \left[0.664 \ Re_{x,c}^{1/2} + 0.037 \left(Re_L^{4/5} - Re_{x,c}^{4/5} \right) \right] Pr^{1/3}$$

or

$$\overline{Nu}_L = \left(0.037 \ Re_L^{4/5} - A \right) Pr^{1/3} \tag{2.174}$$

If the typical, transition Reynolds number is $Re_{x,c} = 5 \times 10^5$

$$\overline{Nu}_L = \left(0.037 \ Re_L^{4/5} - 871 \right) Pr^{1/3} \tag{2.175}$$

Similarly,

$$\overline{C}_{f,L} = \frac{0.074}{Re_L^{1/5}} - \frac{A}{Re_L} \tag{2.176}$$

where $A = 1742$.

If $L \gg x_c$ $(Re_L \gg Re_{x,c})$

$$\overline{Nu}_L = 0.037 \ Re_L^{4/5} \ Pr^{1/3} \quad \text{and} \quad \overline{C}_{f,L} = 0.074 \ Re_L^{-1/5}$$

We know that the fluid properties vary with temperature across the boundary layer and that this variation can certainly influence heat transfer. This influence may be handled by one of the following two ways. In one method all the properties evaluated at a mean boundary layer temperature T_f, termed the film temperature.

$$T_f = \frac{T_w + T_\infty}{2}$$

In the alternate method, evaluate all the properties at T_∞ and multiply the right-hand side of the expression for Nusselt number by an additional parameter. This parameter is commonly of the form $(Pr_\infty/Pr_w)^r$ or $(\mu_\infty/\mu_w)^r$ where w designates *properties at surface temperature.*

REFERENCES

1. H. Blasius, Grenzschichten in Flussigkeiten mit Kleiner Reibung, Z Math Physik, Bd. 56, 1-37, 1908.

2. L. Howarth, On the solution of the laminar boundary layer equations, Proc. Roy. Soc. London A, Vol. 164, 547-579, 1938.

3. H. Schlichting, Boundary Layer Theory. 7th Edition, McGraw-Hill, New York, 1979.

4. E. Pohlhausen, Der Wärmeaustausch zwischen festen Körpen und Flüssigkeiten mit kleiner reibung und kleiner Wärmeleitung, ZAMM, Vol.1, pp 115-121, 1921.

5. E. R. G. Eckert, Survey of Boundary Layer Heat Transfer at High Velocities and High Temperature, WADC Technical Report, pp. 59-62, 1960.

6. K. Pohlhausen, Zur naherungsweisen Intergration der Differentialgleichung der laminaren Reibungsschicht. Zeitschrift für Angewandte Mathematik und Mechanik (Journal of Applied Mathematics and Mechanics) 1, 252-268, 1921.

7. H. Holstein and T. Bohlen, Ein einfaches Verfahren zur Berechnung laminarer Reibungs-schichten, die dem Naherungsverfahren von K. Pohlhausen genugen. Lilienthal-Bericht S. 10,5-16, 1940.

8. A. Walz, Ein neuer Ansatafur das Geschwindigkeitsprofil der laminaren Reibungsschicht Lilienthal-Bericht 141, 8-12, 1941.

9. B. Thwaites, Approximate calculation of the laminar boundary layer, Aeronaut. Q. Vol. 1, pp. 245-280, 1949.

10. T. S. Sarma and S. P. Sukhatme, Local heat transfer from a horizontal cylinder to air in cross flow: Influence of free convection and free stream turbulence, Int. J. Heat Mass Transfer, Vol. 20, pp 51-56, 1977.

11. W. H. Giedt, Investigation of variation of point unit heat transfer coefficient around a cylinder normal to an air stream, Transactions of ASME, Vol. 71 pp. 375-381, 1949.

12. R. Hilpert, Forsch. Geb. Ingenieurwes, Vol. 4, pp 215, 1933.

13. A. Zhukauskas, Heat Transfer from Tubes in Cross Flow, in J. P. Hartnet and T. F. Irvine Eds, Advances in Heat Transfer, Vol. 8, Academic Press, New York, 1972.

14. F. P. Incropera and D. P. Dewitt, Fundamentals of Heat and Mass Transfer, Fifth Edition, John Wiley & Sons, 2002.

EXERCISES

1. Blasius equation for flow over a flat plate must be iterated to find the correct value of $f''(0)$ which causes $f'(\infty)$ to equal 1.0. Use Runge-Kutta method in combination with any root finding method to find (u/U_∞) as a function of η. The parameter $\eta = y/\sqrt{\nu x/U_\infty}$. Having solved the velocity boundary layer equation, solve the thermal boundary layer equation

$$\theta'' + \frac{Pr}{2}f\theta' - \lambda Pr f'\theta = 0$$

   The plate is at a constant temperature T_w and the free stream temperature is T_∞. The non-dimensional temperature is given by $\theta = (T - T_\infty)/(T_w - T_\infty)$. Find θ versus η for different values of Pr. The available boundary condition are: at $\eta = 0, \theta = 1$ and at $\eta = \infty, \theta = 0$. The correct value of $\theta'(0)$ must be obtained to find the variation of θ with η. Write a computer code that is capable of solving the problem for any given value of Prandtl number of the flowing fluid.

2. For the case of flow over a flat heated plate, the thermal boundary layer equation is

$$u\frac{\partial T}{\partial x} + v\frac{\partial T}{\partial y} = \frac{k}{\rho c_p}\left(\frac{\partial^2 T}{\partial y^2}\right)$$

   with

$$u = U_\infty f'(\eta), \ v = -\frac{1}{2}\sqrt{\frac{\nu U_\infty}{x}}\left[f(\eta) - \eta f'(\eta)\right], \ \theta = \frac{T - T_\infty}{T_W - T_\infty}, \ \eta = y/\sqrt{\frac{\nu x}{U_\infty}}$$

   In this problem wall temperature is a function of x while the non-dimensional temperature field and the velocity field are the functions of similarity parameter η. On substitution of the velocity and temperatures in terms of f and θ in the governing equations, separate the groups that are functions of x and η. Show that the thermal boundary layer equation can be written as

$$\theta'' + \frac{Pr}{2}f\theta' - \lambda Pr f'\theta = 0$$

3. After applying the separation of variables technique specified in Problem 2

consider the part that is function of x. We get $(T_w - T_\infty) = Cx^\lambda$. Start with the definition of heat flux at the wall and determine the value of λ that makes the condition of constant wall heat flux on the wall. Use any relation that you have used in Problem 2.

4. The thermal boundary layer equation for constant wall temperature, using similarity variables, can be expressed as

$$\theta'' + \frac{Pr}{2} f\theta' = 0$$

The boundary conditions are $\theta(0) = 1$ and $\theta(\infty) = 0$. Find the solution as $\theta(\eta)$ using appropriate analytical technique.

5. For flow over a wedge-shaped body, the governing equations for mass, momentum and energy are given by

$$\frac{\partial u}{\partial x} + \frac{\partial v}{\partial y} = 0$$

$$u\frac{\partial u}{\partial x} + v\frac{\partial u}{\partial y} = \frac{1}{p}\frac{\partial p}{\partial x} + \nu\frac{\partial^2 \mu}{\partial y^2}$$

$$u\frac{\partial u}{\partial x} + v\frac{\partial T}{\partial y} = \alpha\frac{\partial^2 T}{\partial y^2} + \frac{\mu}{\rho c_p}\left(\frac{\partial u}{\partial y}\right)^2$$

The similarity variables are:

$$\theta(\eta) = \frac{T - T_\infty}{T_w - T_\infty} \text{ and } \eta = y/\sqrt{\frac{\nu x}{U}}$$

The solutions for the main stream velocity are given by

$$u = U f'(\eta) \text{ and } \Psi = \sqrt{\nu u x} \ f(n)$$

$$\text{With } f(\eta) = \int F(\eta)d\eta \text{ and } F(\eta) = F\left[y/\sqrt{\frac{\nu x}{U}}\right]$$

The variables x and y indicate the flow and the normal direction respectively. The temperatures $T_w = T_w(x)$ and T_∞ indicate the surface temperature and the free-stream temperature. Show that the final form of the energy equation is

$$\frac{\theta''}{Pr} + \frac{1}{2} f\theta'(1 + m) - \lambda f'\theta + Ec(f'')^2 = 0$$

The parameter m is related to the wedge angle. Also find out the values of λ for the constant wall temperature and constant wall heat flux conditions.

6. Explain the following:

(a) For a two-dimensional high speed flow over a flat plate, the viscous dissipation effect can be summarized as addition of the term $\mu \left(\frac{\partial u}{\partial y} \right)^2$ on the right-hand side of the energy equation,

$$\rho c_p \left[u \frac{\partial T}{\partial x} + v \frac{\partial T}{\partial y} \right] = k \frac{\partial^2 T}{\partial y^2}$$

(b) For the high speed flow with viscous heating over a flat plate, the wall heat flux is given by

$$q_w'' = -k(T_w - T_\infty) \sqrt{\frac{U_\infty}{\nu x}} \left[1 - \frac{Ec}{2} \phi(0) \right] \tilde{\theta}'(0)$$

where Ec is the Eckert number, k is is the thermal conductivity of the fluid. T_w and T_∞ are the wall temperature and free stream temperature respectively, U_∞ is the free stream velocity, ν is the kinematic viscosity, x is the distance from the leading edge, $\phi \sim \theta/(Ec/2)$. The nondimensional temperature θ and the same without viscous dissipation, $\tilde{\theta}$ are related as

$$\theta = \left[1 - \frac{Ec}{2} \phi(0) \right] \tilde{\theta} + \frac{1}{2} \ Ec \ \phi \ (\eta)$$

The expression for the local Nusselt number, is similar to that of the flow without viscous heating except the reference temperature for calculating the Nu_x is T_{aw} where T_{aw} is the adiabatic wall temperature.

7. Wieghardt derived the integral equation for mechanical energy for the laminar boundary layers. The starting point of this equation was to multiply the equation of motion by u and then integrate from $y = 0$ to $y = h$ (h is greater than the local boundary layer thickness). Can you obtain the equation using only these hints? As with momentum thickness calculations, you may have to define another measure of the boundary layer thickness, known as dissipation energy thickness given by

$$U^3 \delta_3 = \int_0^\infty u \left(U^2 - u^2 \right) dy$$

8. Consider the thermal boundary layer equation for flow over a flat plate as

$$\rho c_p \left(u \frac{\partial T}{\partial x} + v \frac{\partial T}{\partial y} \right) = k \left(\frac{\partial^2 T}{\partial y^2} \right)$$

The plate temperature is T_w (uniform) and the incoming fluid temperature is T_∞. Find the expression for energy integral equation (thermal). Explain how the partial derivative of x is written as the *ordinary derivative* in the final form of energy integral equation.

9. For the flow over a non-zero pressure gradient surface, the boundary layer equation retains the pressure gradient term. The pressure gradient can be expressed as the function of the free stream acceleration (spatial). Show that the momentum integral equation can be expressed in terms of momentum and displacement thicknesses as

$$\frac{d}{dx}(U^2\delta^{**}) + \delta^* U \frac{dU}{dx} = \frac{\tau_w}{\rho}$$

10. The solution of the thermal boundary layer equation on a uniformly heated flat plate is given by

$$\theta(\eta) = \frac{\int_0^\eta (e^{-\frac{Pr}{2}\int_0^\eta f(\eta)d\eta})d\eta}{\int_0^\infty (e^{-\frac{Pr}{2}\int_0^\eta f(\eta)d\eta})d\eta}$$

where $\theta = (T_w - T)/(T_w - T_\infty)$. The value of f can be written (from the Blasius equation) as

$$f = -\frac{2f'''}{f''}$$

On substitution of f in the expression of θ, please show that the average Nusselt number on plate of length L is given by

$$\overline{Nu}_L = 0.664 Pr^{1/3} Re_L^{1/2}$$

Make the assumptions that are appropriate.

APPENDIX

The computer code below uses the Runge-Kutta procedure to solve velocity boundary layer and thermal boundary layer equations over a flat plate via the shooting technique. It employs a fourth order Runge-Kutta method for solving ordinary differential equations and a bisection method for finding roots.

```
# include<stdio.h>
# include<math.h>
//user inputs
# define L 10.0 //length of the domain
# define N 100000 //number of grid points
# define A 0.0 //guess value of H at eta = 0
# define B 1.0 //guess value of H at eta = 0
# define convergence 0.00001 //convergence criteria
# define Pr 1.0

//global variables
float h = L/(N-1); //step size
```

```
//function declaration
float shootingVelocity(float,float[],float[],float[]);
float shootingThermal(float[],float,float[],float[]);
void printVelocityBLData(float[],float[],float[]);
void printThermalBLData(float[],float[]);

//main function
int main()
{
    int iterations = 0;
    float a,b,c_old=1.0,c,c_new,error,f_a,f_b,f_c;
    float f_array[N-1], g_array[N-1],h_array[N-1],Theta[N-1],ThetaDes[N-1];

printf ("********** BOUNDARY LAYER SOLUTION OVER FLAT PLATE
************\n");
printf ("*************** Starting velocity BL calculations ***************\n");

a=A; //These are guessed values of H provided above
b=B;

//bisection method
do
{
iterations++; //counter for number of iterations
   c_new= (a+b) / 2.0; //new guess value

    f_a=shootingVelocity(a,f_array,g_array,h_array);//solve using shooting velocity
technique

   f_b=shootingVelocity(b,f_array,g_array,h_array);
   f_c=shootingVelocity(c_new,f_array,g_array,h_array);

   if(((f_a-1.0)*(f_c-1.0))¿0) //check where root lies
{
   a=c_new;
}
else
{
   b=c_new;
}
   error=fabs((c_new-c_old)/c_old); //calculate error
   c_old=c_new; //update guess value

   printf("Iterations = %d\th(0) = %f\n",iterations,c_new);
```

//output iterations and improved guess to terminal

}while(error>convergence); //iterate up to convergence

printf("Correct guess value of h(0) =%f\n",c_new); //output correct guess value of h(0) to terminal

printf("************* Writing velocity BL data to file *************\n");
//print velocity boundary layer data to file
printVelocityBLData(f_array,g_array,h_array);

iterations =0;

printf("\n");
printf ("*************** Starting thermal BL calculations ***************\n");
printf ("*********************For Prandtl no =%f ****************\n", Pr);

a=A; //Guess value of ThetaDes(0) provided
b=B;

c_old = 1.0;

//bisection method
do
{
   iterations++; //counter for number of iterations
   c_new=(a+b)/2.0; //new guess value

   f_a= shootingThermal(f_array,a,Theta,ThetaDes); //solve using shoootingThermal technique
   f_c= shootingThermal(f_array,c_new ,Theta,ThetaDes);
   f_b= shootingThermal(f_array,b,Theta,ThetaDes);

   if((f_a-1.0)*(f_c-1.0)>0) //check where root lies
{
   a=c_new;
}
else
{
   b=c_new;
}
error=fabs((c_new-c_old)/c_old); //calculate error
c_old = c_new; //update old guess

printf("Iterations =%d\tThetaDes(0) =%f\n",iterations,c_new); //output iterations and improved guess to terminal

```
    }while(error>convergence);

    printf("Correct guess value of ThetaDes(0) =%f\n",c_new); //output correct
guess value of ThetaDes(0) to terminal

    printf("************* Writing Thermal BL data to file *************\n");
//print Thermal boundary layer data to file
printThermalBLData(Theta,ThetaDes);

    printf("******************* Exiting program ********************\n");

    return 0;
}

//shoots using guess h(0) value, returns g(L) value
float shootingVelocity(float h_old,float f_array[],float g_array[],float h_array[])
{
    float
f_old,g_old,f_new,g_new,h_new,k1,k2,k3,k4,l1,l2,l3,l4,m1,m2,m3,m4;
    int i,j;

    f_old=0; //f(0) = 0
    g_old=0; //g(0) = 0

    for(i=0;i<=(N-1);i++) //calculate coefficients of Runge-Kutta method
{
    k1=h*g_old;
    l1=h*h_old;
    m1=h*-0.5*f_old*h_old;

    k2=h*(g_old+0.5*l1);
    l2=h*(h_old+0.5*m1);
    m2=h*-0.5*(f_old+0.5*k1)*(h_old+0.5*m1);

    k3=h*(g_old+0.5*l2);
    l3=h*(h_old+0.5*m2);
    m3=h*-0.5*(f_old+0.5*k2)*(h_old+0.5*m2);

    k4=h*(g_old+l3);
    l4=h*(h_old+m3);
    m4=h*-0.5*(f_old+k3)*(h_old+m3);
```

```
//values of f,g,h for next position
f_new=f_old+(1.0/6.0)*(k1+2.0*k2+2.0*k3+k4);
g_new=g_old+(1.0/6.0)*(l1+2.0*l2+2.0*l3+l4);
h_new=h_old+(1.0/6.0)*(m1+2.0*m2+2.0*m3+m4);

//update f,g,h
f_old=f_new;
g_old=g_new;
h_old=h_new;

//update f,g,h in array
f_array[i] = f_new;
g_array[i] = g_new;
h_array[i] = h_new;

   }
   return g_new;
}

void printVelocityBLData(float f_array[],float g_array[],float h_array[])
{
   int i;
//open output files
FILE*out;
out=fopen("velocityBLData.txt","w");
//fprintf(out,"eta\t\tf\tg\th\n"); //data file variables

for (i=0; i<N;i++) //write to file
{
   fprintf(out,"%f %f %f %f \n",i*h,f_array[i],g_array[i],h_array[i]);
}

//close output file
fclose(out);
}

float shootingThermal(float f_array[],float Y_in,float Theta[],float ThetaDes[])
{
   float Y_old,Y_new,Y1_old,Y1_new,k1,k2,k3,k4,l1,l2,l3,l4;
int i;

   Y_old=Y_in; //Guess value for ThetaDes(0)
   Y1_old=0.0; //Theta(0) = 0

   for(i=0;i<=(N-1);i++) //calculate coefficients of Runge-Kutta method
```

```
{
  k1=-0.5*Pr*f_array[i]*Y_old;
  k2=-0.5*Pr*f_array[i]*(Y_old+(h/2.0)*k1);
  k3=-0.5*Pr*f_array[i]*(Y_old+(h/2.0)*k2);
  k4=-0.5*Pr*f_array[i]*(Y_old+h*k3);

  l1=Y_old;
  l2=Y_old+(h/2.0)*l1;
  l3=Y_old+(h/2.0)*l2;
  l4=Y_old+h*l3;

  //values of Theta, ThetaDes for next position
  Y_new=Y_old+(h/6.0)*(k1+2.0*k2+2.0*k3+k4);
  Y1_new=Y1_old+(h/6.0)*(l1+2.0*l2+2.0*l3+l4);

  //update Theta, ThetaDes
  Y_old=Y_new;
  Y1_old=Y1_new;

  //update Theta, ThetaDes in array
  Theta[i] = Y1_new;
  ThetaDes[i] = Y_new;
}
  return Y1_new;
}

void printThermalBLData(float Theta[],float ThetaDes[])
{
  int i;
//open output files
FILE*out;
out=fopen("thermalBLData.txt","w");
//fprintf(out,"eta\tTheta\tThetaDes\n"); //data file variables

  for (i=0; i<N;i++) //write to file
{
  fprintf(out,"%f %f %f \n",i*h,Theta[i],ThetaDes[i]);
}

//close output file
fclose(out);
}
```

## CHAPTER 3

Internal Flows

3.1   ENTRY FLOW IN DUCT

Let us consider steady flow of uniform velocity entering a duct. The duct is formed by two parallel plates kept $2H$ distance apart. Growth of boundary layers on the side walls has a remarkable influence on flow through constant area ducts.

Boundary layers, initially of zero thickness, are formed on the walls of the duct. The layers grow in size in the downstream until their thickness become equal to half the channel height. More clearly, the boundary-layer thicknesses increase until they meet each other from both sides (Fig. 3.1) at the duct center-line. Before the two boundary-layers meet each other, there prevails a core of fluid which is uninfluenced by viscosity. The volume-flow must be a constant at every section and the boundary-layer thickness increases in the flow direction. Consequently, the inviscid core accelerates, and there is a corresponding fall in pressure.

At small distances from the inlet, the boundary-layers will grow following the pattern of their growth along a flat plate at zero incidence.

For incompressible flow, mass flow rate through a cross section at any x is given by

$$\dot{m} = 2 \int_0^\delta \rho \, u \, dy + \rho \, U(x) \, 2(H - \delta) = \rho \, U_0 \, 2H$$

or

$$\int_0^\delta u \, dy + U(x) \, (H - \delta) = U_0 H$$

or

$$U(x) \left[H - \int_0^\delta \left(1 - \frac{u}{U(x)} \right) dy \right] = U_0 H$$

Using the definition of the displacement thickness δ^* at any section we obtain

$$U(x)[H - \delta^*] = U_0 H$$

The velocity of the potential core is given by

$$U(x) = U_0 \frac{H}{(H - \delta^*)}$$

From the Bernoulli's equation, we can write

$$-\frac{dp}{dx} = \rho \, U(x) \frac{dU(x)}{dx}$$

for the calculation of pressure drop in the entry region of the duct. In order to estimate the boundary-layer thickness, we can use the exact solution of boundary layer equation on a flat plate.

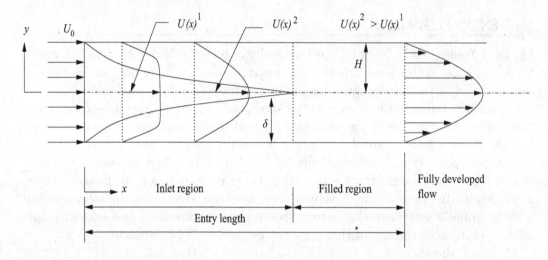

Figure 3.1 Flow development in a parallel plate channel

Owing to the growth of boundary-layer, the inviscid core accelerates. When the boundary-layers merge into each other, the acceleration is not stopped suddenly; rather it continues to persist and is destroyed gradually in the downstream till the fully developed velocity profiles appear ($\partial u / \partial x$ becomes zero) at this stage. The intermediate zone after the merger of the boundary-layers, through which the acceleration of the flow is gradually destroyed is known as the *filled* region. An invariant parabolic velocity profile is obtained beyond this filled zone.

Inlet zone and the filled zone together constitute the entrance region. The pressure gradient in the entrance region takes care of inertia effects and the wall shear and it varies with x. In the fully developed flow, the pressure gradient balances the wall shear stress only and has a constant value. The entry length, which constitutes both the inlet zone and the filled zone, can be estimated through a proper calculation of the filled zone and then by adding it to the inlet zone. Analysis of the filled zone is beyond the scope of this text.

Now we can take recourse to pipe flows. Like the earlier example of flow through a duct, the entrance region exists in the cases of pipe flows, too.

Beyond the entrance region, effect of viscosity extends over the entire cross section and the flow is said to be fully developed. We can define the Reynolds number as

$$Re_D = \frac{\rho u_m D}{\mu} = \frac{\rho D^2 u_m^2}{\mu D u_m} = \frac{\rho D^2 u_m^2}{D^2 \times \mu \frac{u_m}{D}} = \frac{\text{Inertia Force}}{\text{Viscous Force}}$$

where u_m is the mean velocity of the fluid and D is the hydraulic diameter. The critical Reynolds number is usually considered as $Re_{D,C} = 2300$.
Flow is turbulent for $Re_D > 4000$ (generally) and laminar for $Re_D \leq 2300$.
The transition zone lies between them. The estimation of the entrance length (x_e, h) can be done in the following way

$$\left(\frac{x_{e,h}}{D}\right)_{lam} \approx 0.05 Re_D$$

$$10 \leq \left(\frac{x_{e,h}}{D}\right)_{tur} \leq 60$$

We shall assume turbulent fully developed flow for $(x_{e,h}/D) > 10$.

Mean velocity in a circular duct: Because velocity varies over the cross section, it is necessary to work with a mean velocity u_m.

$$\dot{m} = \rho u_m A_c = \rho u_m \pi D^2 / 4, \quad Re_D = \frac{4\dot{m}}{\pi D \mu} \tag{3.1}$$

$$u_m = \frac{\int_{A_c} \rho u(r) dA_c}{\rho A_c} = \frac{2\pi\rho \int_0^{r_0} u(r) r dr}{\rho \pi r_0^2} = \frac{2}{r_0^2} \int_0^{r_0} u(r) r dr \tag{3.2}$$

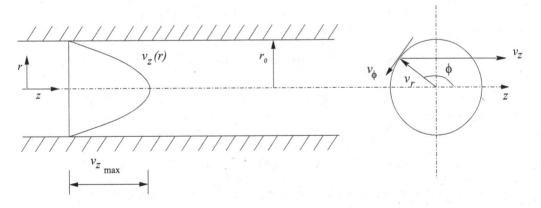

Figure 3.2   Hagen-Poiseullie flow through a pipe

3.2   VELOCITY PROFILE IN FULLY DEVELOPED PIPE FLOW

The fully developed pipe flow is often called Hagen-Poiseullie flow (Fig. 3.2). The general governing equations for flow in a pipe are:

Continuity

$$\frac{\partial v_r}{\partial r} + \frac{v_r}{r} + \frac{1}{r}\frac{\partial v_\phi}{\partial \phi} + \frac{\partial v_z}{\partial z} = 0 \tag{3.3}$$

r momentum

$$\frac{\partial v_r}{\partial t} + v_r \frac{\partial v_r}{\partial r} + \frac{v_\phi}{r}\frac{\partial v_r}{\partial \phi} - \frac{v_\phi^2}{r} + v_z \frac{\partial v_r}{\partial z}$$

$$= -\frac{1}{\rho}\frac{\partial p}{\partial r} + \nu \left(\frac{\partial^2 v_r}{\partial r^2} + \frac{1}{r}\frac{\partial v_r}{\partial r} + \frac{1}{r^2}\frac{\partial^2 v_r}{\partial \phi^2} + \frac{\partial^2 v_r}{\partial z^2} - \frac{v_r}{r^2} - \frac{2}{r^2}\frac{\partial v_\phi}{\partial \phi} \right) \tag{3.4}$$

φ momentum

$$\frac{\partial v_\phi}{\partial t} + v_r \frac{\partial v_\phi}{\partial r} + \frac{v_\phi}{r}\frac{\partial v_\phi}{\partial \phi} + \frac{v_r v_\phi}{r} + v_z \frac{\partial v_\phi}{\partial z}$$

$$= -\frac{1}{\rho}\frac{1}{r}\frac{\partial p}{\partial \phi} + \nu \left(\frac{\partial^2 v_\phi}{\partial r^2} + \frac{1}{r}\frac{\partial v_\phi}{\partial r} + \frac{1}{r^2}\frac{\partial^2 v_\phi}{\partial \phi^2} + \frac{\partial^2 v_\phi}{\partial z^2} - \frac{v_\phi}{r^2} + \frac{2}{r^2}\frac{\partial v_r}{\partial \phi} \right) \tag{3.5}$$

z momentum

$$\frac{\partial v_z}{\partial t} + v_r \frac{\partial v_z}{\partial r} + \frac{v_\phi}{r}\frac{\partial v_z}{\partial \phi} + v_z \frac{\partial v_z}{\partial z}$$

$$= -\frac{1}{\rho}\frac{\partial p}{\partial z} + \nu \left(\frac{\partial^2 v_z}{\partial r^2} + \frac{1}{r}\frac{\partial v_z}{\partial r} + \frac{1}{r^2}\frac{\partial^2 v_z}{\partial \phi^2} + \frac{\partial^2 v_z}{\partial z^2} \right) \tag{3.6}$$

For fully developed flow, v_z is the only non-trivial component. $v_r = v_\phi = 0$. From continuity, $(\partial v_z/\partial z) = 0$; therefore, $v_z = v_z(r, \phi, t)$.

Under steady state, $\partial\,(\text{anything})/\partial t = 0$; for axisymmetric flow, $(\partial\,(\text{any variable})/\partial \phi) = 0$. Therefore,

$$v_z = v_z(r)$$

From r and ϕ momentum equation, we get $p \neq p(r, \phi)$; $p = p(z)$ only.

From z momentum equation, we get

$$\frac{\partial^2 v_z}{\partial r^2} + \frac{1}{r}\frac{\partial v_z}{\partial r} = \frac{1}{\mu}\frac{dp}{dz}$$

or

$$\frac{1}{r}\frac{d}{dr}\left[r \frac{dv_z}{dr} \right] = \frac{1}{\mu}\frac{dp}{dz} \tag{3.7}$$

or

$$r\frac{dv_z}{dr} = \frac{1}{\mu}\frac{dp}{dz}\frac{r^2}{2} + C_1$$

or

$$\frac{dv_z}{dr} = \frac{1}{\mu}\frac{dp}{dz}\frac{r}{2} + \frac{C_1}{r}$$

or

$$v_z = \frac{1}{4\mu}\frac{dp}{dz}r^2 + C_1 \ln r + C_2 \qquad (3.8)$$

The boundary conditions are: at $r = 0, v_z$ is finite $\Rightarrow C_1 = 0$; at $r = r_0, v_z = 0 \Rightarrow$
$0 = \frac{1}{4\mu}\frac{dp}{dz}r_0^2 + C_2$,

$$C_2 = -\frac{1}{4\mu}\frac{dp}{dz}r_0^2$$

$$v_z = -\frac{1}{4\mu}\left(\frac{dp}{dz}\right)r_0^2\left[1 - \left(\frac{r}{r_0}\right)^2\right]$$

Since in most of the applications, we use x as the flow direction and u as the velocity in the flow direction, we propose to substitute z by x and v_z by u.

Using u instead of v_z and x in the place of z, we can write

$$u(r) = -\frac{1}{4\mu}\left(\frac{dp}{dx}\right)r_0^2\left[1 - \left(\frac{r}{r_0}\right)^2\right] \qquad (3.9)$$

$u = u_{max}$ at $r = 0$

$$u_{max} = -\frac{1}{4\mu}\left(\frac{dp}{dx}\right)r_0^2 \qquad (3.10)$$

$$u_m = \frac{2}{r_0^2}\int_0^{r_0} -\frac{r_0^2}{4\mu}\left(\frac{dp}{dx}\right)\left[1 - \left(\frac{r}{r_0}\right)^2\right]rdr$$

or

$$u_m = \frac{2}{r_0^2}\left(\frac{-r_0^2}{4\mu}\frac{dp}{dx}\right)\left[\frac{r^2}{2} - \frac{r^4}{4r_0^2}\right]_0^{r_0} = -\frac{1}{8\mu}\left(\frac{dp}{dx}\right)r_0^2 = \frac{u_{max}}{2} \qquad (3.11)$$

or

$$u_m = \frac{1}{2}u_{max} \quad \text{and} \quad \frac{u(r)}{u_m} = 2\left[1 - \left(\frac{r}{r_0}\right)^2\right] \qquad (3.12)$$

Darcy friction factor is usually defined as

$$h_f = \frac{\Delta p}{\rho g} = \frac{fLu_m^2}{2Dg} \tag{3.13}$$

or

$$f = \frac{(\Delta p/L)D}{\rho u_m^2/2} = \frac{(-dp/dx)D}{(\rho u_m^2/2)} \tag{3.14}$$

Fanning's friction factor is defined as

$$C_f = \frac{\tau_w}{(\rho u_m^2/2)} \tag{3.15}$$

As such,

$$\tau_w = -\mu \frac{\partial u}{\partial r} \mid_{r=r_0}$$

$$\tau_w = -\mu \left(\frac{1}{4\mu} \frac{dp}{dx}\right) 2r_0^2 \frac{r_0}{r_0^2} = -\frac{1}{2} \frac{dp}{dx} r_0 = \frac{1}{2} \left(-\frac{dp}{dx}\right) \frac{D}{2}$$

$$C_f = \frac{(-dp/dx)\,(D/4)}{(\rho u_m^2/2)} \tag{3.16}$$

We have already seen that Eq. (3.14) gives the definition of friction factor as

$$f = \frac{(-dp/dx)\,(D)}{(\rho u_m^2/2)}$$

Therefore, $C_f = f/4$ for a fully developed laminar flow

By substituting f in the expression of mean velocity and defining Reynolds number, we get

$$u_m = -\frac{1}{8\mu} \frac{dp}{dx} r_0^2, \ Re_D = \frac{\rho u_m D}{\mu}$$

we get

$$f = \frac{8\mu u_m D/r_0^2}{\rho u_m^2/2} = \frac{64}{\frac{\rho u_m D}{\mu}} = \frac{64}{Re_D} \tag{3.17}$$

Therefore for fully developed laminar flow, $f = 64/Re_D$.

For fully developed turbulent flow, the analysis is quite involved. At this stage, we may rely on the experimental results, (Fig. 3.3). Figure 3.3 is known as a Moody diagram or Moody chart [1] depicting friction factor as a function of Reynolds number and surface roughness. Correlations that reasonably approximate the smooth surface are:

$$f = 0.316 \, Re_D^{-1/4} \ldots\ldots\ldots Re_D \leq 2 \times 10^4 \tag{3.18}$$

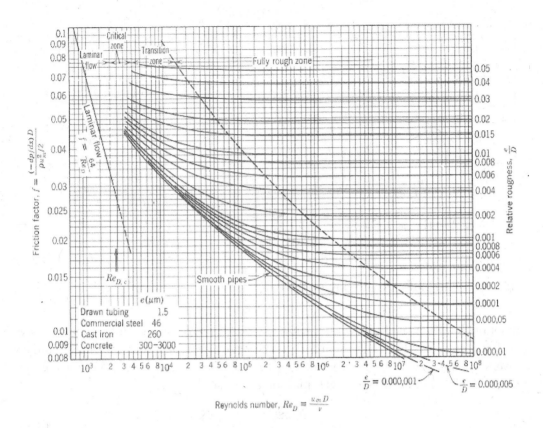

Figure 3.3 Friction factors for pipes

$$f = 0.184 \; Re_D^{-1/5} \ldots \ldots Re_D \geq 2 \times 10^4 \qquad (3.19)$$

Note that f, hence (dp/dx) is constant in fully developed region

$$f = \frac{-(dp/dx)D}{\rho u_m^2/2}$$

$$\Delta p = -\int_{p_1}^{p_2} dp = \int_1^2 f \frac{\rho u_m^2}{2D} dx = \frac{f \rho u_m^2}{2D}(x_2 - x_1) = \frac{f \; \rho u_m^2}{2D} \; L$$

In summary, $f = \frac{64}{Re_D}$ in laminar flow, $f = 0.316 \; Re_D^{-1/4}$ for $Re_D \leq 2 \times 10^4$ (turbulent flow) and $f = 0.184 \; Re_D^{-1/5}$ for $Re_D > 2 \times 10^4$ (for turbulent flow). However, from the above relation we get $\Delta p/\rho g = \frac{fLu_m^2}{2Dg} = h_f$.

3.3 THERMAL CONSIDERATIONS DURING INTERNAL FLOWS

Tube surface conditions require that a uniform wall temperature (UWT) or uniform wall heat flux (UHF) condition prevails. A fully developed temperature profile differs according to the surface conditions. For both surface conditions, however, the

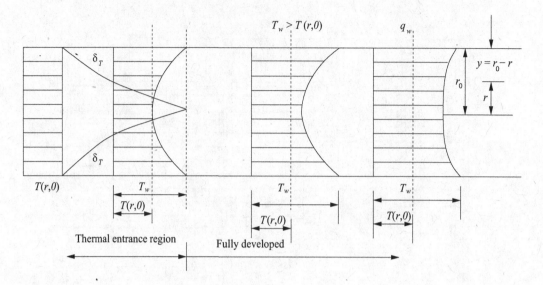

Figure 3.4 Thermally developing duct flow

amount by which the fluid temperature exceeds the entrance temperature increases with increasing x (Fig. 3.4). In order to study the development of thermal boundary layers, we consider a hydrodynamically fully developed flow (Fig. 3.5). For laminar flow, the thermal entrance length may be expressed as

$$\left(\tfrac{x_{e,t}}{D} \right)_{lam} \approx \begin{cases} 0.033 Re_D \ Pr & \text{for UHF} \\ 0.043 Re_D \ Pr & \text{for UWT} \end{cases} \tag{3.20}$$

and we can say, $(x_{e,t/D}) \approx 0.05 \ Re_D \ Pr$.

Comparing this with hydrodynamics boundary layer, it can be said that if $Pr > 1$, the hydrodynamic boundary layer grows more rapidly.

For $Pr > 1$, $x_{e,h} < x_{e,t}$ and $\delta > \delta_T$ at any section; even otherwise $Pr\tfrac{1}{3} \sim (\delta/\delta_T)$. Therefore, for $Pr > 1$, δ has to be $> \delta_T$.

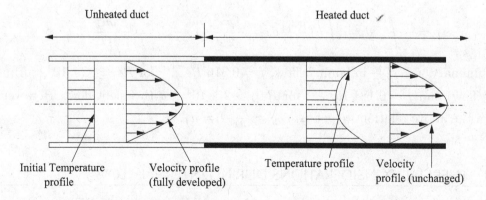

Figure 3.5 Developing temperature field in a fully developed velocity field

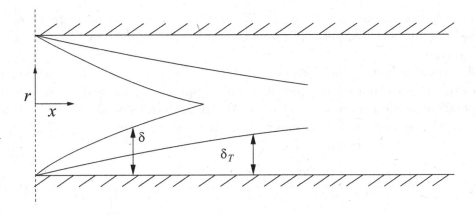

Figure 3.6 Hydrodynamic and thermal boundary layer thickness for $Pr \gg 1$

Inverse is true for $Pr < 1$, $Pr^{1/2} \sim (\delta/\delta_T)$ and consequently δ_T is greater than δ. However, for extremely high Prandtl number fluids (such as oil), $Pr \geq 100$, $x_{e,h} \ll x_{e,t}$. Throughout the thermal entrance region, hydrodynamically fully developed velocity profiles can be assumed (Fig. 3.6).

For turbulent flow, conditions are nearly independent of Prandtl number. We can accept $\left(\frac{x_{e,t}}{D}\right)_{tur} \approx 10$. Bulk mean temperature at any section (x) is given by

$$T_m = \frac{2}{u_m r_0^2} \int_0^{r_0} u(r) \, T(r) \, r dr \qquad (3.21)$$

3.3.1 Newton's Law of Cooling for Internal Flows

$$q_w'' = h(T_w - T_m) \qquad (3.22)$$

The bulk mean temperature T_m is a convenient reference temperature for internal flows. T_m behaves like T_∞ for external flows. Essential difference is T_m varies in the flow direction. T_m increases in the flow direction if heat transfer is from surface to fluid. It is very important to know the variation of T_m under various hydrodynamic and thermal conditions, since the energy transfer is dependent on it.

3.3.2 Fully Developed Thermal Conditions

Heat transfer between the surface and the fluid results in a condition that the fluid temperature must continue to change with x. One might wonder about existence fully developed thermal conditions under such a condition. The situation is different from the hydrodynamic case, for which $(\partial u/\partial x) = 0$ for the fully developed flow. In the presence of heat transfer, (dT_m/dx) as well as $(\partial T/\partial x)$ at any radial location are not zero.

We are actually introducing a new variable $(T_w-T)/(T_w-T_m)$, the condition for which this ratio becomes independent of x is of paramount interest. Although the temperature profile $T(r)$ continues to change with x, the relative shape of the profile does not change and the flow is said to be fully developed. Instead of $(dT_m/dx) = 0$ or $(dT/dx) = 0$, the condition is

$$\frac{d}{dx}\left[\frac{T_w(x) - T(r,x)}{T_w(x) - T_m(x)}\right] = 0 \tag{3.23}$$

The above equation signifies that $(T_w - T)$ changes in the same way as $(T_w - T_m)$ evolves.

Now, we can write

$$\frac{(T_w - T_m)\left(\frac{dT_w}{dx} - \frac{dT}{dx}\right) - (T_w - T)\left(\frac{dT_w}{dx} - \frac{dT_m}{dx}\right)}{(T_w - T_m)^2} = 0$$

or

$$\frac{1}{(T_w - T_m)}\frac{dT_w}{dx} - \frac{dT}{dx}\frac{1}{(T_w - T_m)} - \frac{T_w - T}{(T_w - T_m)^2}\frac{dT_w}{dx} + \frac{(T_w - T)}{(T_w - T_m)^2}\frac{dT_m}{dx} = 0$$

or

$$\frac{dT_w}{dx} - \frac{dT}{dx} - \frac{T_w - T}{T_w - T_m}\frac{dT_w}{dx} + \frac{(T_w - T)}{(T_w - T_m)}\frac{dT_m}{dx} = 0$$

or

$$\frac{dT}{dx} = \frac{dT_w}{dx} - \frac{T_w - T}{T_w - T_m}\frac{dT_w}{dx} + \frac{(T_w - T)}{(T_w - T_m)}\frac{dT_m}{dx} \tag{3.24}$$

Two surface conditions are available: uniform heat flux ($q_w'' = $ constant), and uniform surface temperature, $(dT_w/dx = 0)$. To achieve uniform heat flux, the tube wall should be heated electrically or the outer surface should be uniformly irradiated. If a phase change were taking place at the outer surface, the condition becomes $dT_w/dx = 0$ (uniform surface temperature).

For uniform wall heat flux, we get

$$\frac{dT_w}{dx} = \frac{dT_m}{dx} \tag{3.25}$$

Substituting this condition in the above equation for dT/dx, we get

$$\left.\frac{dT}{dx}\right|_{\text{fully developed}} = \left.\frac{dT_m}{dx}\right|_{\text{fully developed}} \tag{3.26}$$

For uniform wall temperature, $\frac{dT_w}{dx} = 0$

$$\frac{dT}{dx}\bigg|_{fd,t} = \frac{(T_w - T)}{(T_w - T_m)} \frac{dT_m}{dx}\bigg|_{fd,t} \tag{3.27}$$

Some important features of thermally developed flow may be extracted from Eq. (3.23). Since nondimensional temperature profile is independent of x, derivative of the profile with respect to r (temperature gradient from wall to fluid) must also be independent of x. Evaluating this derivative at the tube surface, we obtain

$$\frac{\partial}{\partial r}\left[\frac{T_w - T}{T_w - T_m}\right]_{r=r_0} = \frac{-\partial T/\partial r|_{r=r_0}}{(T_w - T_m)} \neq f(x) \tag{3.28}$$

Again we know,

$$q_w'' = k\frac{\partial T}{\partial r}\bigg|_{r=r_0} = h(T_w - T_m)$$

or

$$\frac{\frac{\partial T}{\partial r}|_{r=r_0}}{(T_w - T_m)} = \frac{h}{k} \tag{3.29}$$

From Eqs. (3.28) and (3.29), we get, $\frac{h}{k} \neq f(x)$ for fully developed flow (Fig. 3.7). Here k is assumed to be constant, hence h is independent of x.

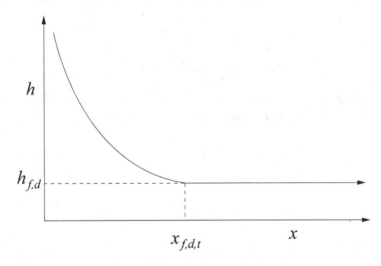

Figure 3.7 Heat transfer coefficient for thermally fully developed flow

Thus we conclude that the local heat transfer coefficient in a thermally fully developed region of pipe or tube does not vary with x.

3.3.3 Energy Balance in Ducted Flows

An energy balance may be applied to a pipe or tube to determine the variation of mean temperature (T_m) with the position along the tube. The total convective energy transfer can be related to the difference in bulk-mean temperatures at the tube outlet and inlet. Figure 3.8 depicts energy balance.

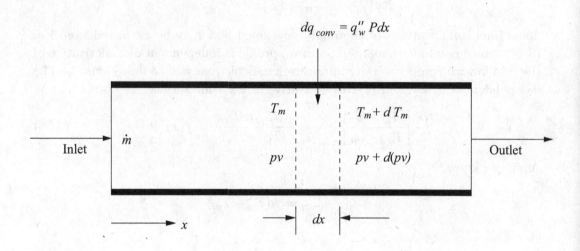

Figure 3.8 Energy balance for flow in a heated tube

$$dq_{conv} + \dot{m}(c_v T_m + pv) - \left[\dot{m}(c_v T_m + pv) + \dot{m}\frac{d(c_v T_m + pv)}{dx}dx \right] = 0$$

or

$$dq_{conv} = \dot{m}\, d(c_v T_m + pv) \tag{3.30}$$

We require finding the exact expression for dq for the compressible as well as incompressible flows. For compressible flows:

$$dq_{conv} = \dot{m}\, d(c_v T_m + pv)$$

or

$$dq_{conv} = \dot{m}\, d(c_v T_m + RT_m)$$

or

$$dq_{conv} = \dot{m}\, d[c_v T_m + (c_p - c_v)T_m]$$

or

$$dq_{conv} = \dot{m}\, c_p dT_m \tag{3.31}$$

For incompressible flows, $dq_{conv} = \dot{m}c_p dT_m$ [since $d(pv) = 0$ and $c_p = c_v$].

Therefore, irrespective of the condition the flow is compressible or incompressible, we can write

$$q_{conv} = \dot{m}c_p(T_{m,0} - T_{m,i}) \tag{3.32}$$

Between any two axial locations, we can write

$$dq_{conv} = \dot{m}c_p dT_m$$

or

$$q_w'' P dx = \dot{m}c_p dT_m \quad [\text{P is the surface perimeter} = \pi D]$$

or

$$\frac{dT_m}{dx} = \frac{q_w'' P}{\dot{m}c_p} = \frac{Ph(T_w - T_m)}{\dot{m}c_p} \tag{3.33}$$

When $T_w > T_m$, heat is transferred to fluid and T_m increases with x. The manner in which Eq. (3.33) varies is required to be understood.

Variation of T_m for uniform wall temperature:
The perimeter, P may vary with x, but it is constant for a pipe. $P/\dot{m}c_p$ is constant. For a fully developed flow, h is constant, although it varies with x in the entrance region. Finally T_w is constant. Therefore, T_m will vary with x except for the trivial case $(T_w = T_m)$ of no heat transfer.

From Eq. (3.33), $\frac{dT_m}{dx} = q_w'' P/\dot{m}c_p = \frac{Ph(T_w-T_m)}{\dot{m}c_p}$, defining $\Delta T = T_w - T_m = $ difference between surface temperature and mean temperature.

$$\frac{dT_m}{dx} = -\frac{d(\Delta T)}{dx} = \frac{Ph(\Delta T)}{\dot{m}c_p} \tag{3.34}$$

or

$$\int_{\Delta T_i}^{\Delta T_0} \frac{d(\Delta T)}{(\Delta T)} = \frac{-P}{\dot{m}c_p} \int_0^L h dx$$

or

$$\ln \frac{(\Delta T_0)}{(\Delta T_i)} = -\frac{PL}{\dot{m}c_p} \cdot \frac{1}{L} \int_0^L h dx$$

or

$$\ln \frac{\Delta T_0}{\Delta T_i} = -\frac{PL}{\dot{m}c_p} \bar{h}_L$$

$$\frac{\Delta T_0}{\Delta T_i} = \frac{T_w - T_{m,0}}{T_w - T_{m,i}} = exp\left(-\frac{PL}{\dot{m}c_p}\bar{h}_L\right) \tag{3.35}$$

If we integrate from the tube inlet to some axial position x within the tube, we shall obtain a more general expression.

$$\frac{T_w - T_m(x)}{T_w - T_{m,i}} = \frac{T_w - T_m(x)}{T_w - T_{m,i}} = exp\left(-\frac{Px}{\dot{m}c_p}\bar{h}\right) \tag{3.36}$$

where $\bar{h}$ is the average value of h from the inlet to the distance x. This result suggests that the temperature difference $(T_w - T_m)$ decays exponentially (also clear from Fig. 3.9) in the flow direction.

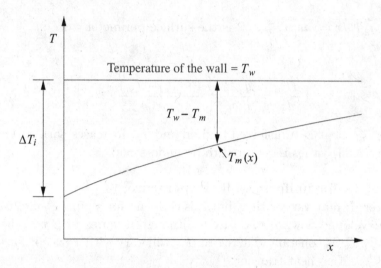

Figure 3.9 Variation of bulk-mean temperature of the fluid in a tube with uniform wall temperature boundary condition

Variation of T_m for uniform wall heat flux:

For uniform wall heat flux, q_w'' is independent of x. Total heat transfer rate is $q_{conv}^{total} = q_w'' PL$. From Eqs. (3.30) through (3.32), $q_w'' PL = \dot{m}c_p(T_{m,0} - T_{m,i})$. Therefore if q_w'', $\dot{m}$, c_p and the geometry is known, it is possible to calculate the fluid temperature rise $(T_{m,0} - T_{m,i})$ or we can write

$$\frac{dT_m}{dx} = \frac{q_w'' P}{\dot{m}c_p} \quad \text{or} \quad T_m = \frac{q_w'' P}{\dot{m}c_p}x + C_1$$

at $x = 0$, $T_m = T_{m,i}$; therefore $C_1 = T_{m,i}$.

$$T_m(x) = T_{m,i} + \frac{q_w'' P}{\dot{m}c_p}x \tag{3.37}$$

We can see the variation $T_m(x)$ with x for constant wall heat flux in Fig. 3.10.

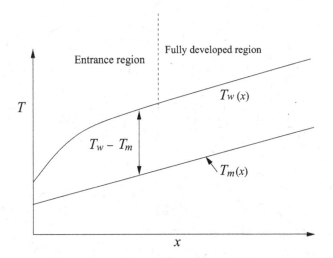

Figure 3.10 Variation of bulk-mean temperature of the fluid in a tube with uniform wall heat flux boundary condition

3.4 LAMINAR FLOW IN CIRCULAR TUBE

The problem of laminar flow in a circular tube has been treated theoretically and the results have been derived to express convection coefficients. We wish to calculate heat transfer under fully developed hydrodynamic and thermal conditions. At any point in the tube the boundary layer approximations may be applied.

3.4.1 Heat Transfer through Circular Tube for Hydrodynamically and Thermally Developed Flow with Uniform Wall Heat Flux (UHF) Condition

The governing energy equation is

$$u\frac{\partial T}{\partial x} + v\frac{\partial T}{\partial r} = \frac{\alpha}{r}\frac{\partial}{\partial r}\left(r\frac{\partial T}{\partial r}\right) \tag{3.38}$$

Solutions of hydrodynamically fully developed flow are $v = 0, \partial u/\partial x = 0$ and $u(r) = 2u_m[1 - (r/r_o)^2]$. Substituting u and v in Eq. (3.38), and applying the thermal condition, we get

$$2u_m\left[1 - \left(\frac{r}{r_0}\right)^2\right]\frac{dT_m}{dx} = \frac{\alpha}{r}\frac{\partial}{\partial r}\left(r\frac{\partial T}{\partial r}\right) \tag{3.39}$$

Note that in the above equation, at any x, $\frac{dT}{dx} = \frac{dT_m}{dx}$. We are solving for T (at any x) with respect to r. At a given x (where dT_m/dx is known), T becomes a function of r only. We can write

$$\frac{1}{r}\frac{d}{dr}\left[r\frac{dT}{dr}\right] = \frac{2u_m}{\alpha}\frac{dT_m}{dx}\left[1 - \left(\frac{r}{r_0}\right)^2\right]$$

or

$$r\frac{dT}{dr} = \frac{2u_m}{\alpha} \frac{dT_m}{dx} \left[\frac{r^2}{2} - \frac{r^4}{4r_0^2} \right] + C_1$$

or

$$\frac{dT}{dr} = \frac{2u_m}{\alpha} \frac{dT_m}{dx} \left[\frac{r}{2} - \frac{r^3}{4r_0^2} \right] + \frac{C_1}{r}$$

or

$$T(r) = \frac{2u_m}{\alpha} \frac{dT_m}{dx} \left[\frac{r^2}{4} - \frac{r^4}{16r_0^2} \right] + C_1 \ln r + C_2 \tag{3.40}$$

At $r = 0$, T is finite; $C_1 = 0$. At $r = r_0$, $T = T_w$

$$C_2 = T_w - \frac{2u_m}{\alpha} \left(\frac{dT_m}{dx} \right) \frac{3r_0^2}{16}$$

$$T(r) = T_w - \frac{2u_m \, r_0^2}{\alpha} \left(\frac{dT_m}{dx} \right) \left[\frac{3}{16} + \frac{1}{16} \left(\frac{r}{r_0} \right)^4 - \frac{1}{4} \left(\frac{r}{r_0} \right)^2 \right] \tag{3.41}$$

Now,

$$T_m = \frac{\int_0^{r_0} u(r) \, T(r) \, r dr}{\int_0^{r_0} u(r) \, r dr}$$

Substituting $(dT_m/dx) = (q_w'' P)/\dot{m}c_p$ in Eq. (3.41), we get from the above equation

$$T_m = \frac{\int_0^{r_0} 2u_m \left(1 - \frac{r^2}{r_0^2} \right) r \left[T_w + \frac{q_w'' r_0}{k} \left\{ \left(\frac{r}{r_0} \right)^2 - \frac{1}{4} \left(\frac{r}{r_0} \right)^4 - \frac{3}{4} \right\} \right] dr}{\int_0^{r_0} 2u_m \left[1 - \left(\frac{r}{r_0} \right)^2 \right] r dr}$$

$$T_m = \frac{2u_m \left[T_w \left(\frac{r_0^2}{2} - \frac{r_0^4}{4r_0^2} \right) + \frac{q_w'' r_0}{k} \int_0^{r_0} \left(r - \frac{r^3}{r_0^2} \right) \left(\frac{r^2}{r_0^2} - \frac{1}{4} \frac{r^4}{r_0^4} - \frac{3}{4} \right) \right] dr}{2u_m \left[\frac{r_0^2}{2} - \frac{r_0^4}{4r_0^2} \right]}$$

$$T_m - T_w = \frac{q_w'' r_0}{k} \left(-\frac{11}{24} \right) \tag{3.42}$$

$$\frac{q_w''}{(T_w - T_m)} \frac{2r_0}{k_f} = Nu_D = \frac{2 \times 24}{11} = 4.36 \tag{3.43}$$

where $q_w'' = h(T_w - T_m) = k\frac{dT}{dr} \big|_{r=r_0}$.

For the UHF boundary condition, the fully developed laminar flow in a tube is $Nu_D = 4.36$ independent of Re and Pr. Finally, the fully developed temperature profile in a tube may be expressed as

$$T(r) = T_w + \frac{q_w'' r_0}{\kappa} \left[\left(\frac{r}{r_0}\right)^2 - \frac{1}{4}\left(\frac{r}{r_0}\right)^4 - \frac{3}{4} \right] \tag{3.44}$$

$$T(r) = T_m + \frac{q_w'' r_0}{\kappa} \left[\left(\frac{r}{r_0}\right)^2 - \frac{1}{4}\left(\frac{r}{r_0}\right)^4 - \frac{7}{24} \right] \tag{3.45}$$

3.4.2 Heat Transfer through Circular Tube for Thermally Fully Developed Laminar Slug Flow with Uniform Wall Temperature (UWT) Condition

There may be a flow situation when unheated entrance length is not short and the thermal diffusivity is more than the momentum diffusivity ($\nu/\propto << 1$); temperature profiles develop more rapidly than the velocity profile. In such a situation, it is appropriate to assume axial velocity to be uniform across the cross-section. However, results developed under such assumption cannot be extended too far downstream.

$$u\frac{\partial T}{\partial x} = \frac{\alpha}{r}\frac{\partial}{\partial r}\left(r\frac{\partial T}{\partial r}\right) \tag{3.46}$$

Slug flow means u is independent of $r, u = U = $ constant

$$U\frac{\partial T}{\partial x} = \alpha\left[\frac{\partial^2 T}{\partial r^2} + \frac{1}{r}\frac{\partial T}{\partial r}\right] \tag{3.47}$$

or

$$\frac{\partial^2 T}{\partial r^2} + \frac{1}{r}\frac{\partial T}{\partial r} = \frac{U}{\alpha}\frac{\partial T}{\partial x} \tag{3.48}$$

T is a function of x and r.

Substituting $\frac{T-T_w}{T_\infty - T_w} = \theta$ and $\frac{r}{r_0} = Y$ and $x = \frac{U r_0^2}{\alpha}X$, we get

$$\frac{\partial^2 \theta}{\partial Y^2} + \frac{1}{Y}\frac{\partial \theta}{\partial Y} = \frac{\partial \theta}{\partial X} \tag{3.49}$$

Assume $\theta = F(X)G(Y)$ where $F = F(X)$ and $G = G(Y)$

$$F\frac{d^2 G}{dY^2} + \frac{F}{Y}\frac{dG}{dY} = G\frac{dF}{dX}$$

or

$$\frac{1}{G}\left[\frac{d^2 G}{dY^2} + \frac{1}{Y}\frac{dG}{dY}\right] = \frac{1}{F}\frac{dF}{dX} = -\beta_n^2$$

The temperature of the fluid must approach that of the wall with increasing X so it is necessary to assume F as a decaying function

$$\frac{1}{F}\frac{dF}{dX} = -\beta_n^2$$

or

$$\frac{dF}{F} = -\beta_n^2 dX$$

or

$$\ln F = -\beta_n^2 X + \ln C \tag{3.50}$$

$F = Ce^{-\beta_n^2 X}$ where C is any arbitrary constant.

Considering the left-hand side of the equation, we get

$$\frac{1}{G}\left[\frac{d^2G}{dY^2} + \frac{1}{Y}\frac{dG}{dY}\right] = -\beta_n^2$$

$$\frac{d^2G}{dY^2} + \frac{1}{Y}\frac{dG}{dY} + G\beta_n^2 = 0$$

$$Y^2\frac{d^2G}{dY^2} + Y\frac{dG}{dY} + Y^2 G\beta_n^2 = 0$$

Substituting $Y^2\beta_n^2 = v^2$, we get

$$\frac{v^2}{\beta_n^2}\frac{d^2G}{d\left(\frac{v}{\beta_n}\right)^2} + \frac{v}{\beta_n}\frac{dG}{d\left(\frac{v}{\beta_n}\right)} + Gv^2 = 0$$

or

$$v^2\frac{d^2G}{dv^2} + v\frac{dG}{dv} + G(v^2 - 0) = 0 \tag{3.51}$$

This is standard Bessel equation of order zero and its general solution is given by the equation

$$G = AJ_0(v) + BY_0(v) \tag{3.52}$$

where $J_0(v)$ and $Y_0(v)$ are Bessel functions of the first kind (order zero) and second kind (order zero).

Short descriptions of Bessel's equation and functions appear in Appendix A at the end of this chapter.

Now we can write

$$\theta = C \cdot e^{-\beta_n^2 X}\{AJ_0(v) + BY_0(v)\}$$

or

$$\theta = \{A_1 \, J_0(v) + B_1 \, Y_0(v)\} \, e^{-\beta_n^2 X} \tag{3.53}$$

or

$$\theta = \{A_1 \, J_0 \, (Y \, \beta_n) \, + \, B_1 \, Y_0 \, (Y\beta_n)\} \, e^{-\beta_n^2 X}$$

Boundary conditions are:
(i) at $r = r_0$, $T = T_w$ for $x > 0$; this means at $Y = 1$, $\theta = 0$ (basically for $v = \beta_n$)
(ii) at $r < r_0$ and $x = 0$, $T = T_\infty$, this leads to at $X = 0$, $\theta = 1$
(iii) at $r = 0, Y = 0, \theta =$ finite.
  Since $Y(0) = -\infty, B_1 = 0$

$$\theta = A_1 \, J_0 \, (Y \, \beta_n) \, e^{-\beta_n^2 X} \tag{3.54}$$

We have already found at $r = r_0, T = T_w$; or at $Y = 1$, $\theta = 0$, this leads to

$$J_0(\beta_n) = 0 \tag{3.55}$$

$\beta_n = 2.4048, 5.5201, 8.6537, 11.7915, 14.93309\ldots$
  For each value of β_n, the problem is satisfied thus far, and each must be considered

$$\theta = \sum_{n=1}^{\infty} A_n \, J_0(\beta_n Y) \exp \left(-\beta_n^2 X\right) \tag{3.56}$$

The remaining boundary condition: $at\,x = 0, T = T_\infty$ or $at\,X = 0, \theta = 1$ demands

$$\sum A_n J_0(\beta_n \, Y) = 1 \tag{3.57}$$

which is a requirement to express in a Fourier series of Bessel functions over the range $0 \leq Y \leq 1$

$$\int_0^1 Y \, J_0 \, (\beta_n Y) \, dY = \int_0^1 A_n \, Y J_0^2 \, (\beta_n Y) \, dY$$

$$A_n = \frac{\int_0^1 Y J_0 \, (\beta_n Y) \, dY}{\int_0^1 Y J_0^2 \, (\beta_n Y) \, dY} = \frac{\int_0^{\beta_n} \frac{v}{\beta_n} \, J_0 \, (v) \, \frac{dv}{\beta_n}}{\frac{1}{2} \, J_1^2 \, (\beta_n)} \tag{3.58}$$

For the above expression, let us refer to the following recurrence relations of Bessel functions:

Relation I

$$\frac{d}{dv} \, [v^n J_n(v)] = v^n J_{n-1}(v)$$

$$\frac{d}{dv}\left[vJ_1(v)\right] = vJ_0(v) \text{ or } \int vJ_0(v)dv = vJ_1(v)$$

Additionally

$$\frac{n}{v}\left[J_n(v)\right] - J_n'(v) = J_{n+1}(v)$$

Which reduces for $n = 0$, to the form, $J_0'(v) = -J_1(v)$

Relation II

$$\int_0^a vJ_n^2(\lambda v)\, dv = \frac{a^2}{2}\left[\left\{J_n'(\lambda a)\right\}^2 + \left(1 - \frac{n^2}{\lambda^2 a^2}\right)J_n^2(\lambda a)\right]$$

or

$$\int_0^a vJ_0^2(\lambda v)\, dv = \frac{a^2}{2}\left[(-J_1(\lambda a))^2 + J_0^2(\lambda a)\right]$$

Now,

$$\int_0^1 YJ_0^2(\beta_n Y)\, dY = \frac{1}{2}\left[(-J_1(\beta_n))^2 + J_0^2(\beta_n)\right]$$

In our case $J_0(\beta_n) = 0$.

Therefore, by making use of Relation I and Relation II, we can write

$$A_n = \frac{\frac{1}{\beta_n^2}\int_0^{\beta_n} vJ_0(v)\, dv}{\frac{1}{2}J_1^2(\beta_n)} = \frac{\frac{1}{\beta_n^2}\left[vJ_1(v)\right]_0^{\beta_n}}{\frac{1}{2}J_1^2(\beta_n)}$$

$$A_n = \frac{2}{\beta_n J_1(\beta_n)}$$

The solution is expressed as

$$\theta = 2\sum_{n=1}^{\infty}\frac{J_0(\beta_n Y)}{\beta_n J_1(\beta_n)}e^{-\beta_n^2 X} \tag{3.59}$$

We know the definition of bulk-mean temperature as

$$T_m = \frac{2}{u_m r_0^2}\int_0^{r_0} u(r)T(r)r\,dr$$

Defining $\theta_m = (T_m - T_w)/(T_\infty - T_w)$, we can write $\frac{T_m - T_w}{T_\infty - T_w} = 2\int_0^1 \frac{u(r)}{u_m}\cdot\frac{T(r) - T_w}{T_\infty - T_w}\cdot$
$\frac{r}{r_o}d\left(\frac{r}{r_o}\right) = 2\int_0^1 \theta Y\,dY$

$$\theta_m = 4\int_0^1 \frac{YJ_0(\beta_n Y)\,dY}{\beta_n J_1(\beta_n)}e^{-\beta_n^2 X}$$

The numerator can be evaluated as

$$\int_0^1 Y\, J_o(\beta_n Y)dY = \int_0^{\beta_n} \frac{v}{\beta_n}J_o(v)\frac{dv}{\beta n} = \frac{1}{\beta_n^2}[vJ_1\,(v)]_o^{\beta n} = \frac{1}{\beta_n^2}[\beta_n\, J_1(\beta_n)]$$

$$\theta_m = 4\sum_{n=1}^{\infty} \frac{e^{-\beta_n^2 X}}{\beta_n^2} \tag{3.60}$$

Again,

$$\frac{d}{dY}\{J_0(\beta_n Y)\} = -\beta_n J_1(\beta_n Y)$$

$$\left.\frac{d\theta}{dY}\right|_{Y=1} = -2\sum_{n=1}^{\infty} e^{(-\beta_n^2 X)} \tag{3.61}$$

The heat flux from the surface may be calculated as

$$h(T_w - T_m) = \left.k\frac{\partial T}{\partial r}\right|_{r=r_0} = \frac{k(T_\infty - T_w)}{r_0}\left.\frac{d\theta}{d\left(\frac{r}{r_o}\right)}\right|_{Y=1} \tag{3.62}$$

$$\frac{hD}{k} = \frac{2\times 2\sum_{n=1}^{\infty}e^{-\beta_n^2 X}}{\theta_m} = \frac{\sum_{n=1}^{\infty}e^{-\beta_n^2 X}}{\sum_{n=1}^{\infty}(e^{-\beta_n^2 X})/\beta_n^2} \tag{3.63}$$

For β_1, $\frac{hD}{k} = \frac{e^{-\beta_1^2 X}\beta_1^2}{e^{-\beta_1^2 X}} = \beta_1^2$ or

$$Nu_D = 5.7831 \tag{3.64}$$

3.4.3 Heat Transfer in Circular Tube for Hydrodynamically and Thermally Developed Flow with Uniform Wall Temperature (UWT) Condition

The governing equation is

$$\rho c_p u\frac{\partial T}{\partial x} = \frac{k}{r}\frac{\partial}{\partial r}\left(r\frac{\partial T}{\partial r}\right) \tag{3.65}$$

$\theta(r) = \frac{T-T_w}{T_c-T_w}$, where T_c is the centerline temperature and T_w is the wall temperature; introducing $\eta = \frac{r}{r_o}, z = \frac{x}{r_0}, U(\eta) = \frac{u}{u_m}$.

We get

$$\frac{\rho c_p u_m U(\eta)}{r_0}\frac{\partial T}{\partial z} = \frac{k}{r_0^2}\frac{1}{\eta}\frac{\partial}{\partial \eta}\left(\eta\frac{\partial T}{\partial \eta}\right)$$

$$\left(\frac{\rho \nu c_p}{k}\right)\frac{u_m r_0}{\nu}U\frac{\partial T}{\partial z} = \frac{1}{\eta}\frac{\partial}{\partial \eta}\left(\eta\frac{\partial T}{\partial \eta}\right)$$

$$\frac{Pr \, Re_D}{2} U \frac{\partial T}{\partial z} = \frac{1}{\eta} \frac{\partial}{\partial \eta} \left(\eta \frac{\partial T}{\partial \eta} \right)$$

Now, $T = T_w + (T_c - T_w)\theta(r)$; $T_c \equiv f_n(z)$

$$\frac{\partial T}{\partial z} = \theta(r) \frac{\partial T_c}{\partial z}$$

$$\frac{\partial T}{\partial r} = (T_c - T_w) \frac{\partial \theta}{\partial r}$$

$$\frac{\partial T}{\partial \eta} = (T_c - T_w) \frac{\partial \theta}{\partial \eta}$$

On substitution,

$$Pe \, \frac{U\theta}{2} \frac{\partial T_c}{\partial z} = \frac{1}{\eta} \frac{\partial}{\partial \eta} \left(\eta \frac{\partial \theta}{\partial \eta} \right) (T_c - T_w)$$

or

$$\left[\frac{1}{T_c(z) - T_w} \right] Pe \, \frac{dT_c}{dz} = \frac{\frac{1}{\eta} \frac{\partial}{\partial \eta} \left(\eta \frac{\partial \theta}{\partial \eta} \right)}{\frac{U(\eta)}{2} \theta(\eta)} = -\lambda^2 \quad \text{(constant)} \qquad (3.66)$$

$$\frac{Pe \, dT_c/dz}{T_c - T_w} = -\lambda^2$$

$$T_c(z) = T_w + Ce^{-\lambda^2 z/Pe}$$

The second equation gives

$$\frac{1}{\eta} \frac{\partial}{\partial \eta} \left(\eta \frac{\partial \theta}{\partial \eta} \right) + \frac{\lambda^2}{2} U\theta = 0$$

where, $\frac{u}{u_m} = 2 \left[1 - \left(\frac{r}{r_0} \right)^2 \right] = 2[1 - \eta^2]$ or

$$\theta'' + \frac{1}{\eta} \theta' + \lambda^2 (1 - \eta^2)\theta = 0 \qquad (3.67)$$

The equation is to be solved using the following boundary conditions, at $r = 0$, $T = T_c$ which means at $\eta = 0$, $\theta = 1$ at $r = r_0$, $T = T_w$ which means at $\eta = 1$, $\theta = 0$. The solution of the above equation is an infinite series.

Let the general solution be

$$\theta = \sum_{n=0}^{\infty} C_n \eta^n$$

Now,

$$\theta = C_0 + C_1\eta + C_2\eta^2 + C_3\eta^3 + \cdots + C_n\eta^n$$

then $\theta' = C_1 + 2\,C_2\eta + 3\,C_3\eta^2 + \cdots + (n+1)\,C_{n+1}\eta^n$

and $\theta'' = 2\,C_2 + 6C_3\eta + \cdots + (n+1)n\,C_{n+1}\,\eta^{n-1} + (n+2)(n+1)C_{n+2}\,\eta^n.$

Plugging in to the original equation, one gets

$$2\,C_2 + 3\cdot 2\,C_3\eta + \cdots + (n+2)(n+1)\,C_{n+2}\eta^n$$

$$+ \quad \frac{C_1}{\eta} + 2\,C_2 + 3\,C_3\eta + \cdots + (n+1)\,C_{n+1}\eta^{n-1} + (n+2)\,C_{n+2}\eta^n$$

$$+ \quad \lambda^2 C_0 + \lambda^2 C_1\eta + \lambda^2 C_2\eta^2 + \cdots + \lambda^2 C_n\eta^n$$

$$= \quad \lambda^2 C_0\eta^2 + \lambda^2 C_1\eta^3 + \lambda^2 C_2\eta^4 + \cdots + \lambda^2 C_n\eta^{n+2}$$

Equating the equal powers of η

$$C_1 = 0$$
$$2.1C_2 + 2C_2 + \lambda^2 C_0 = 0$$
$$3.2C_3 + 3C_3 + \lambda^2 C_1 = 0$$
$$4.3C_4 + 4C_4 + \lambda^2 C_2 = \lambda^2 C_0$$
$$5.4C_5 + 5C_5 + \lambda^2 C_3 = \lambda^2 C_1$$
$$\vdots$$
$$(n+2)(n+1)C_{n+2} + (n+2)C_{n+2} + \lambda^2 C_n = \lambda^2 C_{n-2}$$

As $C_1 = 0, C_3, C_5 \ldots$ all odd coefficients are zero.

Using $n = 2m$,

$$(2m+2)(2m+1)C_{2m+2} + (2m+2)C_{2m+2} + \lambda^2 C_{2m} = \lambda^2 C_{2m-2}$$

$$(2m+2)^2 C_{2m+2} + \lambda^2 C_{2m} = \lambda^2 C_{2m-2}$$

$$C_{2m+2} = \frac{\lambda^2}{(2m+2)^2}(C_{2m-2} - C_{2m})$$

$$C_{2m} = \frac{\lambda^2}{(2m)^2}(C_{2m-4} - C_{2m-2})$$

Therefore, we can write

$$\theta = \sum_{n=0}^{\infty} C_{2n}\eta^{2n} \tag{3.68}$$

where $\quad C_{2n} = \dfrac{\lambda^2}{(2n)^2}[C_{2n-4} - C_{2n-2}] \quad$ for $\quad n \geq 2$

We can also write

$$\theta = \sum_{n=0}^{\infty} C_{2n}\eta^{2n} = C_0 + C_2\eta^2 + C_4\eta^4 + C_6\eta^6 + \cdots$$

Invoking the boundary condition, at $\eta = 0, \theta = 1$ we get

$$C_0 = \theta_{\text{at } \eta=0} = 1$$

We also know at $r = r_0$, i.e., at $\eta = 1, \theta = 0$

$$C_0 + C_2 + C_4 + C_6 + \ldots = 0$$

$$C_0 - \frac{\lambda^2}{4} C_0 + \frac{\lambda^2}{16} \left(1 + \frac{\lambda^2}{4} \right) C_0 - \frac{\lambda^2}{36} \left(\frac{5\lambda^2}{16} + \frac{\lambda^4}{64} \right) C_0 + \cdots = 0$$

$$C_0 \left[1 - \frac{\lambda^2}{4} + \frac{\lambda^2}{16} + \frac{\lambda^4}{64} - \frac{5\lambda^4}{36 \times 16} - \frac{\lambda^6}{36 \times 64} \right] + \cdots = 0$$

Utilizing the fourth power of the series,

$$C_0 \left[1 - \frac{\lambda^2}{4} + \frac{\lambda^2}{16} + \frac{\lambda^4}{64} - \frac{5\lambda^4}{576} \right] = 0$$

Since $C_0 \neq 0$,

$$1 - \frac{\lambda^2}{4} + \frac{\lambda^2}{16} + \frac{\lambda^4}{144} = 0$$

$$\lambda^4 - 27\lambda^2 + 144 = 0$$

Solving numerically by Newton Rapshon method,

$$\lambda = 2.704364 \tag{3.69}$$

3.4.4   Calculation of Nusselt Number for UWT

From the governing equation we get

$$\int_0^{r_0} \rho c_p u \frac{\theta}{r_0} \frac{dT_c}{dz} r \, dr = k r_0 \frac{\partial T}{\partial r} \Big|_{r=r_0}$$

or

$$\int_0^{r_0} k \frac{\nu}{\alpha} \frac{u_m 2 r_0}{\nu} \frac{r}{2r_0^2} \frac{u}{u_m} (T - T_w) \frac{dT_c/dz}{(T_c - T_w)} \, dr = r_0 q_w''$$

$$\int_0^{r_0} \frac{Pr Re_D \frac{dT_c}{dz}}{(T_c - T_w)} k \frac{r}{2r_0^2} \frac{u}{u_m} (T - T_w) \frac{2\pi}{2\pi} dr = r_0 q_w''$$

or

$$\int_0^{r_0} -\frac{\lambda^2 k}{2} \frac{2\pi u r (T - T_w) dr}{2\pi u_m r_0^2} = r_0 q_w''$$

or

$$-\frac{\lambda^2 k}{4} \left\{ \frac{\int_0^{r_0} uT 2\pi r dr}{u_m \pi r_0^2} - \frac{T_w \int_0^{r_0} u 2\pi r dr}{u_m \pi r_0^2} \right\} = r_0 q_w''$$

or

$$-\frac{\lambda^2 k}{4}(T_m - T_w) = r_0 q_w''$$

or

$$\frac{q_w''}{(T_w - T_m)} \frac{2r_0}{k} = \frac{hD}{k} = \frac{\lambda^2}{2}$$

The Nusselt number for hydrodynamically and thermally fully developed flow (subject to uniform wall temperature) is:

$$Nu_D = \frac{\lambda^2}{2} = 3.656 \tag{3.70}$$

For parallel plate channels, Nusselt number for hydrodynamically and thermally fully developed flow is

$$Nu = 7.5407 \text{ (uniform wall temperature)}$$
$$Nu = 8.2352 \text{ (uniform wall heat flux)}$$

In the above two expressions, the Nusselt number has been calculated based on the characteristic length $2b$, where b is the channel height.

3.5 GRAETZ PROBLEM

The problem addresses hydrodynamically developed, thermally developing flow subject to uniform wall heat flux as shown in Fig. 3.11 and is known as the Graetz problem [2]. The problem was further extended by Sellars et al. [3].

For $x < 0$: Both the fluid and the wall have uniform temperature, say T_0.
For $x > 0$: Uniform wall heat flux is applied.

We start with a situation which prevails at a very large x. We refer to hydrodynamically and thermally fully developed flow with uniform wall heat flux.
We have already found the solution as

$$T(x,r) = T_w - \frac{2u_m r_0^2}{\alpha} \left(\frac{dT_m(x)}{dx} \right) \left[\frac{3}{16} + \frac{1}{16} \left(\frac{r}{r_0} \right)^4 - \frac{1}{4} \left(\frac{r}{r_0} \right)^2 \right]$$

After substitution of $\left(\frac{dT_m}{dx} \right)$

$$T(x,r) = T_w + \frac{q_w'' r_0}{k} \left[\left(\frac{r}{r_0} \right)^2 - \frac{1}{4} \left(\frac{r}{r_0} \right)^4 - \frac{3}{4} \right] \tag{3.71}$$

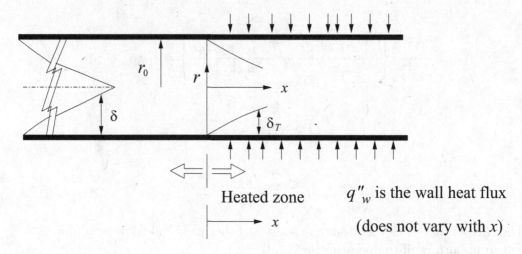

Heated zone q''_w is the wall heat flux

(does not vary with x)

Figure 3.11 Graetz problem with uniform wall heat flux boundary condition

$$T_m(x) = T_w - \frac{11}{48}\frac{2q''_w r_0}{k} \tag{3.72}$$

from Eqs. (3.71) and (3.72)

$$T(x,r) = T_m(x) + \frac{q''_w r_0}{k}\left[\left(\frac{r}{r_0}\right)^2 - \frac{1}{4}\left(\frac{r}{r_0}\right)^4 - \frac{7}{24}\right] \tag{3.73}$$

Also we have found from energy balance

$$T_m(x) = T_{m,i} + \frac{q''_w P}{\dot{m}c_p}x$$

$P = 2\pi r_0; \quad \dot{m} = \rho u_m \pi r_0^2$

$$T_m(x) = T_{m,i} + \frac{2q''_w x}{\rho c_p u_m r_0} \tag{3.74}$$

Substitute Eq. (3.74) in Eq. (3.73) we get

$$T = T_{m,i} + \frac{2q''_w x}{\rho c_p u_m r_0} + \frac{q''_w r_0}{k}\left[\left(\frac{r}{r_0}\right)^2 - \frac{1}{4}\left(\frac{r}{r_0}\right)^4 - \frac{7}{24}\right]$$

$$\frac{T - T_{m,i}}{(q''_w r_0/k)} = \frac{4(x/r_0)}{Re\ Pr} + \left[\left(\frac{r}{r_0}\right)^2 - \frac{1}{4}\left(\frac{r}{r_0}\right)^4 - \frac{7}{24}\right] \tag{3.75}$$

For the present problem, the Eq. (3.75) can be written as

$$\frac{T_{fd} - T_0}{(q''_w r_0/k)} = \frac{4(x/r_0)}{Re\ Pr} + \left[\left(\frac{r}{r_0}\right)^2 - \frac{1}{4}\left(\frac{r}{r_0}\right)^4 - \frac{7}{24}\right] \tag{3.76}$$

T_{fd} is the fully developed temperature and T_0 is the temperature of the flowing fluid at the inlet.

In the developing region, let us define T_d in such a manner

$$T_d = T - T_{fd} \tag{3.77}$$

The approximate form of energy equation (boundary layer approximation has been invoked; also axial conduction is insignificant);

$$u\frac{\partial T}{\partial x} = \frac{\alpha}{r}\frac{\partial}{\partial r}\left(r\frac{\partial T}{\partial r}\right) \tag{3.78}$$

$$u\frac{\partial T_d}{\partial x} = \frac{\alpha}{r}\frac{\partial}{\partial r}\left(r\frac{\partial T_d}{\partial r}\right) \tag{3.79}$$

The boundary conditions are

$$T_d(0,r) = T_0 - T_{fd}(0,r)|_{x=0} \quad \text{at} \quad x = 0$$

$$\left.\frac{\partial T_d}{\partial r}\right|_{r=0} = 0 \quad ; \quad \left.\frac{\partial T_d}{\partial r}\right|_{r=r_0} = 0$$

Let us explain as to why $(\partial T_d/\partial r)$ at $r = r_0 = 0$.

We know

$$\left.k\frac{\partial T}{\partial r}\right|_{r=r_0} = q_w''$$

or

$$k\left[\frac{\partial(T_d + T_{fd})}{\partial r}\right]_{r=r_0} = q_w'' \quad or, \quad \left.k\frac{\partial T_d}{\partial r}\right|_{r=r_0} = q_w'' - \left.k\frac{\partial T_{fd}}{\partial r}\right|_{r=r_0}$$

However,

$$\left.+k\frac{\partial T_{fd}}{\partial r}\right|_{r=r_0} = q_w''$$

The shape of the fully developed temperature profile is the same as the temperature profile of thermally developing flow.

Therefore,

$$\left.\frac{\partial T_d}{\partial r}\right|_{r=r_0} = 0 \quad \text{as well as} \quad \left.\frac{\partial T_d}{\partial r}\right|_{r=0} = 0$$

Keep in mind that $x \to \infty$, $T \to T_{fd}$ and $T_d \to 0$.

We know that,

$$\frac{u}{u_m} = 2\left[1 - \left(\frac{r}{r_0}\right)^2\right] \tag{3.80}$$

From Eqs. (3.79) and (3.80), we get

$$2u_m \left[1 - \left(\frac{r}{r_0} \right)^2 \right] \frac{\partial T_d}{\partial x} = \frac{\alpha}{r} \frac{\partial}{\partial r} \left[r \frac{\partial T_d}{\partial r} \right] \tag{3.81}$$

Non-dimensional parameters are $\tilde{r} = r/r_0$, $\tilde{x} = \frac{x/r_0}{Re\ Pr} = x\alpha/2u_m r_0^2$

$$\theta_d = \frac{T_d}{q_w'' r_0/k} = \frac{T - T_{fd}}{q_w'' r_0/k}$$

Therefore,

$$\frac{\partial T_d}{\partial x} = \frac{\alpha}{2u_m r_0^2} \cdot \frac{q_w'' r_0}{k} \cdot \frac{\partial \theta_d}{\partial \tilde{x}}$$

$$\frac{\partial T_d}{\partial r} = \frac{q_w'' r_0}{k r_0} \frac{\partial \theta_d}{\partial \tilde{r}} = \frac{q_w''}{k} \frac{\partial \theta_d}{\partial \tilde{r}}$$

$$\frac{\partial}{\partial r} \left(r \frac{\partial T_d}{\partial r} \right) = \frac{q_w'' r_0}{k r_0} \frac{\partial}{\partial \tilde{r}} \left(\tilde{r} \frac{\partial \theta_d}{\partial \tilde{r}} \right) = \frac{q_w''}{k} \frac{\partial}{\partial \tilde{r}} \left(\tilde{r} \frac{\partial \theta_d}{\partial \tilde{r}} \right)$$

Substituting these in Eq. (3.81) and simplifying, we finally get

$$(1 - \tilde{r}^2) \frac{\partial \theta_d}{\partial \tilde{x}} = \frac{\partial^2 \theta_d}{\partial \tilde{r}^2} + \frac{1}{\tilde{r}} \frac{\partial \theta_d}{\partial \tilde{r}} \tag{3.82}$$

Boundary conditions are

$$\theta_d(0, \tilde{r}) = \frac{T_0 - T_{fd}(0, \tilde{r})}{q_w'' r_0/k}$$

$$\left. \frac{\partial \theta_d}{\partial \tilde{r}} \right|_{\tilde{r}=0} = \left. \frac{\partial \theta_d}{\partial \tilde{r}} \right|_{\tilde{r}=1} = 0$$

We shall try the method of separation of variables.

Let $\theta_d(\tilde{x}, \tilde{r}) = X(\tilde{x}) R(\tilde{r})$.

Substituting in Eq. (3.82) and rearranging, we get

$$\frac{1}{X} \frac{dX}{d\tilde{x}} = \frac{1}{R(1 - \tilde{r}^2)} \left[\frac{d^2 R}{d\tilde{r}^2} + \frac{1}{\tilde{r}} \frac{dR}{d\tilde{r}} \right] \tag{3.83}$$

Now that the variables are separated, each separated group can be equated to a constant $= -\beta^2$ (say)

$$\frac{1}{X} \frac{dX}{d\tilde{x}} = -\beta^2$$

$$X = A_n e^{-\beta_n^2 \tilde{x}} \tag{3.84}$$

$$\frac{d^2 R}{d\tilde{r}^2} + \frac{1}{\tilde{r}} \frac{dR}{d\tilde{r}} + R\beta_n^2(1 - \tilde{r}^2) = 0 \tag{3.85}$$

or

$$\tilde{r}\frac{d^2 R}{d\tilde{r}^2} + \frac{dR}{d\tilde{r}} + R\tilde{r}\beta_n^2(1 - \tilde{r}^2) = 0 \tag{3.86}$$

The solution of Eq. (3.86) is of the form

$$R = \sum_{n=1}^{\infty} C_n R_n$$

and obtainable for some specific values of β_n (eigenvalues).

The complete solution is given by

$$\theta_d(\tilde{x}, \tilde{r}) = \sum_{n=1}^{\infty} b_n R_n e^{-\beta_n^2 \tilde{x}} \tag{3.87}$$

where R_n = eigenfunctions and β_n = eigenvalues of Eqs. (3.85 or 3.86).

Equations (3.85) or (3.86) are called Sturm–Liouville equation. A Sturm–Liouville system is symbolically stated as

$$\frac{d}{dx}[r(x)\, y'] + [q(x) + \lambda p(x)]\, y = 0$$

Boundary conditions are:

$$y'(a) = h_1\, y(a) \qquad 0 < a < \infty$$
$$y'(b) = h_2\, y(b) \qquad 0 < b < \infty$$

Finite upper and lower bound h_1 and h_2 are real constants. The solution of any function defined in the above domain is given by

$$f(x) = \sum_{n=1}^{\infty} a_n X_n(x)$$

In above the constants a_n are given by

$$a_n = \frac{\int_a^b f(x)\, X_n(x)dx}{\int_a^b X_n^2(x)dx}$$

Now the task is to compare our Eqs. (3.85) or (3.86) with the Sturm–Liouville system and evaluate the constants.

Applying the condition, $x = 0$ in Eq. (3.87) we get

$$\theta_d(0, r) = \sum_{n=1}^{\infty} b_n R_n$$

Substituting $x = 0$ in Eq. (3.76) the result is:

$$\theta_d(0, \tilde{r}) = \frac{T_0 - T_{fd}}{q''_w r_0 / k} = -\left[\tilde{r}^2 - \frac{1}{4} \tilde{r}^4 - \frac{7}{24} \right]$$

Comparing these two equations we get

$$f(\tilde{r}) = \sum_{n=1}^{\infty} b_n R_n = -\left[\tilde{r}^2 - \frac{1}{4} \tilde{r}^4 - \frac{7}{24} \right] \tag{3.88}$$

Now we also obtain

$$b_n = \frac{\int_0^1 \left[(\tilde{r} - \tilde{r}^3) \right] R_n f(\tilde{r}) d\tilde{r}}{\int_0^1 \left[(\tilde{r} - \tilde{r}^3) \right] R_n^2 d\tilde{r}} \tag{3.89}$$

where $\lambda = \beta_n^2$ are the eigenvalues. By integrating Eq. (3.86) we get

$$\int_0^1 \frac{d}{d\tilde{r}} \left(\tilde{r} \frac{dR_n}{d\tilde{r}} \right) d\tilde{r} + \int_0^1 R_n (\tilde{r} - \tilde{r}^3) \lambda \, d\tilde{r} = 0 \qquad \text{for} \quad R = R_n$$

From the boundary conditions, one can write $\frac{dR_n}{d\tilde{r}} = 0$ at $\tilde{r} = 0$ and $\tilde{r} = 1$. Therefore

$$\int_0^1 R_n (\tilde{r} - \tilde{r}^3) \lambda d\tilde{r} = 0 \tag{3.90}$$

We can get λ from Eq. (3.90) and b_n can be valuated as

$$b_n = -\frac{\int_0^1 (\tilde{r} - \tilde{r}^3) R_n (\tilde{r}^2 - \frac{1}{4} \tilde{r}^4 - \frac{7}{24}) d\tilde{r}}{\int_0^1 (\tilde{r} - \tilde{r}^3) R_n^2 d\tilde{r}} \tag{3.91}$$

Going back to the solution, Eq. (3.87)

$$\theta_d = \frac{T - T_{fd}}{q''_w r_0 / k} = \sum_{n=1}^{\infty} b_n e^{-\beta_n^2 \frac{x/r_0}{Re \, Pr}} R_n(\tilde{r})$$

Also, we know T_{fd} from Eq. (3.76) as

$$\frac{T_{fd} - T_0}{q''_w r_0 / k} = \frac{4(x/r_0)}{Re \, Pr} + \left(\frac{r}{r_0} \right)^2 - \frac{1}{4} \left(\frac{r}{r_0} \right)^4 - \frac{7}{24}$$

Combining these two we get

$$T(x, r) = T_0 + \frac{q''_w r_0}{k} \left[\frac{4(x/r_0)}{Re \, Pr} + \left(\frac{r}{r_0} \right)^2 - \frac{1}{4} \left(\frac{r}{r_0} \right)^4 - \frac{7}{24} \right.$$

$$\left. + \sum_{n=1}^{\infty} b_n R_n \, exp \left(\frac{-\beta_n^2 x/r_0)}{Re \, Pr} \right) \right] \tag{3.92}$$

This is the complete solution. The terms b_n and λ (or β_n^2) are to be determined from Eqs. (3.90) and (3.91). The first seven values and functions of Eq. (3.85) were obtained by Siegel, Sparrow and Hallmann [6] and are given in Table 3.1.

Table 3.1
Eigenvalues and eigenfunctions

n	β_n^2	$R_n(1)$	b_n
1	25.6796	-0.492597	0.403483
2	83.8618	0.395508	-0.175111
3	174.167	-0.345872	0.105594
4	296.536	0.314047	-0.732804
5	450.947	-0.291252	0.0550357
6	637.387	0.273808	-0.043483
7	855.850	-0.259852	0.035597

Now we proceed to find the Nusselt number

$$\theta_w = \frac{T_w}{q_w'' r_0/k} \ , \quad \theta_{fd} = \frac{T_{fd}}{q_w'' r_0/k} \ , \quad \theta_0 = \frac{T_0}{q_w'' r_0/k}$$

From Eq. (3.87)

$$\theta_d(\tilde{x},1) = \theta_w - \theta_{fd} = \sum_{n=1}^{\infty} b_n R_n(1) e^{-\beta_n^2 \tilde{x}} \tag{3.93}$$

From Eq. (3.76) we can write

$$(\theta_{fd} - \theta_0) = 4\tilde{x} + \tilde{r}^2 - \frac{1}{4}\tilde{r}^4 - \frac{7}{24} \tag{3.94}$$

Adding Eqs. (3.93) and (3.94) at $\tilde{r} = 1$

$$\theta_w - \theta_0 = 4\tilde{x} + \left(1 - \frac{1}{4} - \frac{7}{24}\right) + \sum_{n=1}^{\infty} b_n R_n(1) e^{-\beta_n^2 \tilde{x}}$$

or

$$(\theta_w - \theta_0) = 4\tilde{x} + \frac{11}{24} + \sum_{n=1}^{\infty} b_n R_n(1) e^{-\beta_n^2 \tilde{x}} \tag{3.95}$$

From energy balance:

$$T_m = T_{mi} + \frac{2q_w'' x}{\rho c_p u_m r_0}$$

Taking $T_{mi} = T_0$

$$(T_m - T_0) = \frac{4q_w'' r_0}{k} \left(\frac{x/r_0}{Re \ Pr}\right)$$

or

$$(\theta_m - \theta_0) = 4\tilde{x} \tag{3.96}$$

from Eqs. (3.95) and (3.96) we get

$$(\theta_w - \theta_m) = \frac{11}{24} + \sum_{n=1}^{\infty} b_n R_n(1) e^{-\beta_n^2 \tilde{x}}$$

By definition,

$$Nu_D = \frac{h 2 r_0}{k} = \frac{q_w''}{(T_w - T_m)} \frac{2 r_0}{k} = \frac{2}{\theta_w - \theta_m}$$

The Nusselt number is:

$$Nu_D = \frac{2}{\frac{11}{24} + \sum_{n=1}^{\infty} b_n R_n(1) e^{-\beta_n^2 \tilde{x}}}$$

For $x \to \infty$

$$Nu_D|_{fd} = \frac{48}{11} = 4.36 \tag{3.97}$$

The result matches the case of hydrodynamically and thermally fully developed flow. The Graetz problem was revisited to include the viscous dissipation [5].

Special Remarks

The equation below represents an alternative way of solving the equation.

$$\tilde{r}\left(\frac{d^2 R}{d\tilde{r}^2}\right) + \left(\frac{dR}{d\tilde{r}}\right) + R \tilde{r} \beta_n^2 (1 - \tilde{r}^2) = 0$$

We can solve this equation with the fourth order Runge–Kutta method by integrating it from $\tilde{r} = 0$ to $\tilde{r} = 1$. However, we need a starting boundary condition on R.

If we closely observe the above equation, we can see that if we multiply the eigenfunction by any arbitrary value Ω, our solution still remains valid. Hence, we can normalize the eigenfunction such that $R_n(0) = 1$, and the solution is still valid. Therefore, the Runge–Kutta formulation appears to be

$$R' = F$$
$$at \ \tilde{r} = 0, \ \ R_n(0) = 1$$
$$F' = -\tilde{r} F + R \beta_n^2$$
$$at \ \tilde{r} \ = 0, \ \ R_n'(0) \ \ or \ \ F(0) = 0$$

We have another boundary condition, that is $R_n'(1) = 0$, which we use as our final target, as we are going to apply shooting technique.

To get approximation for β_n^2, we can refer to the work of Siegel et al. [6], where an approximate solution of β_n^2 is given as $\beta_n^2 = \left\{4n + \frac{4}{3}\right\}^2$. However, this approximation is only near to the actual value of β_n^2 and its value needs to be accurately calculated (after invoking the approximation) by a Newton–Raphson technique. A sample computer code is given in Appendix 3B at the end of this chapter.

3.6 SUMMARY OF SOLUTIONS FOR VARIOUS FLOW AND HEAT TRANSFER SITUATIONS IN PIPE FLOWS

A marvelous compilation of different results due to various investigations are available in Shah [4] and Shah and London [7]. Fig. 3.12 provides the results in a comprehensive manner. Nu_D is plotted against the dimensionless parameter $\frac{x/D}{Re\ Pr}$ or

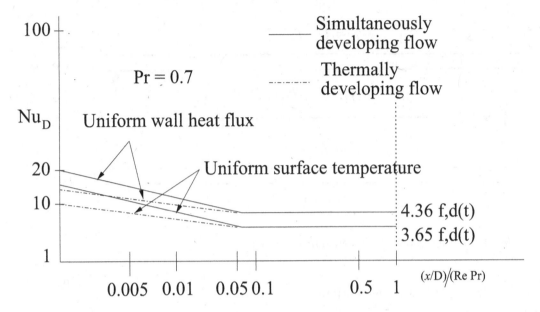

Figure 3.12 Variation of Nusselt number in a circular duct

reciprocal of Graetz number ($G \equiv \frac{D\ Re\ Pr}{x}$). Fully developed values are independent of Prandtl number. Fully developed conditions are reached for $\frac{x/D}{Re\ Pr} \approx 0.05$.

For the constant surface temperature condition, it is desirable to know the average convection coefficient (in the entry length) as $q_{conv} = \overline{h}\ A_w\ \Delta T_{l,m}$. Hausen presents the following correlation for hydrodynamically fully developed flow (laminar) in tubes at constant wall temperature

$$\overline{Nu}_D = 3.66 + \frac{0.0668(D/L)Re_D\ Pr}{1 + 0.04[(D/L)Re_D\ Pr]^{2/3}}$$

Because the above result is for thermally developing flow, it is not generally applicable. For simultaneously developing flow, a suitable correlation due to Sieder and Tate [8] is of the form:

$$\overline{Nu}_D = 1.86 \left[\frac{Re_D\ Pr}{L/D}\right]^{1/3} \left(\frac{\mu}{\mu_w}\right)^{0.14}$$

All properties are evaluated at $T_m = (T_{m,0} + T_{m,i})/2$; μ_w is the viscosity of the fluid at wall temperature.

3.7 HEAT TRANSFER IN COUETTE FLOW

Figure 3.13 shows two infinite parallel plates at a distance of 2H apart. The upper plate moves with a speed v relative to the lower plate. The pressure p is assumed to be constant. The temperature of the upper plate is T_{w1} and the temperature of the lower plate is T_{w0}.

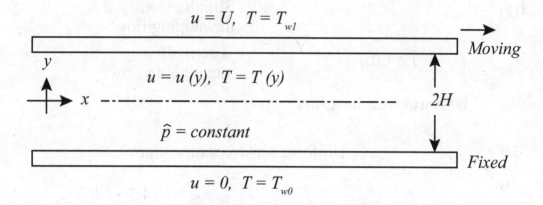

Figure 3.13   Couette flow between parallel plates

For such steady flows, the continuity, momentum and energy equations are

$$\frac{\partial u}{\partial x} = 0 \tag{3.98}$$

$$\mu \frac{d^2 u}{dy^2} = 0 \tag{3.99}$$

$$k \frac{d^2 T}{dy^2} + \mu \left(\frac{du}{dy} \right)^2 = 0 \tag{3.100}$$

Also, $u = u(y)$ and $T = T(y)$. Eq. (3.99) can be integrated twice to obtain

$$u = C_1 y + C_2$$

The boundary conditions at $y = (-H), u = 0$ and at $y = (+H), u = U$. Invoking the boundary conditions we get $C_1 = U/2H$ and $C_2 = U/2$.

The velocity distribution becomes

$$u = \frac{U}{2} \left(1 + \frac{y}{H} \right) \tag{3.101}$$

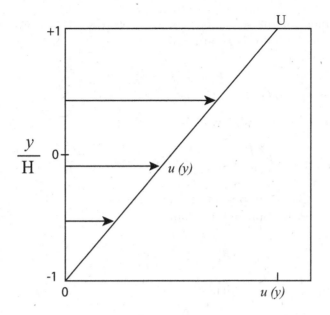

Figure 3.14 Velocity distribution in Couette flow

This is plotted in Fig. 3.14 and observed to be a straight line. Shear stress at any point in the flow is given by

$$\tau = \mu \frac{du}{dy} = \frac{\mu U}{2H} \tag{3.102}$$

The skin friction at the wall is

$$C_f = \frac{\tau_w}{\frac{1}{2}\rho U^2} = \frac{\mu}{\rho U H} = \frac{1}{Re_H} \tag{3.103}$$

Also one can define Poiseuille number as

$$P_0 = C_f Re_H = \frac{2H\tau_w}{\mu U} = 1 \tag{3.104}$$

From (3.101), we get $\frac{du}{dy} = \frac{U}{2H}$. From (3.100) we can write

$$\frac{d^2T}{dy^2} = -\frac{\mu}{k}\frac{U^2}{4H^2}$$

$$T = -\frac{\mu U^2}{4kH^2}\frac{y^2}{2} + C_3 y + C_4 \tag{3.105}$$

The thermal conditions require at $y = (-H), T = T_{w0}$ and at $y = (+H), T = T_{w1}$. We obtain $C_3 = \frac{(T_{w1}-T_{w0})}{2H}$ and $C_4 = \frac{(T_{w1}+T_{w0})}{2} + \frac{\mu U^2}{8k}$.

The final solution becomes

$$T = \left(\frac{T_{w1}+T_{w0}}{2} + \frac{T_{w1}-T_{w0}}{2}\frac{y}{H}\right) + \frac{\mu U^2}{8k}\left(1 - \frac{y^2}{H^2}\right) \tag{3.106}$$

T can be nondimensionalized by $(T_{w1} - T_0)$. From the second term of the RHS of Eq. (3.106) we get the following parameter

$$\frac{\mu U^2}{k(T_{w1} - T_{w0})} = \frac{\mu c_p}{k} \cdot \frac{U^2}{c_p(T_{w1} - T_{w0})} = PrEc \tag{3.107}$$

The rate of heat transfer at the walls is

$$q_w = k\frac{\partial T}{\partial y}\big|_{\pm H} = \frac{k}{2H}(T_{w1} - T_{w0}) \pm \frac{\mu U^2}{4H} \tag{3.108}$$

The symbol $\pm H$ refers to upper and tower surfaces.
The heat transfer coefficient may be defined as

$$h = \frac{q_w}{T_{w1} - T_{w0}} \tag{3.109}$$

The Nusselt number may be calculated as

$$Nu_{2H} = \frac{h(2H)}{k} = 1 \pm \frac{PrEc}{2} \tag{3.110}$$

In the absence of strong viscous dissipation, $Nu_{2H} = 1$.

3.8   CONVECTION CORRELATIONS FOR NON-CIRCULAR TUBES

To a first approximation, many of the circular tube results may be applied by using hydraulic diameter as characteristic length which is

$$D_h = \frac{4Ac}{P}$$

Here, Ac is flow cross sectional area and P is wetted perimeter. However, in a non-circular tube, the convection coefficients vary around the periphery, approaching zero in the corners.

For the turbulent flows (i.e., for $Re_{D_h} > 2300$) often it is reasonable to use the modified $(D \rightarrow D_h)$ circular tube correlations. However, for laminar flow, the use of circular tube correlations is less accurate. For such cases Nusselt numbers corresponding to fully developed conditions may be obtained from Shah and London [7]. Some results are shown in Table 3.2.

Table 3.2
Nusselt numbers for fully developed laminar flow in tubes of differing cross sections

Cross section	$\dfrac{b}{a}$	$Nu_D = (hD_h)/k$	
		uniform q''_w	uniform T_w
○	–	4.36	3.66
▢ (a, b = 1.0)	1.0	3.61	2.98
▭ (a, b)	1.43	3.73	3.08
▭ (a, b)	2.0	4.12	3.39
▭ (a, b)	3.0	4.79	3.96
"	4.0	5.33	4.44
"	8.0	6.49	5.60
(parallel plate)	∞	8.23	7.54
△		3.00	2.35

REFERENCES

1. L. F. Moody, 1944, Friction factors for pipe flow, ASME Trans, Vol. 66, pp. 671-684.

2. L. Graetz, 1885, Uber die Warmeleitungs fahigkeit von Flussigkeiten, Part 1, Ann. Phys. Chem., Vol. 18, 1883, pp. 79-94, Part 2, Ann. Phys, Chem, Vol. 25, pp. 337-357.

3. Sellars, J. R., Tribus, M., and Klein, J. S., 1956, Heat Transfer to Laminar Flow in a Round Tube or Flat Conduit: The Graetz Problem Extended, Trans ASME, Vol. 78, pp. 441-448.

4. Shah, R. K., 1975, Laminar Flow Friction and Forced Convection Heat Transfer in Ducts of Arbitrary Geometry, Int. J. Heat Mass Transfer, Vol. 18, pp. 849-862.

5. Basu, T., and Roy, D. N., 1985, Laminar Heat Transfer in a Tube with Viscous Dissipation, Int. J. Heat Mass Transfer, Vol. 28, pp. 699-701.

6. Siegel, R., Sparrow, E. M., and Hallman, T. M., 1958, Steady Laminar Flow Heat Transfer in a Circular Tube with a Prescribed Wall Heat Flux, Appl. Scient. Res., A7, pp. 386-392.

7. Shah, R. K., and London, A. L., 1978, Advances in Heat Transfer, Laminar Flow Forced Convection in Ducts, Academic Press, New York.

8. Sieder, E. N., and Tate, G. E., 1936, Heat Transfer and Pressure Drop of Liquids in Tubes, Ind. Eng. Chem., Vol. 28, pp. 1429-1436.

EXERCISES

1. In a hydrodynamically developed and thermally developed flow through a tube of radius r_0, constant wall heat flux boundary condition is applied. The governing equation is given by

$$u\frac{\partial T}{\partial x} + v\frac{\partial T}{\partial r} = \frac{\alpha}{r}\frac{\partial}{\partial r}\left(r\frac{\partial T}{\partial r}\right)$$

It is known that $\frac{dT_m}{dx} = \frac{q_w'' P}{\dot{m}c_p}$ where T_m is the bulk mean temperature, q_w'' is the wall heat flux. P is the perimeter of the duct, $\dot{m}$ is the mass flow rate and c_p is the specific heat and constant pressure. It is also known that at $r = 0, T$ is finite and at $r = r_0, T = T_w$. Show that the Nusselt number based on the duct diameter at any x is 4.36

Furthermore show that the temperature distribution is given by

$$T(r) = T_m + \frac{q_w R}{k}\left[\left(\frac{r}{r_0}\right)^2 - \frac{1}{4}\left(\frac{r}{r_0}\right)^4 - \frac{7}{24}\right]$$

2. A viscous fluid is flowing through a straight pipe of inner radius r_0. Due to viscous dissipation, the fluid tends to warm up as it flows through the pipe. In order to control this effect, cooling is provided all along the pipe wall, which is isothermal (T_w= constant). The flow is hydrodynamically and thermally fully developed. The energy equation for the fluid with constant properties reduces in this case to

$$k\frac{1}{r}\frac{d}{dr}\left(r\frac{\partial T}{\partial r}\right) + \mu\phi = 0$$

where ϕ is the viscous dissipation term $\left(\phi = \left(\frac{dv_z}{dr}\right)^2\right)$ and where $v_z(r)$ is the fully developed velocity profile given by $v_z(r) = \frac{C_o}{4\mu}(r^2 - r_o^2)$ and $C_o = \frac{dp}{dz}$.

(a) Determine the temperature distribution $T(r)$ in the fluid.
(b) Calculate the total heat transfer rate, if the pipe length is L.

3. Consider a hydrodynamically and thermally fully developed flow through a tube of radius R. The velocity profile is parabolic. The governing equation for the radial variation is given by $\theta'' + (\theta'/\eta) + \lambda^2(1 - \eta^2)\theta = 0$. A constant temperature boundary condition $(T = T_w)$ is maintained at the wall. The non-dimensional temperature is given by $\theta = (T - T_w)/(T_c - T_w)$. T_c is the centerline temperature. Show that the radial variation of the temperature is given by

$$\theta = \Sigma_{n=0}^{\infty} \quad C_{2n} \, \eta^{2n}$$

where $C_{2n} = \frac{\lambda^2}{(2n)^2}[C_{2n-4} - C_{2n-2}]$, $C_2 = -\frac{\lambda^2}{4}C_0, C_0 = 1$, and $\eta = r/R$. The parameter, $-\lambda^2$ is the separation constant, used while applying the separation of variables technique.

4. Consider a concentric tube annulus. Focus your attention in the annular region. The outer surface of the tube is insulated (heat flux is zero). The inner surface has a uniform heat flux. Hydrodynamicallly and thermally fully developed laminar flow may be assumed to exist.
   (a) Determine the velocity profile, $u(r)$ in the annular region.
   (b) Determine the temperature profile, $T(r)$.

   Hints: Start with the governing equation for flow

$$\frac{\mu}{r}\frac{d}{dr}\left(r\frac{du}{dr}\right) = \frac{dp}{dx}$$

   The boundary conditions are

$$u(r_i) = 0, \quad u(r_0) = 0$$

   Calculate the velocity profile. The normal component of velocity is zero. The energy equation for this problem can be written as

$$\frac{1}{r}\frac{\partial}{\partial r}\left(r\frac{\partial T}{\partial r}\right) = \frac{u}{\alpha}\frac{dT_m}{dx}$$

   The thermal boundary conditions are: $q_o'' = 0$ and $q_i'' = $ constant, also $T(r_i) = T_{w,i}$.

5. Consider slug flow through a heated tube of radius R. The flow is thermally fully developed. The constant wall temperature condition $(T_w =$ constant$)$ is maintained at the tube wall. The temperature of the fluid at the entrance is given by T_∞. You are required to find the solution, i.e., temperature distribution $T(x,r)$ at any distance, x from the entry.

(a) First find out the expression for the solution in terms of the unknown constants.

(b) Clearly indicate the boundary conditions you need to determine the constants.

(c) Which boundary condition will lead to an eigenvalue problem and which one would necessitate the use of orthogonal relations?

6. In a hydrodynamically developed and thermally developing flow through a tube of radius R constant wall heat flux boundary condition is applied. The solution of the thermally fully developed flow field is

$$T(r) = T_w - \frac{2u_m R^2}{\alpha} \left(\frac{dT_m}{dx}\right) \left[\frac{3}{16} + \frac{1}{16}\left(\frac{r}{R}\right)^4 - \frac{1}{4}\left(\frac{r}{R}\right)^2\right]$$

An unheated length (velocity is fully developed) precedes the heated section. The quantity

$$\frac{dT_m}{dx} = \frac{q_w'' P}{\dot{m} c_p}$$

Show that

$$T(r) = T_w + \frac{q_w'' R}{k} \left[\left(\frac{r}{R}\right)^2 - \frac{1}{4}\left(\frac{r}{R}\right)^4 - \frac{7}{24}\right]$$

Also from the energy balance in a tube, one can obtain $T_m(x) = T_{m,i} + \frac{q_w'' P}{\dot{m} c_p} x$, where $P = 2\pi R$, $\dot{m} = \rho u_m \pi R^2$. Find a relation between the fully developed temperature, T_{fd} in the far downstream and the bulk mean temperature T_o at a location where the unheated initial length prevails. Derive the boundary condition for T_d at the tube wall. Given, $T_d = (T - T_{fd})$.

7. Slug flow is an idealized pipe flow for which velocity profile does not change in the flow direction and it is uniform over the entire pipe cross-section. For the case of a laminar slug flow through a pipe with the uniform heat flux boundary condition, determine the temperature profile $T(r)$ of the fluid in a pipe with radius R when the flow is thermally fully developed. Also find the Nusselt number based on the pipe diameter Nu_D.

8. Consider a cylindrical rod (heating element) of length L and diameter D that is enclosed with a concentric tube. Water flows through the annular region between the rod and the tube at a rate $\dot{m}$. The outer surface of the tube is insulated. Heat generation occurs within the rod, and the volumetric generation rate is known to vary with the distance along the element. The variation is given by

$$\dot{q}(x) = \dot{q}_o(x/L)^2 \text{ where } \dot{q}_o(W/m^3) \text{ is a constant}$$

A convection coefficient h exists between the surface of the rod and the water.

(a) Obtain an expression for the local heat flux, $q''(x)$ and the total heat transfer, Q from the heating element to the water. (b) Obtain an expression for the variation of bulk mean temperature, $T_m(x)$ of the water with distance x along the tube. (c) Obtain an expression for the variation of the surface temperature of the heating element, $T_w(x)$ with distance x along the tube.

9. In Problem 3, the value of λ can be determined by a Newton–Raphson technique. Let us assume that its value is $\lambda = 2.7043$ (need not prove). The governing equation may be represented by its integral form as

$$\int_0^R \rho c_p u \frac{\theta}{R} \frac{dT_c}{dz} r dr = Rk \frac{\partial T}{\partial r}\big|_{r=R}$$

Show that the Nusselt number at any x, based on the tube diameter is 3.56.

10. Refer to Problem 5. The solution of the temperature field is given by

$$\theta = 2 \sum_{n=1}^{\infty} \frac{J_0(\beta_n Y)}{\beta_n J_1(\beta_n)} e^{-\beta_n X}$$

Where β_n are the eigenvalues, J_o is the Bessel function of the first kind (order zero) and X is the non-dimensional axial distance from the entry.

Find the bulk mean temperature at any distance from the entry. Show that the Nusselt number for such flows is 5.7831 (given β_1 2.4048).

APPENDIX 3A

Bessel's equation is given by

$$x^2 \frac{d^2y}{dx^2} + x \frac{dy}{dx} + (x^2 - n^2)y = 0 \tag{3.111}$$

in which n is a constant known as Bessel's equation of order n; its solutions are known as Bessel functions. Bessel used them in the 19th century to solve a problem of dynamic astronomy. The two solutions of Bessel's equation are

(i) Bessel function of first kind (Order n) given by

$$J_n(x) = \left(\frac{x}{2}\right)^n \Sigma_{m=0}^{\infty} \frac{(-1)^m}{m!\Gamma(m+n+1)} \left(\frac{x}{2}\right)^{2m} \tag{3.112}$$

(ii) Bessel function of second kind (order n) given by

$$Y_n(x) = \frac{2}{\pi}\left\{\ell n\left(\frac{x}{2}\right) + \gamma\right\} J_n(x) - \frac{1}{\pi} \Sigma_{m=0}^{n-1} \frac{(n-m-1)!}{m!} \left(\frac{x}{2}\right)^{2m-n} \tag{3.113}$$

where $\gamma = 0.5772156$ is Euler's constant. When $n = 0$, the first solution of Bessel's equation is given by

$$J_0(x) = \Sigma_{m=0}^{\infty} \frac{(-1)^m}{m!\Gamma(m+1)} \left(\frac{x}{2}\right)^{2m} = 1 - \frac{x^2}{2^2} + \frac{x^4}{2^2 4^2} - \frac{x^6}{2^2 4^2 6^2} + \cdots \tag{3.114}$$

Similarly the second solution becomes

$$Y_0(x) = \frac{2}{\pi}\left\{\ell n\left(\frac{x}{2}\right) + \gamma\right\} J_0(x) - V_0(x) \cdots \tag{3.115}$$

$Y_0(x)$ is called the Bessel function of the second kind of order zero and when x is small, we may write

$$Y_0(x) = \frac{2}{\pi}(\ln x - (\ln 2 - \gamma)) \tag{3.116}$$

The complete solution of Bessel's equation of order zero is

$$y = AJ_0(x) + BY_0(x) \tag{3.117}$$

Both $J_o(x)$ and $Y_o(x)$ are oscillatory functions and their graphs are shown in Fig. 3.15. It should be noted that both $J_0(x)$ and $Y_0(x)$ vanish at an infinite sequence of values of x. In this respect they behave similarly to the trigonometrical functions $cos x$ and $sin x$ which vanish respectively when $x = \left(m + \frac{1}{2}\right)\pi$ and $x = m\pi$.

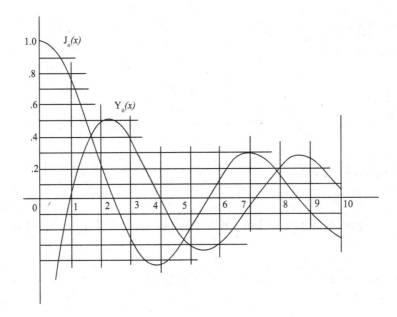

Figure 3.15 Bessel's functions of order zero

APPENDIX 3B

The following program solves the Sturm–Liouville equation as a solution of the Graetz problem. The method of the solution aims to employ Runge–Kutta method of order 4 to solve the Sturm–Liouville equation. As the obtained solution should satisfy the boundary conditions of governing differential equations, a shooting technique is used to evaluate the eigenvalues. To speed up the computation the eigenvalue is first approximated from the analytical expression, and then the exact value of the eigenvalue is obtained using Newton–Raphson method.

```
*/
# include <stdio.h>
# include <stdlib.h>
# include <conio.h>
# include <string.h>
int main()
{

// declaring the variable to be used
double f,rn,beta,r,fs,rns,rs,k1,l1,k2,l2,k3,l3,k4,l4,h,fm,betam,betan,zz,bn,
deno,deno1,num,num1;
int i,n;
FILE *fp;
// creating the file to write the final output. Note that this is not a required step
fp = fopen("eigenvalues and eigenfunctions.txt","w+");
```

```
//RK4 computation

f = 0.0;
rn = 1.0;
h = 0.000001;
beta = 80.0;

//f(1) at i-1
for (n = 1; n < 40; n++)
{
```

//general first approximation of beta which will be modified using Newton-Raphson method

```
beta = ((4.0*n) + (4.0/3.0))*((4.0*n) + (4.0/3.0));
for(i = 1; i > 0; i++)


    {
fm = f;
f = 0.0;
rn = 1.0;

for(r = 0.000001; r < 1.000001; r = r + 0.000001)
{

fs = f;
rns = rn;
rs = r;

k1 = h*fs;
l1 = h*(-(rns*beta*(1.0 - (rs*rs))) - (fs/rs));

fs = f + (l1*0.5);
rns = rn + (k1*0.5);
rs = r + (h*0.5);

k2 = h*fs;
l2 = h*(-(rns*beta*(1.0 - (rs*rs))) - (fs/rs));

fs = f + (l2*0.5);
rns = rn + (k2*0.5);
rs = r + (h*0.5);

k3 = h*fs;
```

```
l3 = h*(-(rns*beta*(1.0 - (rs*rs))) - (fs/rs));

fs = f + (l3);
rns = rn + (k3);
rs = r + (h);

k4 = h*fs;
l4 = h*(-(rns*beta*(1.0 - (rs*rs))) - (fs/rs));

rn = rn + (1.0/6.0)*(k1 + 2.0*k2 + 2.0*k3 + k4);
f = f + (1.0/6.0)*(l1 + 2.0*l2 + 2.0*l3 + l4);


       }
zz = f - 0.0;
if(zz ¡ 0){zz = zz*(-1.0);}
if(zz > 0.00001)
       {
if(i == 1 || i == 2){betan = beta + 0.5; betam = beta; beta = betan;}
else{
if(i == 3){fm = 0.0;}
betan = beta - ((f)*(beta - betam)/(f - fm));
betam = beta;
beta = betan;}
       }
else{break;}
       }
//values for beta obtained
f = 0.0;
rn = 1.0;
num1 = 0;
deno1 = 0;
num = 0;
deno = 0;
for(r = 0.000001; r < 1.000001; r = r + 0.000001)
{
fs = f;
rns = rn;
rs = r;

k1 = h*fs;
l1 = h*(-(rns*beta*(1.0 - (rs*rs))) - (fs/rs));

fs = f + (l1*0.5);
rns = rn + (k1*0.5);
```

```
rs = r + (h*0.5);

k2 = h*fs;
l2 = h*(-(rns*beta*(1.0 - (rs*rs))) - (fs/rs));

fs = f + (l2*0.5);
rns = rn + (k2*0.5);
rs = r + (h*0.5);

k3 = h*fs;
l3 = h*(-(rns*beta*(1.0 - (rs*rs))) - (fs/rs));

fs = f + (l3);
rns = rn + (k3);
rs = r + (h);

k4 = h*fs;
l4 = h*(-(rns*beta*(1.0 - (rs*rs))) - (fs/rs));

rn = rn + (1.0/6.0)*(k1 + 2.0*k2 + 2.0*k3 + k4);
f = f + (1.0/6.0)*(l1 + 2.0*l2 + 2.0*l3 + l4);

num1 = (r*r - (0.25*pow(r,4.0)) - (7.0/24.0))*(r - (r*r*r))*rn*0.000001;
num = num + num1;

deno1 = (r - (r*r*r))*(rn*rn)*0.000001;
deno = deno + deno1;

    }
bn = -num/deno;

    //printing the eigenvalues and eigenfunction
printf("%d %lf %lf %lf  n",n,beta,rn,bn);
fprintf(fp,"%d %lf %lf %lf  n",n,beta,rn,bn);

    }
}
```

## CHAPTER 4

Solution of Complete Navier–Stokes and Energy Equations for Incompressible Internal Flows

4.1 INTRODUCTION

Different basic flow configurations are observed in different parts of the heat exchangers, for example, the fluid may flow between closely spaced plates that effectively form the configuration of flow through a duct. Although laminar duct flows do not occur as extensively as turbulent duct flows, they do occur in a number of important situations. In the case of plate-fin heat exchangers, the fin spacing is so small and the mean velocity range is such that the flows are often laminar. Conventionally, it is usual to assume that a higher heat transfer rate is achieved with turbulent flow than with laminar flow. However, sometimes a design that involves laminar flow is more efficient from the viewpoint of energy transfer since associated pressure penalty is less.

In this chapter, we shall consider hydrodynamically developing flow in a plane channel. Flow and heat transfer in a duct are considered in this chapter. The flow situation pertaining to its application is shown in Fig. 4.1. Even though computation techniques related to convective heat transfer in a duct are discussed here, the analysis for the pipe flow is a straightforward extension.

4.2 SOLUTION OF NAVIER STOKES EQUATIONS IN CARTESIAN COORDINATE

In Cartesian coordinates, weak conservative forms of the governing equations for incompressible three-dimensional flows are

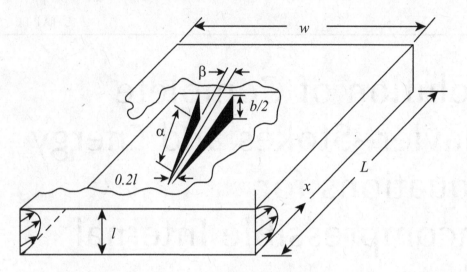

Figure 4.1 Flow in a narrow channel with built-in obstacle [1]

$$\frac{\partial u}{\partial x} + \frac{\partial v}{\partial y} + \frac{\partial w}{\partial z} = 0 \tag{4.1}$$

$$\frac{\partial u}{\partial t} + \frac{\partial (u^2)}{\partial x} + \frac{\partial (uv)}{\partial y} + \frac{\partial (uw)}{\partial z} = -\frac{\partial p}{\partial x} + \frac{1}{Re}\left[\frac{\partial^2 u}{\partial x^2} + \frac{\partial^2 u}{\partial y^2} + \frac{\partial^2 u}{\partial z^2}\right] \tag{4.2}$$

$$\frac{\partial v}{\partial t} + \frac{\partial (uv)}{\partial x} + \frac{\partial (v^2)}{\partial y} + \frac{\partial (vw)}{\partial z} = -\frac{\partial p}{\partial y} + \frac{1}{Re}\left[\frac{\partial^2 v}{\partial x^2} + \frac{\partial^2 v}{\partial y^2} + \frac{\partial^2 v}{\partial z^2}\right] \tag{4.3}$$

$$\frac{\partial w}{\partial t} + \frac{\partial (uw)}{\partial x} + \frac{\partial (vw)}{\partial y} + \frac{\partial (w^2)}{\partial z} = -\frac{\partial p}{\partial z} + \frac{1}{Re}\left[\frac{\partial^2 w}{\partial x^2} + \frac{\partial^2 w}{\partial y^2} + \frac{\partial^2 w}{\partial z^2}\right] \tag{4.4}$$

We aim at solving the complete Navier–Stokes equations. The procedure described in this section is based on finite difference discretization and on the solution of a Poisson equation to determine the pressure. The method uses primitive variables u, v, w and p as function of x, y, z, t and Re.

4.2.1 Staggered Grid

The major difficulty encountered during solution of incompressible flow is the non-availability of any obvious equation for pressure. Different grid-mesh for each of the dependent variables may be conceptualized. Such a staggered grid for the dependant variables in a flow field was first used by Harlow and Welch [2], in their very well known MAC (marker and cell) method. Since then, it has been used by many researchers. Specifically, SIMPLE (semi-implicit method for pressure-linked equations) procedure of Patankar and Spalding [3] is very popular. Fig. 4.2 shows a two dimensional staggered grid where independent variables ($u_{i,j}, v_{i,j}$ and $p_{i,j}$) with the identical indices are staggered to one another. Computational domain can be divided into a number of cells, which are shown as "main control volume" in

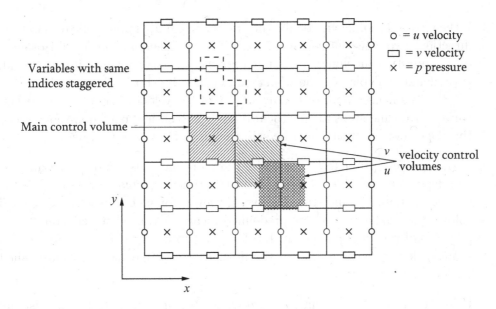

Figure 4.2 Staggered grid

Fig. 4.2. The velocity components are located at the center of the cell faces to which they are normal. In such cases the pressure difference between the two adjacent cells appears as the driving force for the velocity component located at the interface of these cells. The finite difference approximation is now physically meaningful and the pressure field accepts a reasonable pressure distribution for a plausible velocity field.

Another important advantage is that transport rates across the faces of the control volumes can be computed without interpolation of velocity components. The detailed outline of the solution procedures for the full Navier–Stokes equations with primitive variables using staggered grid will be discussed in the subsequent sections.

4.2.2 Introduction to MAC method

The MAC method of Harlow and Welch [2] is one of the earliest and most useful methods for solving the Navier–Stokes equations for incompressible flows. A modified version of the original MAC method due to Hirt and Cook [4] has been deployed by the researchers to solve a variety of flow and heat transfer problems.

The text discusses the modified MAC method and highlights the salient features of the solution algorithm so that the readers are confident when they write a computer program easily. The MAC algorithm consists of the following:

1. Unsteady Navier–Stokes equations for incompressible flows in weak conservative form and the continuity equation are the governing equations.

2. The mathematical character of the equations is elliptic in space and parabolic in time. The solution proceeds in the time direction. At each time step, a converged solution in space is obtained but any converged solution at a given time step is the intermediate solution of the physical problem.

3. If the physical problem is steady, then after some finite number of steps in time direction, two consecutive time-steps will show identical solutions. However, in a computing machine this is not possible and a very small upper bound, is predefined. Typically, the upper bound may be chosen between 10^{-3} and 10^{-5}. If the maximum discrepancy of any of the velocity components for two consecutive time steps at any location over the entire space does not exceed the upper bound, then it can be said that the steady solution has evolved.

4. If the physical problem is unsteady, the aforesaid maximum discrepancy of any dependent variable for two consecutive time steps may not fall below the predefined upper bound stated above. However, for such a situation, specified velocity components at some predefined locations can be stored over a long duration of time and plots of such velocity component with time (often called signals) depict the character of the flow. At this stage such flows are called unstready flows.

5. Using the unsteady momentum equations, we compute explicitly a provisional value of the velocity components for the next time step.

Consider the weak conservative form of the nondimensional momentum equation in the x direction:

$$\frac{\partial u}{\partial t} + \frac{\partial (u^2)}{\partial x} + \frac{\partial (uv)}{\partial y} + \frac{\partial (uw)}{\partial z} = -\frac{\partial p}{\partial x} + \frac{1}{Re}\nabla^2 u \qquad (4.5)$$

It is assumed that at $t = n^{th}$ level, we have a converged solution. Then for the next time step we can write

$$\frac{\tilde{u}_{i,j,k}^{n+1} - u_{i,j,k}^{n}}{\delta t} = [\text{CONDIFU-DPDX}]_{i,j,k}^{n} \qquad (4.6)$$

or

$$\tilde{u}_{i,j,k}^{n+1} = u_{i,j,k}^{n} + \delta t \ [\text{CONDIFU-DPDX}]_{i,j,k}^{n} \qquad (4.7)$$

$[CONDIFU - DPDX]_{i,j,k}^{n}$ consists of discretized convective and diffusive terms, and the discretized form of the pressure gradient. In a similar manner the provisional values for $\tilde{v}_{i,j,k}^{n+1}$ and $\tilde{w}_{i,j,k}^{n+1}$ can be explicitly computed. These explicitly advanced velocity components may not constitute a realistic flow field. Now, with these provisional $\tilde{u}_{i,j,k}^{n+1}$, $\tilde{v}_{i,j,k}^{n+1}$ and $\tilde{w}_{i,j,k}^{n+1}$ values, continuity equation is evaluated in each cell. The non-zero value of $(\nabla \cdot \mathbf{V})$ in any cell indicates mass accumulation or annihilation in that cell which is not physically possible. Therefore the pressure at any cell is directly linked with the value of the $(\nabla \cdot \mathbf{V})$ of that cell. On one hand the pressure has to be corrected with the help of the nonzero divergence value and on the other, the velocity components have to be adjusted. The correction procedure continues through an iterative cycle till the divergence free velocity field is ensured in every cell. A Poisson equation for pressure can be formed for this purpose.

6. Boundary conditions are to be applied after the explicit evaluation of the velocity components using the Navier–Stokes equations through each time step. Since the governing equations are elliptic in space, boundary conditions on all confining surfaces are required. The boundary conditions are applied after every pressure-velocity iteration. The five principal boundary conditions to be considered are rigid no-slip walls, free-slip walls, inflow and outflow boundaries, and periodic (repeating) boundaries.

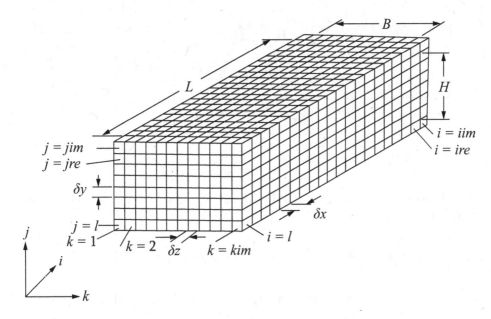

Figure 4.3 Domain discretization of a three-dimensional domain

Formulation of Problem

The computational domain is divided into a set of small cells having edge lengths $\delta x, \delta y$ and δz (Fig. 4.3). In each cell, velocity components are located at the center of the cell faces to which they are normal and pressure and temperature are defined at the center of the cell. Cells are labeled with an index (i,j,k) which denotes the cell number as counted from the origin in the x, y and z directions respectively. Also $p_{i,j,k}$ is the pressure at the center of the (i,j,k) cell, while $u_{i,j,k}$ is the x-direction velocity at the center of the face between cells (i,j,k) and (i+1, j, k) and so on (Fig. 4.4). Similarly $v_{i,j,k}$ is the y-direction velocity located at the center of the face between the cells (i,j,k) and (i,j+1,k) and so on (Fig 4.4). The velocities are not defined at the nodal points, but whenever required, they are to be found by interpolation. For example, with uniform grids the x-direction velocity at the center of (i,j,k) cell as given by $u_{i-1/2,j,k}$ and we can write $u_{i-1/2,j,k} = \frac{1}{2}[u_{i-1,j,k} + u_{i,j,k}]$. Where a product or square of such quantities appears, it is to be interpolated first and then the product is to be formed. Convective terms are discretized using a weighted averaged of second upwind and space centered scheme (Hirt et al. [4]). Diffusive terms are discretized by a central differencing scheme. Let us consider the discretized terms of the x-momentum equation (Fig. 4.4):

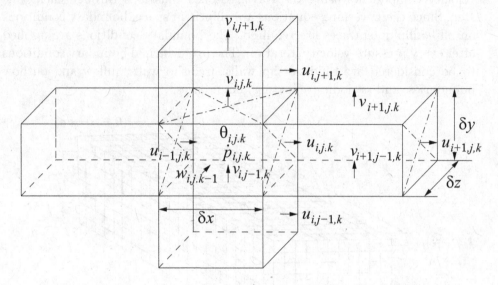

Figure 4.4 Three-dimensional staggered grid showing the locations of the discretized variables

$$\frac{\partial(u^2)}{\partial x} = \frac{1}{4\delta x}[(u_{i,j,k} + u_{i+1,j,k})(u_{i,j,k} + u_{i+1,j,k})$$

$$+ \quad \alpha|(u_{i,j,k} + u_{i+1,j,k})|(u_{i,j,k} - u_{i+1,j,k})$$

$$- \quad (u_{i-1,j,k} + u_{i,j,k})(u_{i-1,j,k} + u_{i,j,k})$$

$$- \quad \alpha|(u_{i-1,j,k} + u_{i,j,k})|(u_{i-1,j,k} - u_{i,j,k})] \tag{4.8}$$

The coefficient α in Eq. (4.8) gives the desired amount of upstream (donor cell) differencing. When α is zero, the difference equation is centered in space. When α is equal to unity, the difference equation reduces to the full upstream or donor cell form. If α is small (between 0.2 and 0.3) the discretized equation tends toward centered in space. Introduction of modulus sign keeps the discretized equation valid irrespective of the direction (positive or negative) of the local u−velocity. Similarly,

$$\frac{\partial(uv)}{\partial y} = \frac{1}{4\delta y}[(v_{i,j,k} + v_{i+1,j,k})(u_{i,j,k} + u_{i,j+1,k})$$

$$+ \quad \alpha|(v_{i,j,k} + v_{i+1,j,k})|(u_{i,j,k} - u_{i,j+1,k})$$

$$- \quad (v_{i,j-1,k} + v_{i+1,j-1,k})(u_{i,j-1,k} + u_{i,j,k})$$

$$- \quad \alpha|(v_{i,j-1,k} + v_{i+1,j-1,k})|(u_{i,j-1,k} - u_{i,j,k})] \tag{4.9}$$

$$\frac{\partial(uw)}{\partial z} = \frac{1}{4\delta z}[(w_{i,j,k} + w_{i+1,j,k})(u_{i,j,k} + u_{i,j,k+1})$$

$$+ \alpha|(w_{i,j,k} + w_{i+1,j,k})|(u_{i,j,k} - u_{i,j,k+1})$$

$$- (w_{i,j,k-1} + w_{i+1,j,k-1})(u_{i,j,k-1} + u_{i,j,k})$$

$$- \alpha|(w_{i,j,k-1} + w_{i+1,j,k-1})|(u_{i,j,k-1} - u_{i,j,k})] \qquad (4.10)$$

The pressure gradient term in the x−momentum equation may be discretized as

$$\frac{\partial p}{\partial x} = \frac{p_{i+1,j,k} - p_{i,j,k}}{\delta x} \equiv \text{DPDX} \qquad (4.11)$$

The diffusive terms of x−momentum equation are discretized as

$$\frac{\partial^2 u}{\partial x^2} = \frac{u_{i+1,j,k} - 2u_{i,j,k} + u_{i-1,j,k}}{(\delta x)^2} \qquad (4.12)$$

$$\frac{\partial^2 u}{\partial y^2} = \frac{u_{i,j+1,k} - 2u_{i,j,k} + u_{i,j-1,k}}{(\delta y)^2} \qquad (4.13)$$

$$\frac{\partial^2 u}{\partial z^2} = \frac{u_{i,j,k+1} - 2u_{i,j,k} + u_{i,j,k-1}}{(\delta z)^2} \qquad (4.14)$$

Therefore, the term CONDIFU may be expressed as

$$\text{CONDIFU} = \left[-\left(\frac{\partial(u^2)}{\partial x} + \frac{\partial(uv)}{\partial y} + \frac{\partial(uw)}{\partial z} \right) + \frac{1}{Re}\left(\frac{\partial^2 u}{\partial x^2} + \frac{\partial^2 u}{\partial y^2} + \frac{\partial^2 u}{\partial z^2} \right) \right] \qquad (4.15)$$

As mentioned earlier, the quantity $\tilde{u}_{i,j,k}^{n+1}$ is now evaluated explicitly from the discretized form of Eq. (4.2) as

$$\tilde{u}_{i,j,k}^{n+1} = u_{i,j,k}^n + \delta t\, [\text{CONDIFU-DPDX}]_{i,j,k}^n \qquad (4.16)$$

Similarly, we can evaluate

$$\tilde{v}_{i,j,k}^{n+1} = v_{i,j,k}^n + \delta t\, [\text{CONDIFV-DPDY}]_{i,j,k}^n \qquad (4.17)$$

$$\tilde{w}_{i,j,k}^{n+1} = w_{i,j,k}^n + \delta t\, [\text{CONDIFW-DPDZ}]_{i,j,k}^n \qquad (4.18)$$

The terms

$$\frac{\partial p}{\partial y} = \frac{p_{i,j+1,k} - p_{i,j,k}}{\partial y} \equiv \text{DPDY} \qquad (4.19)$$

and

$$\frac{\partial p}{\partial z} = \frac{p_{i,j,k+1} - p_{i,j,k}}{\partial z} \equiv \text{DPDZ} \qquad (4.20)$$

The terms CONDIFV and CONDIFW in Eqs. (4.17) and (4.18) contain the discretized convection-diffusion contributions of $y-$momentum and $z-$momentum equations respectively.

The explicitly advanced tilde velocities may not necessarily lead to a flow field with zero mass divergence in each cell. This implies that at this stage the pressure distribution is not correct. Pressure in each cell has to be corrected in such a way that there is no net mass flow in or out of the cell. In the original MAC method, the corrected pressures were obtained from the solution of a Poisson equation for pressure. A related technique developed by Chorin [5] involved a simultaneous iteration on pressure and velocity components. Vieceli [6] showed that the two methods as applied to MAC are equivalent. We shall make use of the iterative correction procedure of Chorin [5] in order to obtain a divergence-free velocity field. We will also discuss the mathematical method for this iterative procedure.

The relationship between the explicitly advanced velocity component and velocity at the previous time step may be written as

$$\tilde{u}_{i,j,k}^{n+1} = u_{i,j,k}^{n} + \delta t \left[\frac{p_{i,j,k}^{n} - p_{i+1,j,k}^{n}}{\delta x} \right] + [\text{CONDIFU}]_{i,j,k}^{n} \delta t \qquad (4.21)$$

where $[\text{CONDIFU}]_{i,j,k}^{n}$ is only the contribution from convection and diffusion terms. On the other hand, the corrected velocity component (unknown) is related to the corrected pressure (also unknown) in the following way:

$$u_{i,j,k}^{n+1} = u_{i,j,k}^{n} + \delta t \left[\frac{p_{i,j,k}^{n+1} - p_{i+1,j,k}^{n+1}}{\delta x} \right] + [\text{CONDIFU}]_{i,j,k}^{n} \delta t \qquad (4.22)$$

From Eqs. (4.21) and (4.22)

$$u_{i,j,k}^{n+1} - \tilde{u}_{i,j,k}^{n+1} = \delta t \left[\frac{p_{i,j,k}^{c} - p_{i+1,j,k}^{c}}{\delta x} \right] \qquad (4.23)$$

where the pressure correction may be defined as

$$p_{i,j,k}^{c} = p_{i,j,k}^{n+1} - p_{i,j,k}^{n} \qquad (4.24)$$

Neither the pressure correction nor $u_{i,j,k}^{n+1}$ are known explicitly at this stage. Hence, one cannot be calculated without the help of the other. Calculations are done in an iterative cycle and we write

Corrected Velocity = Estimated Velocity ± Correction

$$u_{i,j,k}^{n+1} = \tilde{u}_{i,j,k}^{n+1} + \delta t \left[\frac{p_{i,j,k}^{c} - p_{i+1,j,k}^{c}}{\delta x} \right]$$

In a similar way, we can formulate the following array:

$$u_{i,j,k}^{n+1} = \tilde{u}_{i,j,k}^{n+1} + \delta t \left[\frac{p_{i,j,k}^{c} - p_{i+1,j,k}^{c}}{\delta x} \right] \qquad (4.25)$$

$$u_{i-1,j,k}^{n+1} = \tilde{u}_{i-1,j,k}^{n+1} - \delta t \left[\frac{p_{i,j,k}^c - p_{i-1,j,k}^c}{\delta x} \right] \tag{4.26}$$

$$v_{i,j,k}^{n+1} = \tilde{v}_{i,j,k}^{n+1} + \delta t \left[\frac{p_{i,j,k}^c - p_{i,j+1,k}^c}{\delta y} \right] \tag{4.27}$$

$$v_{i,j-1,k}^{n+1} = \tilde{v}_{i,j-1,k}^{n+1} - \delta t \left[\frac{p_{i,j,k}^c - p_{i,j-1,k}^c}{\delta y} \right] \tag{4.28}$$

$$w_{i,j,k}^{n+1} = \tilde{w}_{i,j,k}^{n+1} + \delta t \left[\frac{p_{i,j,k}^c - p_{i,j,k+1}^c}{\delta z} \right] \tag{4.29}$$

$$w_{i,j,k-1}^{n+1} = \tilde{w}_{i,j,k-1}^{n+1} - \delta t \left[\frac{p_{i,j,k}^c - p_{i,j,k-1}^c}{\delta z} \right] \tag{4.30}$$

The correction is done through the continuity equation. Plugging-in the above relationship into the continuity equation (Eq. (4.1)) yields

$$\left[\frac{u_{i,j,k}^{n+1} - u_{i-1,j,k}^{n+1}}{\delta x} + \frac{v_{i,j,k}^{n+1} - v_{i,j-1,k}^{n+1}}{\delta y} + \frac{w_{i,j,k}^{n+1} - w_{i,j,k-1}^{n+1}}{\delta z} \right]$$

$$= \left[\frac{\tilde{u}_{i,j,k}^{n+1} - \tilde{u}_{i-1,j,k}^{n+1}}{\delta x} + \frac{\tilde{v}_{i,j,k}^{n+1} - \tilde{v}_{i,j-1,k}^{n+1}}{\delta y} + \frac{\tilde{w}_{i,j,k}^{n+1} - \tilde{w}_{i,j,k-1}^{n+1}}{\delta z} \right]$$

$$- \delta t \left[\frac{p_{i+1,j,k}^c - 2p_{i,j,k}^c + p_{i-1,j,k}^c}{(\delta x)^2} + \frac{p_{i,j+1,k}^c - 2p_{i,j,k}^c + p_{i,j-1,k}^c}{(\delta y)^2} \right.$$

$$\left. + \frac{p_{i,j,k+1}^c - 2p_{i,j,k}^c + p_{i,j,k-1}^c}{(\delta z)^2} \right]. \tag{4.31}$$

or

$$\left[\frac{u_{i,j,k}^{n+1} - u_{i-1,j,k}^{n+1}}{\delta x} + \frac{v_{i,j,k}^{n+1} - v_{i,j-1,k}^{n+1}}{\delta y} + \frac{w_{i,j,k}^{n+1} - w_{i,j,k-1}^{n+1}}{\delta z} \right]$$

$$= \left[\frac{\tilde{u}_{i,j,k}^{n+1} - \tilde{u}_{i-1,j,k}^{n+1}}{\delta x} + \frac{\tilde{v}_{i,j,k}^{n+1} - \tilde{v}_{i,j-1,k}^{n+1}}{\delta y} + \frac{\tilde{w}_{i,j,k}^{n+1} - \tilde{w}_{i,j,k-1}^{n+1}}{\delta z} \right]$$

$$+ \frac{2\delta t(p_{i,j,k}^c)}{\delta x^2} + \frac{2\delta t(p_{i,j,k}^c)}{\delta y^2} + \frac{2\delta t(p_{i,j,k}^c)}{\delta z^2}$$

In deriving the above expression, it is assumed that the pressure corrections in the neighboring cells are zero. Under such an approximation, we can write

$$0 = (Div)_{i,j,k} + p_{i,j,k}^c \left[2\delta t \left(\frac{1}{\delta x^2} + \frac{1}{\delta y^2} + \frac{1}{\delta z^2} \right) \right] \tag{4.32}$$

or

$$p_{i,j,k}^c = \frac{-(Div)_{i,j,k}}{\left[2\delta t \left(\frac{1}{\delta x^2} + \frac{1}{\delta y^2} + \frac{1}{\delta z^2}\right)\right]}$$

In order to accelerate the calculation, the pressure correction equation is modified as

$$p_{i,j,k}^c = \frac{-\omega_0 (Div)_{i,j,k}}{\left[2\delta t \left(\frac{1}{\delta x^2} + \frac{1}{\delta y^2} + \frac{1}{\delta z^2}\right)\right]} \tag{4.33}$$

where ω_0 is the overrelaxation factor. A value of $\omega_0 = 1.7$ is commonly used. The value of ω_0 giving most rapid convergence should be determined by numerical experimentation. After calculating $p_{i,j,k}$, the pressure in the cell (i,j,k) is adjusted as

$$p_{i,j,k}^{n+1} \longrightarrow p_{i,j,k}^n - p_{i,j,k}^c \tag{4.34}$$

Now the pressure and velocity components for each cell are corrected through an iterative procedure in such a way that for the final pressure field, the velocity divergence in each cell vanishes. The process is continued till a divergence-free velocity is reached with a prescribed upper bound; here a value of 0.0001 is recommended. Finally we discuss another important observation. If the velocity boundary conditions are correct and a divergence-free converged velocity field has been obtained, eventually correct pressure will be determined in all the cells at the boundary. Thus, this method avoids the application of pressure boundary conditions. This typical feature of modified MAC method has been discussed in more detail by Peyret and Taylor [7]. However, it was also shown by Brandt, Dendy and Ruppel [8] that the aforesaid pressure-velocity iteration procedure of correcting pressure is equivalent to the solution of Poisson equation for pressure. One can formulate the Poisson equation from pressure-correction from Eq. (4.31), as

$$\frac{p_{i+1,j,k}^c - 2p_{i,j,k}^c + p_{i-1,j,k}^c}{(\delta x)^2} + \frac{p_{i,j+1,k}^c - 2p_{i,j,k}^c + p_{i,j-1,k}^c}{(\delta y)^2} + \frac{p_{i,j,k+1}^c - 2p_{i,j,k}^c + p_{i,j,k-1}^c}{(\delta z)^2}$$

$$= \frac{1}{\delta t}\left[\frac{\tilde{u}_{i,j,k}^{n+1} - \tilde{u}_{i-1,j,k}^{n+1}}{\delta x} + \frac{\tilde{v}_{i,j,k}^{n+1} - \tilde{v}_{i,j-1,k}^{n+1}}{\delta y} + \frac{\tilde{w}_{i,j,k}^{n+1} - \tilde{w}_{i,j,k-1}^{n+1}}{\delta z}\right] \tag{4.35}$$

Varying i from 2 to ire, j from 2 to jre and k from 2 to kre, and applying boundary conditions, one can obtain a system of equations that can be written in the form of $\mathbf{A}\,\tilde{x} = b$. $\mathbf{A}$ is a septa-diagonal matrix and the system of equations can be solved by using the linear equation solver.

Boundary Conditions

The boundary conditions are imposed by setting appropriate velocities in the fictitious cells surrounding the physical domain (Fig. 4.5).

Consider, for example, the bottom boundary of the computational domain. If this boundary has to represent a rigid no-slip wall, the normal velocity on the wall must be zero and the tangential velocity components should also be zero. Here we consider a stationary wall. With reference to the Fig. 4.5, we have

$$\left.\begin{aligned} v_{i,1,k} &= 0 \\ u_{i,1,k} &= -u_{i,2,k} \\ w_{i,1,k} &= -w_{i,2,k} \end{aligned}\right\} \quad \begin{aligned} &\text{for } i = 2 \text{ to ire} \\ &\text{and } k = 2 \text{ to kre} \end{aligned} \qquad (4.36)$$

If the left side-wall the wall is a free-slip (vanishing shear) boundary, the normal velocity must be zero and the tangential velocities should have no normal gradient.

$$\left.\begin{aligned} w_{i,j,1} &= 0 \\ u_{i,j,1} &= u_{i,j,2} \\ v_{i,j,1} &= v_{i,j,2} \end{aligned}\right\} \quad \begin{aligned} &\text{for } i = 2 \text{ to ire} \\ &\text{and } j = 2 \text{ to jre} \end{aligned} \qquad (4.37)$$

If the front plane is to be provided with inflow boundary conditions, the desired velocity profile may be recommended. Generally, normal velocity components are set to zero and a uniform or parabolic axial velocity may be deployed. Hence with reference to Fig. 4.5, we can write

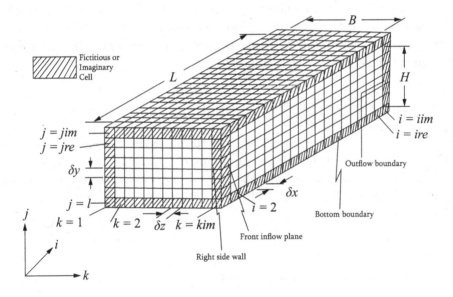

Figure 4.5 Boundary conditions and fictitious boundary cells

$$\left.\begin{aligned} v_{1,j,k} &= -v_{2,j,k} \\ w_{1,j,k} &= -w_{2,j,k} \\ u_{1,j,k} &= 1.0 \quad \text{or} \\ u_{1,j,k} &= 1.5\left[1 - ((j_m - j)/j_m)^2\right] \end{aligned}\right\} \quad \begin{aligned} &\text{for } j = 2 \text{ to jre} \\ &\text{and } k = 2 \text{ to kre} \end{aligned} \qquad (4.38)$$

where j_m is the horizontal midplane.

Continuative or outflow boundaries are major concerns for low-speed calculations, because it can affect the entire flow upstream. We need a prescription that permits fluid to flow out of the grid-mesh with a minimum of upstream influence. The common equations for such a boundary is $\nabla \mathbf{V} \cdot \mathbf{n} = 0$, where $\mathbf{n}$ is the unit normal vector.

The boundary condition that has more generality at the outflow is described by Orlanski [9]. This condition allows changes inside the flow field to be transmitted outward, but not vice-versa:

$$\frac{\partial \Psi}{\partial t} + U_{av}\frac{\partial \Psi}{\partial x} = 0 \tag{4.39}$$

where U_{av} is the average velocity at the outflow plane and Ψ represents u, v, w or any dependent variable.

Numerical Stability Considerations

For accuracy, the mesh size must be chosen small enough to resolve the expected spatial variations in all dependent variables.

Once a grid-mesh has been chosen, the choice of the time increment is governed by two restrictions, namely, the Courant-Fredrichs-Lewy (CFL) condition and the restriction on the basis of grid-Fourier numbers. According to the CFL condition, material cannot move through more than one cell in one time step, because the difference equations assume fluxes only between the adjacent cells. Therefore the time increment must satisfy the inequality.

$$\delta t < \min\left\{\frac{\delta x}{|u|}, \frac{\delta y}{|v|}, \frac{\delta z}{|w|}\right\} \tag{4.40}$$

where the minimum is with respect to every cell in the mesh. Typically, δt is chosen equal to one-fourth to one-third of the minimum cell transit time. When the viscous diffusion terms are more important, the condition necessary to ensure stability is dictated by the restriction on the grid Fourier numbers, which results in

$$\nu \delta t < \frac{1}{2} \cdot \left(\frac{\delta x^2 \delta y^2 \delta z^2}{\delta x^2 + \delta y^2 + \delta z^2}\right) \tag{4.41}$$

in dimensional form. After nondimensionilization, this leads to

$$\delta t < \frac{1}{2} \cdot \left(\frac{\delta x^2 \delta y^2 \delta z^2}{\delta x^2 + \delta y^2 + \delta z^2}\right) Re \tag{4.42}$$

The final δt for each time increment is the minimum of the δt's obtained from Eqs. (4.40) and (4.42) The last quantity needed to ensure numerical stability is the upwind parameter α. In general, α should be slightly larger than the maximum value of $|u\delta t/\delta x]$ or $|v\delta t/\delta y]$ occurring in the mesh, that is,

$$\max\left\{\left|\frac{u\delta t}{\delta x}\right|, \left|\frac{v\delta t}{\delta y}\right|, \left|\frac{w\delta t}{\delta z}\right|\right\} \le \alpha < 1 \tag{4.43}$$

As a ready prescription, a value between 0.2 and 0.4 can be used for α. If α is too large, an unnecessary amount of numerical diffusion (artificial viscosity) may be introduced.

Higher Order Upwind Differencing

More accurate solutions are obtained if the convective terms are discretized by higher order schemes. One significant improvement is known as third-order upwind differencing (see Kawamura et al. [10]). The following example illustrates the essence of this discretization scheme.

$$\left(u\frac{\partial u}{\partial x} \right)_{i,j} = u_{i,j} \left[\frac{-u_{i+2,j} + 8(u_{i+1,j} - u_{i-1,j}) + u_{i-2,j}}{12\Delta x} \right]$$

$$+ |u_{i,j}| \left[\frac{u_{i+2,j} - 4u_{i+1,j} + 6u_{i,j} - 4u_{i-1,j} + u_{i-2,j}}{4\Delta x} \right] \quad (4.44)$$

Higher order upwinding is an emerging area of research in computational fluid dynamics. However, no unique resolution for solving a wide variety of problems has yet appeared. Interested readers are referred to Vanka et al. [11] for more stimulating information on related topics.

One of the most widely used higher order schemes is known as QUICK (Leonard [12]). The QUICK scheme may be written in a compact manner in the following way

$$f\frac{\partial u}{\partial x}|_i = f_i \left[\frac{u_{i-2} - 8u_{i-1} + 8u_{i+1} - u_{i+2}}{12\Delta x} \right]$$

$$+ f_i \left\{ \frac{(\Delta x)^2}{24} \right\} \left[\frac{-u_{i-2} + 2u_{i-1} - 2u_{i+1} + u_{i+2}}{(\Delta x)^3} \right]$$

$$+ |f_i| \left\{ \frac{(\Delta x)^3}{16} \right\} \left[\frac{u_{i-2} - 4u_{i-1} + 6u_i - 4u_{i+1} + u_{i+2}}{(\Delta x)^4} \right] \quad (4.45)$$

Sample Results

The isolines for the axial velocity behind a delta-winglet placed inside a channel are shown in Fig. 4.6. These results were obtained by Biswas et. al., [13] who used MAC to solve for a three-dimensional flow field in a channel containing a delta-winglet as a vortex generator. Excellent comparisons between the predicted results experimental observations are revealed through this figure. Experiments were conducted using hot-wire probes. The MAC algorithm has been extensively used by the researchers to solve flows in complex geometry. Robichaux, Tafti and Vanka [14] deployed the MAC algorithm for large eddy simulation (LES) of turbulent channel flows. Of course, they performed the time integration of the discretized equations by using a fractional step method (Kim and Moin [15]). More about the implementation techniques of higher order schemes in MAC algorithm is available elsewhere [16].

4.3   SOLUTION OF ENERGY EQUATION IN CARTESIAN COORDINATE

The energy for incompressible flows, neglecting mechanical work and gas radiation, may be written in dimensional form as

$$\rho c_p \left[\frac{\partial T}{\partial t^*} + u^* \frac{\partial T}{\partial x^*} + v^* \frac{\partial T}{\partial y^*} + w^* \frac{\partial T}{\partial z^*} \right] = k\nabla^2 T + \mu\phi^* \tag{4.46}$$

where ϕ^* is the viscous dissipation given as

$$\phi^* = 2\left[\left(\frac{\partial u^*}{\partial x^*}\right)^2 + \left(\frac{\partial v^*}{\partial y^*}\right)^2 + \left(\frac{\partial w^*}{\partial x^*}\right)^2 \right] + \left\{ \frac{\partial u^*}{\partial x^*} + \frac{\partial v^*}{\partial y^*} \right\}^2$$
$$+ \left\{ \frac{\partial w^*}{\partial y^*} + \frac{\partial v^*}{\partial z^*} \right\}^2 + \left\{ \frac{\partial w^*}{\partial x^*} + \frac{\partial u^*}{\partial z^*} \right\}^2$$

Equation (4.46) may be nondimensionalized in the following way:

$$u = \frac{u^*}{U_\infty}, \quad v = \frac{v^*}{U_\infty}, \quad w = \frac{w^*}{U_\infty}, \quad \theta = \frac{T - T_\infty}{T_w - T_\infty}$$
$$x = \frac{x^*}{L}, \quad y = \frac{y^*}{L}, \quad z = \frac{z^*}{L}, \quad t = \frac{t^*}{L/U_\infty}$$

Substituting the above variables in Eq. (4.46) we obtain

$$\frac{\rho c_p U_\infty (T_w - T_\infty)}{L} \left[\frac{\partial \theta}{\partial t} + u\frac{\partial \theta}{\partial x} + v\frac{\partial \theta}{\partial y} + w\frac{\partial \theta}{\partial z} \right]$$
$$= \frac{(T_w - T_\infty)k}{L^2} \left[\frac{\partial^2 \theta}{\partial x^2} + \frac{\partial^2 \theta}{\partial y^2} + \frac{\partial^2 \theta}{\partial z^2} \right] + \frac{\mu U_\infty^2}{L^2}\phi \tag{4.47}$$

where ϕ is the nondimensional form of ϕ^*. Finally, the normalized energy equation becomes

$$\frac{\partial \theta}{\partial t} + u\frac{\partial \theta}{\partial x} + v\frac{\partial \theta}{\partial y} + w\frac{\partial \theta}{\partial z} = \frac{1}{Pe}\left[\frac{\partial^2 \theta}{\partial x^2} + \frac{\partial^2 \theta}{\partial y^2} + \frac{\partial^2 \theta}{\partial z^2} \right] + \frac{Ec}{Re}\phi \tag{4.48}$$

where Pe, the Peclet number is given as

$$\frac{1}{Pe} = \frac{(T_w - T_\infty)k}{L^2} \cdot \frac{L}{\rho c_p U_\infty (T_w - T_\infty)}$$

or

$$\frac{1}{Pe} = \frac{k}{L\rho c_p U_\infty} = \frac{k}{\mu c_p} \cdot \frac{\mu}{\rho L U_\infty} = \frac{1}{Pr} \cdot \frac{1}{Re}$$

Further, the ratio of Eckert number and Reynolds number can be written as

$$\frac{Ec}{Re} = \frac{\mu U_\infty^2}{L^2} \cdot \frac{L}{\rho c_p U_\infty (T_w - T_\infty)} = \frac{U_\infty^2}{c_p(T_w - T_\infty)} \cdot \frac{1}{\rho U_\infty L/\mu}$$

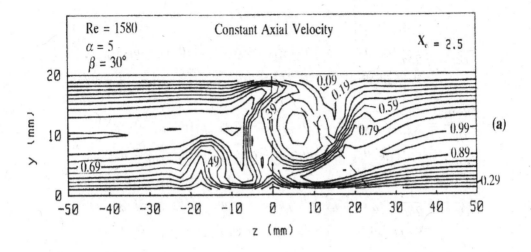

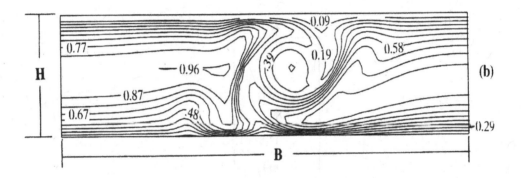

Figure 4.6 The isolines for the axial velocity behind a delta winglet: (a) experimental results and (b) numerical simulation

4.3.1 Solution Procedure

The steady state energy equation, neglecting the dissipation term, may be written in the following conservative form as

$$\frac{\partial(u\theta)}{\partial x} + \frac{\partial(v\theta)}{\partial y} + \frac{\partial(w\theta)}{\partial z} = \frac{1}{Pe}\left[\frac{\partial^2\theta}{\partial x^2} + \frac{\partial^2\theta}{\partial y^2} + \frac{\partial^2\theta}{\partial z^2}\right] \qquad (4.49)$$

Eq. (4.49) may be written as

$$\nabla^2\theta = Pe[CONVT]_{i,j,k}^m \qquad (4.50)$$

where $[CONVT]_{i,j,k}^m$ is the discretized convective terms on the left-hand side of Eq. (4.49) and m stands for the iterative counter. The discretized terms of CONVT can be expressed as (refer to Fig. 4.7)

$$\frac{\partial(u\theta)}{\partial x} = \frac{1}{2\delta x} \left[u_{i,j,k} \left(\theta_{i,j,k} + \theta_{i+1,j,k} \right) + \alpha |u_{i,j,k}| \left(\theta_{i,j,k} - \theta_{i+1,j,k} \right) \right.$$
$$\left. -u_{i-1,j,k} \left(\theta_{i-1,j,k} + \theta_{i,j,k} \right) - \alpha |u_{i-1,j,k}| \left(\theta_{i-1,j,k} - \theta_{i,j,k} \right) \right]$$

$$\frac{\partial(v\theta)}{\partial y} = \frac{1}{2\delta y} \left[v_{i,j,k} \left(\theta_{i,j,k} + \theta_{i,j+1,k} \right) + \alpha |v_{i,j,k}| \left(\theta_{i,j,k} - \theta_{i,j+1,k} \right) \right.$$
$$\left. -v_{i,j-1,k} \left(\theta_{i,j-1,k} + \theta_{i,j,k} \right) - \alpha |v_{i,j-1,k}| \left(\theta_{i,j-1,k} - \theta_{i,j,k} \right) \right]$$

$$\frac{\partial(w\theta)}{\partial z} = \frac{1}{2\delta z} \left[w_{i,j,k} \left(\theta_{i,j,k} + \theta_{i,j,k+1} \right) + \alpha |w_{i,j,k}| \left(\theta_{i,j,k} - \theta_{i,j,k+1} \right) \right.$$
$$\left. -w_{i,j,k-1} \left(\theta_{i,j,k-1} + \theta_{i,j,k} \right) - \alpha |w_{i,j,k-1}| \left(\theta_{i,j,k-1} - \theta_{i,j,k} \right) \right]$$

To start with, we can assume any guess value of θ throughout the flow field. Since u, v, w are known from the solution of momentum equation (Eq. (4.49)) is now a

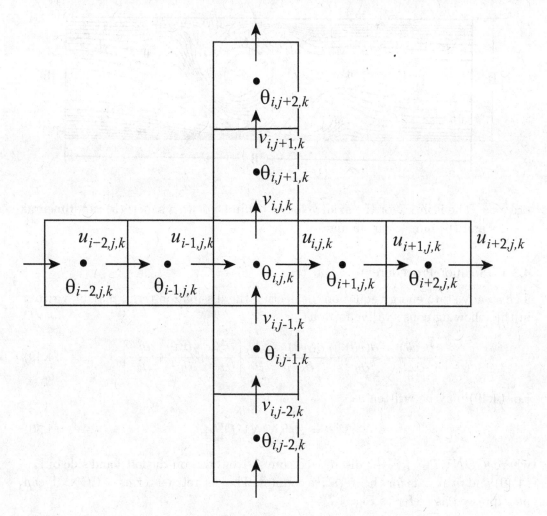

Figure 4.7 Neighboring cells of the variable $\theta_{i,j,k}$

linear equation. However, from the guess value of θ and known correct values of u, v and w the left-hand side of Eq. (4.49) has been evaluated using a weighted average scheme or the QUICK scheme [12] may be adapted for discretization of the convective terms too. After discretizing and evaluating right-hand side of Eq. (4.50) we obtain a Poisson equation for temperature with a source term on the right-hand side. Now, we shall follow successive over relaxation (SOR) technique for solving discretized form of Eq. (4.50). Consider a discretized equation as

$$\frac{\theta_{i+1,j,k} - 2\theta_{i,j,k} + \theta_{i-1,j,k}}{(\delta x)^2} + \frac{\theta_{i,j+1,k} - 2\theta_{i,j,k} + \theta_{i,j-1,k}}{(\delta y)^2}$$

$$+ \frac{\theta_{i,j,k+1} - 2\theta_{i,j,k} + \theta_{i,j,k-1}}{(\delta z)^2} = S^{*m} \quad (4.51)$$

where

$$S^{*m} \equiv Pe[CONVT]_{i,j,k}^m \quad \text{is a known quantity}$$

From Eq. (4.51) we can write

$$A^{*m} - \theta_{i,j,k}\left(\frac{2}{\delta x^2} + \frac{2}{\delta y^2} + \frac{2}{\delta z^2}\right) = S^{*m}$$

or

$$\theta_{i,j,k} = \frac{A^{*m} - S^{*m}}{\left(\frac{2}{\delta x^2} + \frac{2}{\delta y^2} + \frac{2}{\delta z^2}\right)} \quad (4.52)$$

where

$$A^{*m} = \frac{\theta_{i+1,j,k}^m + \theta_{i-1,j,k}^m}{(\delta x)^2} + \frac{\theta_{i,j+1,k}^m + \theta_{i,j-1,k}^m}{(\delta y)^2} + \frac{\theta_{i,j,k+1}^m + \theta_{i,j,k-1}^m}{(\delta z)^2}$$

$\theta_{i,j,k}$ in Eq. 4.52 may be assumed to be the most recent value and it may be written as $\theta_{i,j,k}^{m'}$. In order to accelerate the speed of computation we introduce an overrelaxation factor ω. Thus

$$\theta_{i,j,k}^{m+1} = \theta_{i,j,k}^m + \omega\left[\theta_{i,j,k}^{m'} - \theta_{i,j,k}^m\right] \quad (4.53)$$

where $\theta_{i,j,k}^m$ is the previous value, $\theta_{i,j,k}^{m'}$ the most recent value and $\theta_{i,j,k}^{m+1}$ the calculated better guess. The procedure will continue till the required convergence is achieved. This is equivalent to Gauss-Siedel procedure for solving a system of linear equations.

4.3.2 Flow Chart

A sample flow chart for computation is presented as Fig. 4.8.

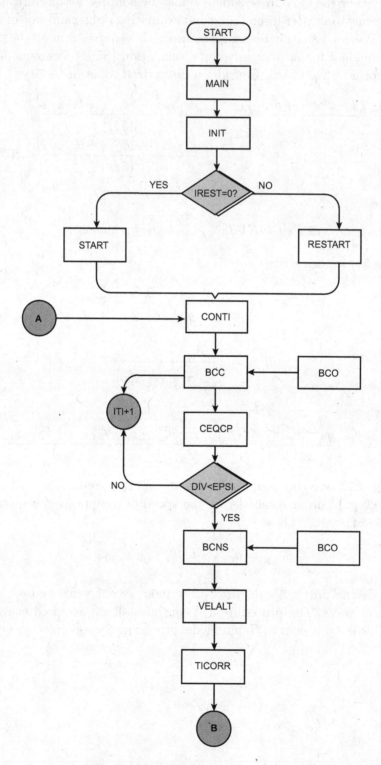

Figure 4.8 The flowchart and its description

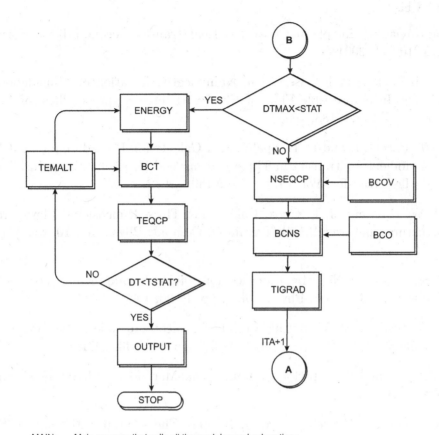

MAIN: Main program that calls all the modules and subroutines.
CONTI: Module for solving continuity equation
BCC: Boundary conditions for the confining surfaces
BCO: Boundary conditions for obstacle
CEQCP: Continuity equation for constant property cases
DIV: Velocity divergence in each cell
EPSI: Pre-defined small numerical value
ITI: Iterative counter for the continuity equation
BCNS: Boundary conditions for the Navier-Stokes equations
VELALT: Changes the converged velocity arrays, i.e., makes the converged values
        at n+1 th time step as n th time level to start the next time step
TICORR: Calculates δt
DTMAX: Maximum discrepancy of the dependant variables between two time steps
STAT: Predefined small numerical value
NSEQCP: Discretized Navier-Stokes equations for constant property
BCOV: Special boundary conditions for obstacles
TIGRAD: Calculation of change in dependant variables between two time steps
ITA: Iterative counter for calculating the Navier-Stokes equations
ENERGY: Module for calculation of energy equation
BCT: Boundary conditions for temperatures
TEQCP: Discretized energy equation for constant property cases
DT: Maximum discrepancy of temperature in any cell between two iterative steps
TSTAT: Predefined small numerical value
TEMALT: Changes the temperature arrays from (m+1) iterative level to mth
        iterative level to start the next iteration

Figure 4.8 (Cont.) The flowchart and its description

REFERENCES

1. M. Kaviany, Principles of Convective Heat Transfer, Second Edition, Springer, pp. 210-211, 2001.

2. F. H. Harlow, and J. E. Welch, Numerical Calculation of Time-dependent Viscous Incompressible Flow of Fluid with Free Surfaces, Phys. of Fluids, Vol. 8, pp. 2182-2188, 1965.

3. S. V. Patankar, and D. B. Spalding, A Calculation Procedure for Heat Mass and Momentum Transfer in Three Dimensional Parabolic Flows, Int. J. Heat and Mass Transfer, Vol. 15, pp. 1787-1805, 1972.

4. C. W. Hirt, and J. L. Cook, Calculating Three Dimensional Flows around Structures and over Rough Terrain, J. Comput. Phys., Vol. 10, pp. 324-340, 1972.

5. A. J. Chorin, A Numerical Method for Solving Incompressible Viscous Flow Problems, J. Comput. Phys., Vol. 2, pp. 12-26, 1967.

6. J. A. Viecelli, A Computing Method for Incompressible Flows Bounded by Moving Walls, J. Comput. Phys., Vol. 8, pp. 119-143, 1971.

7. P. Peyret, and T. D. Taylor, Computational Methods for Fluid Flow, Springer, New York, 1983.

8. A. Brandt, J. E. Debdy, and H. Ruppel, The Multigrid Method for Semi-Implicitly Hydrodynamic Codes, J. Comput. Phys., Vol. 34, pp. 348-370, 29-1980.

9. I. Orlanski, A Simple Boundary Condition for Unbounded Flows, J. Comput. Phys., Vol. 21, pp. 251-269, 1976.

10. T. Kawamura, H. Takami and K. Kuwahara, Computation of High Reynolds Number Flow around a Circular Cylinder with Surface Roughness. Fluid Dynamics Research, Vol. 1, pp. 145-162, 1986.

11. S. P. Vanka, B. C.-J. Chen, and W. T. Sha, A Semi-Implicit Calculation Procedure for Flow Described in Body-Fitted Coordinate Systems, Num. Heat Trans., Vol. 3, pp. 1-19, 1980.

12. B. P. Leonard, A Stable and Accurate Convective Modeling Procedure Based on Quadratic Upstream Interpolation, Comp. Methods Appl. Mech. Engr., Vol. 19, pp. 59-98, 1979.

13. G. Biswas, K. Torii, D. Fujii . and K. Nishino, Numerical and Experimental Determination of Flow Structure and Heat Transfer Effects of Longitudinal Vortices in a Channel Flow. Int. J. Heat Mass Trans., Vol. 39, pp. 3441-3451, 1996.

14. J. Robichaux, D. K. Tafti, and S. P. Vanka, Large Eddy Simulations of Turbulence on CM-2, Num. Heat Trans., Part B, Vol. 21, pp. 267-388, 1992.

15. J. Kim, and P. Moin, Application of a Fractional Step Method to Incompressible Navier–Stokes Equations, J. Comput. Phys., Vol. 59, pp. 308-323, 1985.

16. G. Biswas, Solution of Navier–Stokes Equations for Incompressible Flows using MAC and SIMPLE Algorithms, pp. 229-274, in Computational Fluid Flow and Heat Transfer, eds K. Muralidhar and T. Sundararajan, Narosa Publishing House, 2003.

EXERCISES

1. Consider the following nonlinear equation

$$u\frac{\partial u}{\partial x} = \mu\frac{\partial^2 u}{\partial y^2}$$

   Is this equation in conservative form? If not, suggest a conservative form. Consider a domain in $x..(x = 0$ to $x = L)$ and in $y..(y = 0$ to $y = H)$ and assume that all the values of the dependent variable are known at $x = 0$ (from $y = 0$ to $y = H$) at every Δy interval. Develop an implicit expression for determining u at all the points along $(y = 0$ to $y = H)$ at the next $(x+\Delta x)$ location.

2. Write the expression for the finite difference quotient for the convective term of the Burger's equation given by

$$\frac{\partial \omega}{\partial t} + u\frac{\partial \omega}{\partial x} = \mu\frac{\partial^2 \omega}{\partial x^2}$$

   Use upwind differencing on a weak conservative form of the equation. The upwind differencing is known to retain the transportive property. Show that the formulation preserves the conservative property of the continuum as well.

3. Consider the two-dimensional heat conduction equation

$$\frac{\partial T}{\partial t} = \alpha\left(\frac{\partial^2 T}{\partial x^2} + \frac{\partial^2 T}{\partial y^2}\right)$$

   While formulating the numerical scheme for solution, you plan to follow ADI method in two special dimensions. Determine the expressions that would be used to predict $T_{i,j}^{n+1/2}$ from $T_{i,j}^n$ and then $T_{i,j}^{n+1}$. Show that the temporal accuracy of such a scheme is of second order.

4. In a simplified marker and cell (MAC) method for a two-dimensional flow, the provisional velocities for the next time step are predicted explicitly from the expression

$$\bar{u}_{i,j}^{n+1} = u_{i,j}^n + \delta t\frac{(p_{i,j}^n - p_{i+1,j}^n)}{\delta x} + \delta t[CONDIFU]_{i,j}^n$$

The term CONDIFU signifies discretized convection and diffusion terms of $x-$ momentum equation. In the absence of $p_{i,j}^{n+1}$, the pressures of nth time step are used in the calculation. Hence we get $\tilde{u}_{i,j}^{n+1}$ instead of $u_{i,j}^{n+1}$. Find an expression for correcting the $u-$velocity in each cell. Also derive an expression for $v-$velocity correction. From these two expressions and the expressions of the predicted correct velocities, find an implicit equation for pressure correction. Show that this equation can be called a Poisson equation involving pressure-correction as the variable.

5. Assume you are using MAC-family of algorithms for solving two-dimensional Navier–Stokes (N-S) and energy equations in a heated channel. Having solved the N-S equations, you know the velocity field. Outline a scheme for solving non-dimensional form of energy equation involving SOR scheme. In a staggered grid arrangement, how would you apply the temperature boundary condition (say, $\theta = \theta_w$) if you want to implement it on the bottom wall?

## CHAPTER 5

Fluid Flow Solutions in Complex Geometry

5.1 INTRODUCTION

For most fluid mechanics problems, the geometry of the problem can not be represented by a Cartesian mesh. Instead it is common for the boundaries to be curved in space. Some typical examples are turbine-blade passages, heat-exchangers, combustion chambers, aircrafts, vehicles, mixing vessels, flow around large structures like building, cooling towers, air-conditioning systems. The need for the full Navier-Stokes simulation of complex fluid flows arises in numerous engineering problems. Maliska and Raithby [1] described an economical method of solving the equations of motion for two and three dimensional problems using non-orthogonal boundary-fitted mesh. In this chapter, the finite volume discretization of the transport equation on a non-orthogonal mesh is discussed for steady, two-dimensional, incompressible and laminar flows. The first order upwind and QUICK schemes are implemented. Pressure-velocity coupling is achieved via the SIMPLE algorithm as described by Patankar [2].

5.2 BOUNDARY-FITTED MESHES

Structured meshes are characterized by regular connectivity, i.e., the points of the grid can be indexed (by two indices in two dimensions and three indices in three dimensions). Structured mesh generation techniques concentrate on meshing domains with irregular boundaries, e.g., flow and heat transfer in turbine blades, flow in blood vessels, flow inside 2-D planer channel. Generally the meshes are generated so that they fit the boundaries, with one coordinate surface forming the boundary. This gives accurate solutions near the boundary and enables the use of fast and accurate solvers. For fluid flow these grids allow the easy application of turbulence models, which usually require the grid to be aligned with the boundary.

5.3   MAPPING

Mapping is done to convert the regions having irregular shape (physical domain, Fig. 5.1(a)) into the computational domain (Fig. 5.1(b)) where the geometry becomes regular with a suitable transformation. A curvilinear mesh is generated over the physical domain such that one member of each family of curvilinear coordinate lines is coincident with the boundary contour of the physical domain. Mapping may be done in any coordinate system, i.e., Cartesian, polar, or spherical coordinates. Depending on the choice of the values of ξ, η along the boundary segments of the physical region, a variety of other acceptable configurations can be generated in the computational domain as shown in Fig. 5.1.

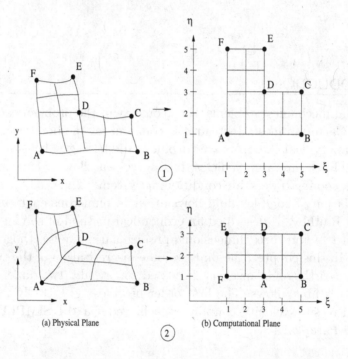

(a) Physical Plane   (b) Computational Plane

Figure 5.1   (1) Mapping the L-shaped irregular region into an L-shaped regular region; (2) mapping the L-shaped irregular region into a rectangular region [3]

5.4   COORDINATE TRANSFORMATION RELATIONS

The partial differential equation has been given in x, y independent variables. The transformation of this partial equation is needed from the x, y to the ξ, η independent variables. The coordinate (x, y) and (ξ, η) represents the coordinates in the physical domain and computational domain respectively. The transformation from the x, y to the ξ, η variables can be expressed as

$$\xi \equiv \xi(x, y), \qquad \eta \equiv \eta(x, y) \qquad (5.1)$$

and the inverse transformation is given by

$$x \equiv x(\xi, \eta), \qquad y \equiv y(\xi, \eta) \tag{5.2}$$

The expression of the Jacobian of the transformation J is

$$J = J\left(\frac{x, y}{\xi, \eta}\right) = \begin{vmatrix} x_\xi & y_\xi \\ x_\eta & y_\eta \end{vmatrix} = x_\xi y_\eta - x_\eta y_\xi \neq 0 \tag{5.3}$$

where

$$x_\xi = \frac{\partial x}{\partial \xi} \qquad x_\eta = \frac{\partial x}{\partial \eta} \qquad y_\xi = \frac{\partial y}{\partial \xi} \qquad y_\eta = \frac{\partial y}{\partial \eta} \tag{5.4}$$

Jacobian is the ratio of the volumes in the transformed and physical spaces. The transformation relations between the coordinates of physical domain and computational domain are as follows

$$\xi_x = \frac{1}{J} y_\eta \qquad \xi_y = -\frac{1}{J} x_\eta \qquad \eta_x = -\frac{1}{J} y_\xi \qquad \eta_y = \frac{1}{J} x_\xi \tag{5.5}$$

The derivatives ξ_x, ξ_y, η_x and η_y are known as metrics and the derivatives x_ξ, x_η, y_ξ and y_η are known as computational derivatives.

5.5 GOVERNING EQUATIONS IN BOUNDARY-FITTED COORDINATE SYSTEMS

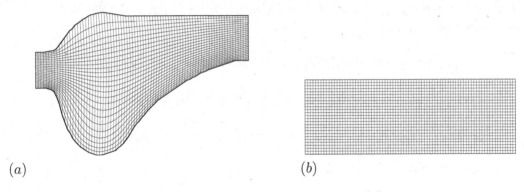

(a)   (b)

Figure 5.2 (a) Physical plane and (b) computational plane [4]

The differential equations of motion are required as the starting point for the solution in the transformed space where ξ and η are the independent variables. Figure 5.2 illustrates the relationship of the physical and transformed domains. The transformation of the physical domain (x, y) to the computational domain (ξ, η) is achieved via transformation metrics which are related to the physical coordinates as follows.

$$\begin{bmatrix} \xi_x & \xi_y \\ \eta_x & \eta_y \end{bmatrix} = \frac{1}{J} \begin{bmatrix} y_\eta & -x_\eta \\ -y_\xi & x_\xi \end{bmatrix}$$

where J is the Jacobian of the transformation given by

$$J = x_\xi\, y_\eta - x_\eta\, y_\xi \tag{5.6}$$

In Cartesian coordinates, the governing equations for incompressible two-dimensional steady laminar flow are

Continuity equation:

$$\frac{\partial u}{\partial x} + \frac{\partial v}{\partial y} = 0 \tag{5.7}$$

x-momentum equation:

$$\frac{\partial (u^2)}{\partial x} + \frac{\partial (uv)}{\partial y} = -\frac{\partial p}{\partial x} + \frac{1}{Re}\left(\frac{\partial^2 u}{\partial x^2} + \frac{\partial^2 u}{\partial y^2}\right) \tag{5.8}$$

y-momentum equation:

$$\frac{\partial (uv)}{\partial x} + \frac{\partial (v^2)}{\partial y} = -\frac{\partial p}{\partial y} + \frac{1}{Re}\left(\frac{\partial^2 v}{\partial x^2} + \frac{\partial^2 v}{\partial y^2}\right) \tag{5.9}$$

Energy equation:

$$\frac{\partial (uT)}{\partial x} + \frac{\partial (vT)}{\partial y} = \frac{1}{Re\,Pr}\left(\frac{\partial^2 T}{\partial x^2} + \frac{\partial^2 T}{\partial y^2}\right) \tag{5.10}$$

The transformation of Eqs. (5.7) to (5.10) is given by the continuity equation:

$$U_\xi + V_\eta = 0 \tag{5.11}$$

The momentum equations can be rewritten in ξ, η coordinates as follows:
ξ-momentum equation:

$$\frac{1}{J}\left[y_\eta(uu)_\xi - y_\xi(uu)_\eta\right] + \frac{1}{J}\left[-x_\eta(uv)_\xi + x_\xi(uv)_\eta\right] = -\frac{1}{J}\left[y_\eta p_\xi - y_\xi p_\eta\right]$$
$$+\frac{1}{Re}\left[\frac{1}{J}\left\{\frac{1}{J}y_\eta\left[(y_\eta u)_\xi - (y_\xi u)_\eta\right] - \frac{1}{J}x_\eta\left[-(x_\eta u)_\xi + (x_\xi u)_\eta\right]\right\}_\xi\right.$$
$$\left.+\frac{1}{J}\left\{-\frac{1}{J}y_\xi\left[(y_\eta u)_\xi + (y_\xi u)_\eta\right] + \frac{1}{J}x_\xi\left[-(x_\eta u)_\xi + (x_\xi u)_\eta\right]\right\}_\eta\right] \tag{5.12}$$

η-momentum equation:

$$\frac{1}{J}\left[y_\eta(uv)_\xi - y_\xi(uv)_\eta\right] + \frac{1}{J}\left[-x_\eta(vv)_\xi + x_\xi(vv)_\eta\right] = -\frac{1}{J}\left[-x_\eta p_\xi + x_\xi p_\eta\right]$$
$$+\frac{1}{Re}\left[\frac{1}{J}\left\{\frac{1}{J}y_\eta\left[(y_\eta v)_\xi - (y_\xi v)_\eta\right] - \frac{1}{J}x_\eta\left[-(x_\eta v)_\xi + (x_\xi v)_\eta\right]\right\}_\xi\right.$$
$$\left.+\frac{1}{J}\left\{-\frac{1}{J}y_\xi\left[(y_\eta v)_\xi + (y_\xi v)_\eta\right] + \frac{1}{J}x_\xi\left[-(x_\eta v)_\xi + (x_\xi v)_\eta\right]\right\}_\eta\right] \tag{5.13}$$

Energy equation:

$$\frac{1}{J}\left[y_\eta(uT)_\xi - y_\xi(uT)_\eta\right] + \frac{1}{J}\left[-x_\eta(vT)_\xi + x_\xi(vT)_\eta\right]$$
$$= \frac{1}{Re\,Pr}\left[\frac{1}{J}\left\{\frac{1}{J}y_\eta\left[(y_\eta T)_\xi - (y_\xi T)_\eta\right] - \frac{1}{J}x_\eta\left[-(x_\eta T)_\xi + (x_\xi T)_\eta\right]\right\}_\xi\right.$$
$$\left.+\frac{1}{J}\left\{-\frac{1}{J}y_\xi\left[(y_\eta T)_\xi + (y_\xi T)_\eta\right] + \frac{1}{J}x_\xi\left[-(x_\eta T)_\xi + (x_\xi T)_\eta\right]\right\}_\eta\right] \tag{5.14}$$

Both Eqs. (5.12) and (5.13) can be written in general form for general variable ϕ as

$$(U\phi)_\xi + (V\phi)_\eta = S^\phi(\xi,\eta) + \left\{\frac{\Gamma^\phi}{J}(\alpha\phi_\xi - \beta\phi_\eta)\right\}_\xi + \left\{\frac{\Gamma^\phi}{J}(-\beta\phi_\xi + \gamma\phi_\eta)\right\}_\eta \quad (5.15)$$

where

$$\alpha = x^2{}_\eta + y^2{}_\eta \qquad \beta = x_\xi x_\eta + y_\xi y_\eta \qquad \gamma = x^2{}_\xi + y^2{}_\xi \qquad (5.16)$$

$$S^u(\xi,\eta) = -y_\eta\, p_\xi + y_\xi\, p_\eta \quad \text{and} \quad \Gamma^u = \frac{1}{Re} \qquad \text{for} \qquad \phi = u$$

$$S^v(\xi,\eta) = x_\eta\, p_\xi - x_\xi\, p_\eta \quad \text{and} \quad \Gamma^v = \frac{1}{Re} \qquad \text{for} \qquad \phi = v$$

$$S^T(\xi,\eta) = 0 \qquad\qquad\quad \text{and} \quad \Gamma^T = \frac{1}{RePr} \quad \text{for} \qquad \phi = T$$

The relationships between the Cartesian and contravariant velocity components are

$$U = y_\eta u - x_\eta v \qquad\qquad V = x_\xi v - y_\xi u \qquad\qquad (5.17)$$

5.6 DISCRETIZATION

The governing equations are discretized on a structured grid. The velocity components and the scalar variables (pressure, temperature) are located on the grid in a staggered manner, as shown in Fig. 5.3(a). Control volumes of pressure and velocity in computational plane are shown in Fig. 5.4. The governing equations written in the generalized boundary-fitted coordinates are integrated and discretized over the control volume whose dimensions in the computational domain are given by $\Delta\xi \times \Delta\eta$ as shown in Fig. 5.3(b). The discretized form of the continuity equation

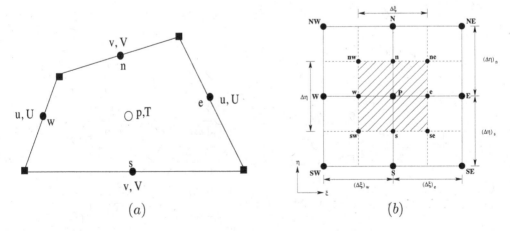

Figure 5.3 (a) Staggered grid and (b) grid for variable ϕ at point P

can be written as

$$(U_e - U_w)\Delta\eta + (V_n - V_s)\Delta\xi = 0 \qquad\qquad (5.18)$$

The discretized form of generalized momentum Eq. (5.15) is as follows

$$[(U\phi)_e - (U\phi)_w]\,\Delta\eta + [(V\phi)_n - (V\phi)_s]\,\Delta\xi = \text{Source}^\phi(\xi,\eta)$$

$$+ \left\{ \left[\frac{\Gamma^\phi}{J}(\alpha\phi_\xi - \beta\phi_\eta)\right]_e - \left[\frac{\Gamma^\phi}{J}(\alpha\phi_\xi - \beta\phi_\eta)\right]_w \right\}\Delta\eta$$

$$+ \left\{ \left[\frac{\Gamma^\phi}{J}(-\beta\phi_\xi + \gamma\phi_\eta)\right]_n - \left[\frac{\Gamma^\phi}{J}(-\beta\phi_\xi + \gamma\phi_\eta)\right]_s \right\}\Delta\xi \qquad (5.19)$$

For ξ momentum equation ($\phi = u$)

$$\text{Source}^u(\xi,\eta) = -(y_\eta)_e(p_E - p_P)\Delta\eta + (y_\xi)_e\left\{\frac{1}{4}(p_{NE} + p_N - p_{SE} - p_S)\right\}\Delta\xi$$

For η momentum equation ($\phi = v$)

$$\text{Source}^v(\xi,\eta) = (x_\eta)_n\left\{\frac{1}{4}(p_{NE} + p_E - p_{NW} - p_W)\right\}\Delta\eta - (x_\xi)_n(p_N - p_P)\Delta\xi$$

For energy equation ($\phi = T$)

$$\text{Source}^T(\xi,\eta) = 0$$

The two terms on the left-hand side of Eq. (5.19) are the convective fluxes at the control volume faces; the first term on the right-hand side is the source term, and the last two are the diffusion fluxes. The non-linear convective terms can be discretized using a variety of convection schemes. Now, the first order upwind scheme is implemented for treating the convective fluxes. Using first order upwind scheme for convective terms, the Eq. (5.19) can be written as

$$\{\phi_P[|0, F_e|] - \phi_E[|0, -F_e|]\} - \{\phi_W[|0, F_w|] - \phi_P[|0, -F_w|]\} + \{\phi_P[|0, F_n|]$$

$$-\phi_N[|0, -F_n|]\} - \{\phi_S[|0, F_s|] - \phi_P[|0, -F_s|]\} = \text{Source}^\phi(\xi,\eta)$$

$$+D_e(\phi_E - \phi_P) - D_w(\phi_P - \phi_W) + D_n(\phi_N - \phi_P) - D_s(\phi_P - \phi_S)$$

$$+\{-G_e(\phi_N + \phi_{NE} - \phi_S - \phi_{SE}) + G_w(\phi_N + \phi_{NW} - \phi_S - \phi_{SW})$$

$$-G_n(\phi_E + \phi_{NE} - \phi_W - \phi_{NW}) + G_s(\phi_E + \phi_{SE} - \phi_W - \phi_{SW})\} \qquad (5.20)$$

where

$$F_e = U_e\Delta\eta \quad D_e = \frac{\Gamma^\phi\alpha_e\Delta\eta}{J_e(\Delta\xi)_e} \quad G_e = \frac{\Gamma^\phi\beta_e}{4J_e}$$

$$F_w = U_w\Delta\eta \quad D_w = \frac{\Gamma^\phi\alpha_w\Delta\eta}{J_w(\Delta\xi)_w} \quad G_w = \frac{\Gamma^\phi\beta_w}{4J_w}$$

$$F_n = V_n\Delta\xi \quad D_n = \frac{\Gamma^\phi\gamma_n\Delta\xi}{J_n(\Delta\eta)_n} \quad G_n = \frac{\Gamma^\phi\beta_n}{4J_n} \qquad (5.21)$$

$$F_s = V_s\Delta\xi \quad D_s = \frac{\Gamma^\phi\gamma_s\Delta\xi}{J_s(\Delta\eta)_s} \quad G_s = \frac{\Gamma^\phi\beta_s}{4J_s}$$

Finally, the Eq. (5.20) can be written in the following form after rearrangement

$$a_P\,\phi_P = a_E\,\phi_E + a_W\,\phi_W + a_N\,\phi_N + a_S\,\phi_S + \text{Source}^\phi(\xi,\eta) + G^\phi$$

$$= \sum a_{nb}\,\phi_{nb} + \text{Source}^\phi(\xi,\eta) + G^\phi \qquad (5.22)$$

where

$$a_E = D_e + [|0, -F_e|] \quad a_W = D_w + [|0, F_w|]$$

$$a_N = D_n + [|0, -F_n|] \quad a_S = D_s + [|0, F_s|]$$

$$a_P = a_E + a_W + a_N + a_S \tag{5.23}$$

$$G^\phi = -G_e(\phi_N + \phi_{NE} - \phi_S - \phi_{SE}) + G_w(\phi_N + \phi_{NW} - \phi_S - \phi_{SW})$$

$$-G_n(\phi_E + \phi_{NE} - \phi_W - \phi_{NW}) + G_s(\phi_E + \phi_{SE} - \phi_W - \phi_{SW})$$

Now using deferred QUICK scheme for convective terms, the Eq. (5.19) can be written in final form as

$$a_P \, \phi_P^n = a_E \, \phi_E^n + a_W \, \phi_W^n + a_N \, \phi_N^n + a_S \, \phi_S^n + b_E \, \phi_E^{n-1} + b_W \, \phi_W^{n-1}$$

$$+ b_N \, \phi_N^{n-1} + b_S \, \phi_S^{n-1} + b_{EE} \, \phi_{EE}^{n-1} + b_{WW} \, \phi_{WW}^{n-1} + b_{NN} \, \phi_{NN}^{n-1}$$

$$+ b_{SS} \, \phi_{SS}^{n-1} + b_P \, \phi_P^{n-1} + \text{Source}^\phi(\xi, \eta) + G^\phi \tag{5.24}$$

where

$$a_E = D_e + [|0, -F_e|] \quad a_W = D_w + [|0, F_w|]$$

$$a_N = D_n + [|0, -F_n|] \quad a_S = D_s + [|0, F_s|]$$

$$a_P = a_E + a_W + a_N + a_S$$

$$b_E = \frac{1}{8}\{-3\,[|0, F_e|] - 2\,[|0, -F_e|] + [|0, -F_w|]\}$$

$$b_W = \frac{1}{8}\{-3\,[|0, -F_w|] - 2\,[|0, F_w|] + [|0, F_e|]\}$$

$$b_N = \frac{1}{8}\{-3\,[|0, F_n|] - 2\,[|0, -F_n|] + [|0, -F_s|]\}$$

$$b_S = \frac{1}{8}\{-3\,[|0, -F_s|] - 2\,[|0, F_s|] + [|0, F_n|]\} \tag{5.25}$$

$$b_{EE} = -\frac{1}{8}[|0, -F_e|] \quad b_{WW} = -\frac{1}{8}[|0, F_w|]$$

$$b_{NN} = -\frac{1}{8}[|0, -F_n|] \quad b_{SS} = -\frac{1}{8}[|0, F_s|]$$

$$b_P = b_E + b_W + b_N + b_S + b_{EE} + b_{WW} + b_{NN} + b_{SS}$$

$$G^\phi = -G_e(\phi_N + \phi_{NE} - \phi_S - \phi_{SE})^{n-1} + G_w(\phi_N + \phi_{NW} - \phi_S - \phi_{SW})^{n-1}$$

$$-G_n(\phi_E + \phi_{NE} - \phi_W - \phi_{NW})^{n-1} + G_s(\phi_E + \phi_{SE} - \phi_W - \phi_{SW})^{n-1}$$

5.7   STEADY CALCULATION USING PSEUDO-TRANSIENT

For steady state problems the SIMPLE coupling scheme can be considered as a pseudo-transient process, each iteration of the scheme corresponding to a pseudo time step. In pseudo-transient process, the discretized momentum equation for general variable ϕ using deferred QUICK scheme [5] can be written as

$$a_P\, \phi_P^n = a_E\, \phi_E^n + a_W\, \phi_W^n + a_N\, \phi_N^n + a_S\, \phi_S^n + a_P^0 \phi_P^{n-1} + b_E\, \phi_E^{n-1} + b_W\, \phi_W^{n-1}$$

$$+ b_N\, \phi_N^{n-1} + b_S\, \phi_S^{n-1} + b_{EE}\, \phi_{EE}^{n-1} + b_{WW}\, \phi_{WW}^{n-1} + b_{NN}\, \phi_{NN}^{n-1}$$

$$+ b_{SS}\, \phi_{SS}^{n-1} + b_P\, \phi_P^{n-1} + \text{Source}^\phi(\xi, \eta) + G^\phi \tag{5.26}$$

where

$$a_P^0 = J_P \frac{\Delta\xi\Delta\eta}{\Delta\tau} \tag{5.27}$$

$$a_P = a_E + a_W + a_N + a_S + a_P^0 \tag{5.28}$$

5.8 NON-ORTHOGONAL PRESSURE CORRECTION EQUATION

The pressure correction equation is based on the SIMPLE algorithm developed for incompressible flows and Cartesian coordinates by Patankar [2]. The algorithm has been extended to boundary-fitted curvilinear coordinates in order to handle arbitrarily shaped flow boundaries. It is assumed that u^* and v^* are the velocity components that satisfy the momentum equations with a given distribution of p^*. Then the contravariant velocities, U^* and V^*, are calculated from Cartesian velocities, u^* and v^*. Since U^* and V^* in general will not satisfy the continuity equation, the pressure p^* must be corrected. In the SIMPLE procedure, the correct pressure (p) and velocities (u, v) can be obtained by adding a correction as follows

$$p = p^* + p' \quad u = u^* + u' \quad v = v^* + v' \tag{5.29}$$

And the corrections in contravariant velocities are given by

$$U' = U - U^* = y_\eta(u - u^*) - x_\eta(v - v^*) \tag{5.30}$$

$$V' = V - V^* = x_\xi(v - v^*) - y_\xi(u - u^*) \tag{5.31}$$

The equations for u_e and v_e are obtained by integration of the ξ and η momentum equation over the u control volume as shown in Fig. 5.4. These equations can be obtained by putting $\phi = u_e$ and $\phi = v_e$ in Eq. (5.22)

$$a_e u_e = \sum a_{nb} u_{nb} - \left[(y_\eta)_e (p_E - p_P)\Delta\eta - \right.$$

$$\left. (y_\xi)_e \left\{ \frac{1}{4}(p_{NE} + p_N - p_{SE} - p_S) \right\} \Delta\xi \right] + G^u \tag{5.32}$$

$$a_e v_e = \sum a_{nb} v_{nb} - \left[-(x_\eta)_e (p_E - p_P)\Delta\eta + \right.$$

$$\left. (x_\xi)_e \left\{ \frac{1}{4}(p_{NE} + p_N - p_{SE} - p_S) \right\} \Delta\xi \right] + G^v \tag{5.33}$$

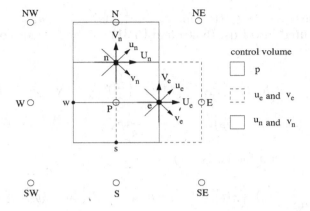

Figure 5.4 A grid for which Cartesian and contravariant velocities are not aligned [1]

The momentum Eqs. (5.32) and (5.33) can be written for guess values of pressure and velocities as follows

$$a_e u^*_e = \sum a_{nb} u^*_{nb} - \Big[(y_\eta)_e (p^*_E - p^*_P)\Delta\eta -$$

$$(y_\xi)_e \Big\{\frac{1}{4}(p^*_{NE} + p^*_N - p^*_{SE} - p^*_S)\Big\}\Delta\xi\Big] + G^{u^*} \qquad (5.34)$$

$$a_e v^*_e = \sum a_{nb} v^*_{nb} - \Big[-(x_\eta)_e (p^*_E - p^*_P)\Delta\eta +$$

$$(x_\xi)_e \Big\{\frac{1}{4}(p^*_{NE} + p^*_N - p^*_{SE} - p^*_S)\Big\}\Delta\xi\Big] + G^{v^*} \qquad (5.35)$$

Similar equations can be written for u_n, v_n, u_w, v_w, u_s, v_s, u^*_n, v^*_n, u^*_w, v^*_w, u^*_s and v^*_s. If Eqs. (5.34) and (5.35) are subtracted from Eqs. (5.32) and (5.33) respectively, the connection between the pressure and velocity correction appears as follows

$$a_e u'_e = a_e\{u_e - u^*_e\} = \sum a_{nb} u'_{nb} - \Big[(y_\eta)_e (p'_E - p'_P)\Delta\eta$$

$$-(y_\xi)_e \Big\{\frac{1}{4}(p'_{NE} + p'_N - p'_{SE} - p'_S)\Big\}\Delta\xi\Big] + G^{u'} \qquad (5.36)$$

$$a_e v'_e = a_e\{v_e - v^*_e\} = \sum a_{nb} v'_{nb} - \Big[-(x_\eta)_e (p'_E - p'_P)\Delta\eta$$

$$+(x_\xi)_e \Big\{\frac{1}{4}(p'_{NE} + p'_N - p'_{SE} - p'_S)\Big\}\Delta\xi\Big] + G^{v'} \qquad (5.37)$$

Now, neglecting all the terms on the right-hand sides of Eqs. (5.36) and (5.37) except the pressure terms and rearranging, we get

$$u_e = u^*_e - \frac{1}{a_e}\Big[(y_\eta)_e (p'_E - p'_P)\Delta\eta -$$

$$(y_\xi)_e \Big\{\frac{1}{4}(p'_{NE} + p'_N - p'_{SE} - p'_S)\Big\}\Delta\xi\Big] \qquad (5.38)$$

$$v_e = v^*_e - \frac{1}{a_e}\Big[-(x_\eta)_e (p'_E - p'_P)\Delta\eta +$$

$$(x_\xi)_e \Big\{\frac{1}{4}(p'_{NE} + p'_N - p'_{SE} - p'_S)\Big\}\Delta\xi\Big]. \qquad (5.39)$$

Similar expressions can be written for $u_n - u^*_n, v_w - v^*_w, u_s - u^*_s$ and $v_s - v^*_s$. These are substituted into Eqs. (5.30) and (5.31) to obtain equations for U and V in terms of U^*, V^* and p'.

$$U_e = U^*_e - \frac{(y_\eta)_e}{a_e}\left[(y_\eta)_e\ (p'_E - p'_P)\Delta\eta - (y_\xi)_e\left\{\frac{1}{4}(p'_{NE} + p'_N - p'_{SE} - p'_S)\right\}\Delta\xi\right]$$
$$+\frac{(x_\eta)_e}{a_e}\left[-(x_\eta)_e\ (p'_E - p'_P)\Delta\eta + (x_\xi)_e\left\{\frac{1}{4}(p'_{NE} + p'_N - p'_{SE} - p'_S)\right\}\Delta\xi\right] \quad (5.40)$$

Similarly, the expression for V_n is

$$V_n = V^*_n + \frac{(y_\xi)_n}{a_n}\left[(y_\eta)_n\left\{\frac{1}{4}(p'_{NE} + p'_E - p'_{NW} - p'_W)\right\}\Delta\eta - (y_\xi)_e(p'_N - p'_P)\Delta\xi\right]$$
$$-\frac{(x_\xi)_n}{a_n}\left[-(x_\eta)_n\left\{\frac{1}{4}(p'_{NE} + p'_E - p'_{NW} - p'_W\right\}\Delta\eta + (x_\xi)_n\ (p'_N - p'_P)\Delta\xi\right] \quad (5.41)$$

The expression for U_w and V_s can be written in similar manner. Substituting the contravariant velocity components into the continuity Eq. (5.18) and rearranging, we obtain the pressure correction equation

$$A_P\ p'_P = A_E\ p'_E + A_W\ p'_W + A_N\ p'_N + A_S\ p'_S + A_{NE}\ p'_{NE}$$
$$+A_{NW}\ p'_{NW} + A_{SE}\ p'_{SE} + A_{SW}\ p'_{SW} + B \quad (5.42)$$

where

$$A_E = \frac{\alpha_e(\Delta\eta)^2}{a_e} - \frac{\beta_n\Delta\xi\Delta\eta}{4a_n} + \frac{\beta_s\Delta\xi\Delta\eta}{4a_s}$$

$$A_W = \frac{\alpha_w(\Delta\eta)^2}{a_w} + \frac{\beta_n\Delta\xi\Delta\eta}{4a_n} - \frac{\beta_s\Delta\xi\Delta\eta}{4a_s}$$

$$A_N = \frac{\gamma_n(\Delta\xi)^2}{a_n} + \frac{\beta_w\Delta\xi\Delta\eta}{4a_w} - \frac{\beta_e\Delta\xi\Delta\eta}{4a_e}$$

$$A_S = \frac{\gamma_s(\Delta\xi)^2}{a_s} - \frac{\beta_w\Delta\xi\Delta\eta}{4a_w} + \frac{\beta_e\Delta\xi\Delta\eta}{4a_e}$$

$$A_{NE} = -\frac{\beta_e\Delta\xi\Delta\eta}{4a_e} - \frac{\beta_n\Delta\xi\Delta\eta}{4a_n}$$

$$A_{SE} = \frac{\beta_e\Delta\xi\Delta\eta}{4a_e} + \frac{\beta_s\Delta\xi\Delta\eta}{4a_s} \quad (5.43)$$

$$A_{NW} = \frac{\beta_w\Delta\xi\Delta\eta}{4a_w} + \frac{\beta_n\Delta\xi\Delta\eta}{4a_n}$$

$$A_{SW} = -\frac{\beta_w\Delta\xi\Delta\eta}{4a_w} - \frac{\beta_s\Delta\xi\Delta\eta}{4a_s}$$

$$A_p = A_E + A_W + A_N + A_S$$

$$A_{NE} + A_{SE} + A_{NW} + A_{SW} = 0$$

$$B = -[(U^*_e - U^*_w)\Delta\eta + (V^*_n - V^*_s)\Delta\xi]$$

The term B represents the mass imbalance in the main (p) control volume. The coefficients α, β and γ are defined in Eq. (5.16), evaluated at the points in Fig. 5.4

denoted by their subscripts. a_e is the central coefficient in the u_e or v_e equation (Eqs. (5.40) and (5.41)); similarly a_n, a_w, a_s are the central coefficients in the equations for u_n or v_n, u_w or v_w and u_s or v_s.

After calculating p', the pressure is updated from the following relation.

$$p = p^* + \alpha_p \, p' \tag{5.44}$$

Then the Cartesian velocities (u, v) are corrected from the relation of Eqs. (5.40) and (5.41) and contravariant velocities (U, V) are calculated using the new values of u and v from Eq. (5.17). On a boundary fitted coordinate system both the Cartesian and the contravariant velocities are corrected after each iteration.

Solution Procedure

The iterative process involves the following steps:

1. Guess the pressure field p^*.
2. Calculate coefficients for all equations using the best available velocities.
3. Solve the momentum Eqs. (5.34) and (5.35) to obtain u^*, v^*.
4. Calculate the corresponding contravariant velocities U^*, V^* from $U^* = y_\eta u^* - x_\eta v^*$, $V^* = x_\xi v^* - y_\xi u^*$.
5. Solve the p' equation from Eq. (5.42).
7. Correct pressure p from Eq. (5.44).
8. Correct u and v using the velocity correction Eqs. (5.38) and (5.39).
9. Calculate U and V using $U = y_\eta u - x_\eta v$, $V = x_\xi v - y_\xi u$.
10. Treat the corrected pressure as a new guessed pressure p^*.
11. Return to step 2 and repeat the steps if convergence has not been achieved.

When the solution is converged, the discretization equation for other scaler variable like temperature is solved, if the particular variable does not influence the flow field.

5.9   BOUNDARY CONDITIONS

The boundary conditions for the field equation should be transformed from the x, y to the ξ, η coordinates. Three types of boundary conditions are considered generally as described below.

First Kind Boundary Conditions

If the value of field variable ϕ is prescribed at the boundary in the physical domain, then no transformation is needed, because the values of ϕ specified along the boundaries in the physical domain remain the same at the corresponding grid locations along the boundaries in the computational domain. This is called Dirichlet boundary condition.

Second Kind Boundary Conditions

If the derivative of the field variable ϕ normal to the boundary is prescribed in the physical domain, then the normal derivative $\frac{\partial \phi}{\partial n}$ should be expressed in terms of the ξ, η independent variables of the computational domain. This is also called Neumann boundary condition. The normal derivatives can be written in the following form (Fig. 5.5).

$$\text{Top wall } S_1 \quad \frac{\partial \phi}{\partial n_1} = \frac{1}{J\sqrt{\gamma}}(\gamma\varphi_\eta - \beta\varphi_\xi)$$

$$\text{Right wall } S_2 \quad \frac{\partial \phi}{\partial n_2} = \frac{1}{J\sqrt{\alpha}}(\alpha\varphi_\xi - \beta\varphi_\eta)$$

$$\text{Bottom wall } S_3 \quad \frac{\partial \phi}{\partial n_3} = \frac{-1}{J\sqrt{\gamma}}(\gamma\varphi_\eta - \beta\varphi_\xi) \qquad (5.45)$$

$$\text{Left wall } S_4 \quad \frac{\partial \phi}{\partial n_4} = \frac{-1}{J\sqrt{\alpha}}(\alpha\varphi_\xi - \beta\varphi_\eta)$$

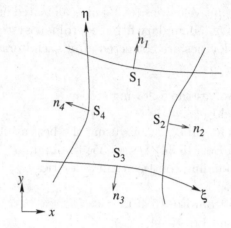

Figure 5.5 Outward drawn unit normal vectors to $\xi = $ constant and $\eta = $ constant lines

Third Kind Boundary Conditions

This type of boundary condition is a combination of the function ϕ and its normal derivative specified on the boundary. This is also called Robin's boundary condition.

$$a\phi + b\frac{\partial \phi}{\partial n} = h \qquad (5.46)$$

The transformation of Eq. 5.46 to the ξ, η variables in the computational domain is all we need to transfer the normal derivative $\frac{\partial \varphi}{\partial n}$ according to the relations given by Eq. (5.45).

In this chapter, we have discussed a solution procedure for solving two-dimensional Navier-Stokes equations in curvilinear coordinate systems. The readers may refer to the work of Demirdžić et al. [6] to solve some problems using the algorithms discussed in this chapter. The readers are encouraged to study the methods for solving three-dimensional Navier-Stokes equations using unstructured grids (Dalal et al. [7]).

REFERENCES

1. C. R. Maliska and G. D. Raithby, 1984, A Method for Computing Three Dimensional Flows Using Non-Orthogonal Boundary-Fitted Co-Ordinates, Int. J. Num. Meth. in Fluids, Vol. 4, pp. 519-537.

2. S. V. Patankar, 1980, Numerical Heat Transfer and Fluid Flow, Hemisphere Publishing.

3. M. Necati Ozisik, 1994, Finite Difference Methods in Heat Transfer, CRC Press.

4. W. Shyy, S. S. Tong and S. M. Correa, 1985, Numerical Recirculating Flow Calculation Using a Body-Fitted Coordinate System, Num. Heat Trans., Vol. 8, pp. 99-113.

5. T. Hayase, J. C. Humphrey and R. Greif, 1992, A Consistently Formulated QUICK Scheme for Fast and Stable Convergence Using Finite-Volume Iterative Calculation Procedures, J. Comp. Physics, Vol. 98, pp. 180-118.

6. Demirdžić I., Ž. Lilek, and Perić M., 1992, Fluid flow and heat transfer test problems for non-orthogonal grids: Bench-mark solutions, Int. J. Num. Meth. in Fluids, Vol. 15(3), pp. 329–354.

7. A. Dalal, V. Eswaran, and G. Biswas, 2008, A Finite Volume Method for Navier-Stokes Equations on Unstructured Meshes, Num. Heat Trans.: Part B, Vol. 54(3), pp. 238-259.

EXERCISES

1. Consider the steady flow inside an inclined cavity whose upper lid is moving at constant velocity $u = 1$ as shown in Fig. 5.6. This problem has become a benchmark to test algorithms for flows in complex geometries (Demirdžić et al. [6]). The flow domain is a rhombus with angle $\beta = 30^o$ and $\beta = 45^o$, and cavity side, $L = 1$, in both cases (see Fig. 5.6). Solve the problems using curvilinear grid for $Re = 400$, 1000 and different side angles. Draw the streamlines and compare the u velocity along the vertical centerline and v velocity along the horizontal centerline of the cavity with the solutions of Demirdžić et al. [6].

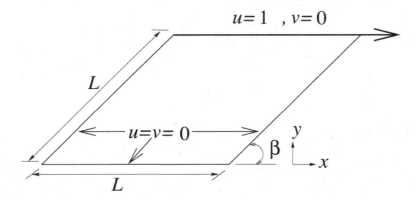

Figure 5.6   Geometry of lid-driven cavity with inclined side wall

2. Consider a differentially heated cavity with inclined side walls as studied by Demirdžić et al. [6]. Figure 5.7 shows the geometry and boundary conditions of the problem of interest. For this study, the inclination angle is $\beta = 45^o$, Rayleigh number is 10^6, and Prandtl numbers are 0.1 and 10. Solve the problem using

the curvilinear grid and show the streamlines and isotherms for different cases. Compare the local Nusselt number on the cold wall for both the Prandtl numbers with the results of Demirdžić et al. [6].

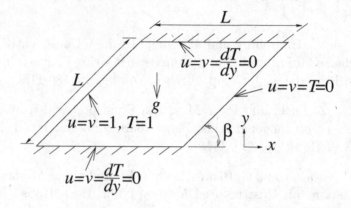

Figure 5.7 Geometry of differentially heated cavity with inclined side walls

3. Consider flow over a heated cylinder placed eccentrically on the vertical center-line of a square duct, as studied by Demirdžić et al. [6]. In this case, the cylinder is kept at a temperature of $T_H = 1$, the vertical side walls are maintained at $T_C = 0$, and the horizontal side walls are assumed adiabatic. Consider only one-half of the geometry due to the existence of a vertical plane of symmetry as shown in Fig. 5.8. The vertical shift of the cylinder center from the duct center is $\Delta y_c = 0.1$, the radius of the cylinder is $R = 0.2$, and the duct dimension

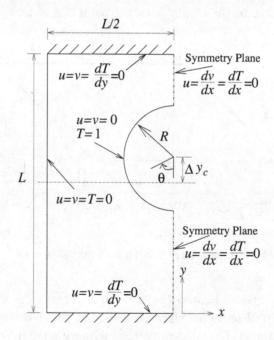

Figure 5.8 Geometry and boundary conditions

is $L = 1$. The geometry and boundary conditions are shown in Fig. 5.8. Flows at $Ra = 10^6$ are studied for two values of the Prandtl number ($Pr = 0.1$ and 10). Solve the problem using the curvilinear grid and show the streamlines and isotherms for both the Prandtl numbers. Plot the local variations of the Nusselt number along the cold and hot walls and compare with the results of Demirdžić et al. [6]. Calculate the maximum Nusselt number on the cold and hot walls and the average Nusselt number on the cold wall for different Prandtl numbers.

## CHAPTER 6

Turbulent Flow and Heat Transfer

6.1 INTRODUCTION

The following are the characteristics of turbulent motion

- **Irregularity**
  Complex variations of velocity and temperature with space and time (fluctuations) are the dominant characteristics of a turbulent flow. The irregular motion is generated due to random fluctuations (Fig. 6.1). It is postulated that the fluctuations inherently come from disturbances (such as, roughness of the solid surface) and they may be either reduced due to viscous damping or may grow by drawing energy from the free stream. At a Reynolds number less than the critical, the kinetic energy of flow is not enough to sustain the random fluctuations against the viscous damping and in such cases laminar flow continues to exist. At a Reynolds number somewhat higher than critical, the kinetic energy of flow supports the growth of fluctuations and transition to turbulence is induced.

- **Strong mixing**
  The fluctuating turbulent motion promotes higher level of transfer of momentum, heat and mass; - is practically the most important feature. Turbulent flows appear to be random. Turbulent flows are not always free of coherent structures. The coherent structures component is periodic or at least repeatable.

- **Three-dimensional turbulent motion**
  For a parallel flow, the axial velocity component is

$$u(y, t) = \overline{u}(y) + u'(\Gamma, t)$$

where y is the normal direction and Γ is any space variable.

Even if the bulk motion is parallel, the fluctuation u' being random varies in all directions. Now let us look at the continuity equation

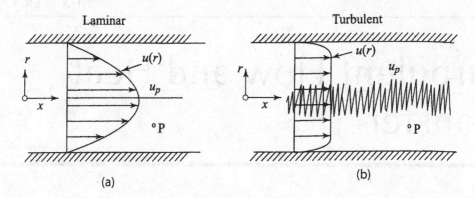

Figure 6.1 Presence of turbulence is appreciated from the velocity signal (u_p) at any point P in the flow field

$$\frac{\partial \overline{u}}{\partial x} + \frac{\partial u'}{\partial x} + \frac{\partial v}{\partial y} + \frac{\partial w}{\partial z} = 0$$

Since $\frac{\partial u'}{\partial x} \neq 0$, the above equation depicts that y and z components of velocity exist even for the parallel flow if the flow is turbulent. We can write

$$u(y,t) = \overline{u}(y) + u'\,(\Gamma, t)$$
$$v = 0 + v'\,(\Gamma, t)$$
$$w = 0 + w'\,(\Gamma, t)$$

Turbulence is a three-dimensional process in which vortex stretching and tilting are essential mechanisms for energy exchange between different scales of motion.

- **Turbulent motion carries vorticity and consists of interacting eddies**
  Wide spectrum of eddy sizes and corresponding fluctuation frequencies are shown in Fig. 6.2. In a turbulent flow, energy is fed from large scales mainly to smaller scales by nonlinear processes.

- The term *homogeneous turbulence* implies that the velocity fluctuations in the system are random.

- The average turbulent characteristics are independent of the position of the fluid, i.e., invariant to axis translation.

- If the velocity fluctuations are independent of the axis of reference (invariant to axis rotation and reflection), the restriction leads to *isotropic turbulence*, which by definition is always homogeneous.

Turbulence is generally damped by viscosity. One of the effects of viscosity on turbulence is to make the flow more homogeneous and less dependent on direction. **If the turbulence has the same structure quantitatively in**

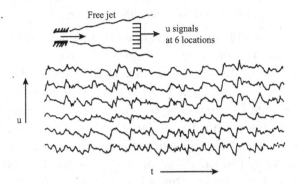

Figure 6.2 Wide spectrum of eddy sizes and corresponding frequencies

all parts of the flow field, the turbulence is said to be homogeneous. The turbulence is called isotropic if its statistical features have no directional preference and perfect disorder persists. Its velocity fluctuations are independent of the axis of reference, i.e., invariant to axis rotation and reflection. Isotropic turbulence is by its definition always homogeneous. In such a situation, the gradient of the mean velocity does not exist. The mean velocity is either zero or constant throughout. However, when the mean velocity has a gradient the turbulence is called anisotropic.

A little more discussion on homogeneous and isotropic turbulence is needed at this stage. The term *homogeneous turbulence* implies that the velocity fluctuations in the system are random. The average turbulent characteristic are independent of the position in the field, i.e., invariant to axis translation.

Consider the root mean square (rms) velocity fluctuations:

$$u' = \sqrt{\overline{u^2}}, \quad v' = \sqrt{\overline{v^2}}, \quad w' = \sqrt{\overline{w^2}}$$

In homogeneous turbulence, the rms values of u', v' and w' can all be different, but each value must be uniform over the entire turbulent field. Even if the rms fluctuation of any component, say u's are constant over the entire field, the instantaneous values of u may differ from point to point at any instant.

In addition to its homogeneous nature, if the velocity fluctuations are independent of the axis of reference, i.e., invariant to axis rotation and reflection, the situation leads to isotropic turbulence, which by definition as mentioned earlier, is always homogeneous.

In isotropic turbulence, fluctuations are independent of the direction of reference and

$$\sqrt{\overline{u^2}} = \sqrt{\overline{v^2}} = \sqrt{\overline{w^2}}$$

or

$$u' = v' = w'$$

Again, it is of relevance to say that even if the rms fluctuations at any point are same, their instantaneous values may differ from each other at any instant.

Turbulent flow is also diffusive. In general, turbulence brings about better mixing of a fluid and produces an additional diffusive effect. The term *eddy diffusion* is often used to distinguish this effect from molecular diffusion. The effects caused by mixing are as if the viscosity is increased by a factor of 100 or more. At a large Reynolds number there exists a continuous transport of energy from the free stream to the large eddies. Then smaller eddies are formed continuously from the large eddies. Then from the large eddies smaller eddies are continuously formed. Near the wall, the smallest eddies dissipate energy and destroy themselves (Tennekes and Lumley [1]).

- **Length scales of turbulence**
  The large scale motions with length scale L are unstable for large Reynolds numbers. The scale L is commensurate with the dimension of the flow domain. We shall try to understand smallest length scales of turbulent flows. At the smallest length scale, viscosity can be effective in smoothing out the velocity fluctuations. As mentioned, turbulence is generally damped by the viscosity ν and the main contribution to energy dissipation ϵ can be attributed to very small length scales of the order of the Kolmogorov-scale, l_s given by $(\nu^3/\epsilon)^{1/4}$. The parameters that govern the small scale motions include dissipation rate per unit mass, ϵ. The dimenssion of ϵ is $(m^2 s^{-3})$. The dimension of kinematic viscosity, ν is (m^2/s). Making use of these parameters, it is possible to form length scale, time scale and velocity scale as

$$l_s = \left(\frac{\nu^3}{\epsilon}\right)^{1/4}, \quad t_s = (\nu/\epsilon)^{1/2} \quad and \quad v_s = (\nu\epsilon)^{1/4}$$

These are known as Kolmogorov scales of length, time and velocity. The Reynolds number formed by l_s and $v_s (l_s\, v_s/\nu) = 1$.

In the case of homogeneous turbulence, at high Reynolds numbers, the dissipation rate, ϵ equals the production-rate of turbulence and is of the order of u_k^3/L_k where u_k is similar to root-mean-square velocity of turbulence and L_k is the integral length scale of turbulence.

For the smaller length scales within the intertial range (where kinetic energy production rates due to external forcing and viscous dissipation rates are small) the turbulent motion becomes more and more homogeneous and isotropic even if the large scale motion is anisotropic. It is worth mentioning that ratio of largest to smallest length scales is L/L_s, which is proportional to $(Re)^{3/4}$.

6.2   CLASSICAL IDEALIZATION OF TURBULENT FLOWS

Since detailed descriptions of turbulent fluctuations are usually not of interest to engineers we take a statistical approach by averaging turbulence data. Different

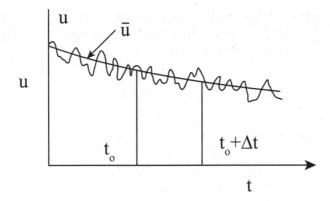

Figure 6.3   Mean motion and fluctuations

trends of variation of the the mean and fluctuating components are shown in Figs. 6.3 and 6.4. Fig 6.4(b) reveals an unstready mean motion. The mean velocity has a time period t_2 and a large time period signifies low frequency oscillation of the mean motion. The high frequency oscillations have a time period of t_1.

Statistical quantities may be calculated as

$$u_i = \bar{u}_i + u_i', \ p = \bar{p} + p', \ \bar{u} = \frac{1}{\Delta t} \int_{to}^{to+\Delta t} u \, dt \tag{6.1}$$

However the fluctuating components do not bring about the bulk displacement of a fluid element. The instantaneous displacement is $u' \, dt$ and if that is indeed not responsible for the bulk motion, we can conclude that

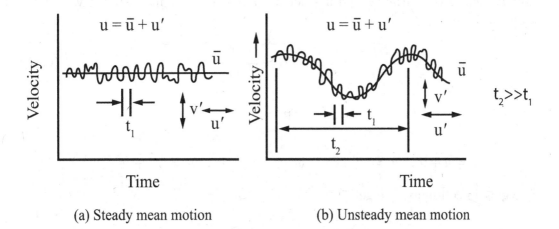

Figure 6.4   Steady and unsteady mean motions in a turbulent flow

$$\int_{to}^{to+\Delta t} u' \, dt = 0 \tag{6.2}$$

Due to the interaction of fluctuating components, macroscopic momentum transport takes place. Therefore, interaction effect between two fluctuating components over long period is nonzero and this yields

$$\int_{to}^{to+\Delta t} u' \, v' \, dt \neq 0 \tag{6.3}$$

We take time average of these two integrals and write

$$\overline{u'} = \frac{1}{\Delta t} \int_{to}^{to+\Delta t} u' \, dt = 0 \tag{6.4}$$

and

$$\overline{u'v'} = \frac{1}{\Delta t} \int_{to}^{to+\Delta t} u' \, v' \, dt \neq 0 \tag{6.5}$$

Now, we can make a general statement with any two fluctuating parameters, say, with f' and g' as (f' and g' can be vectors or passive scalars)

$$\overline{f'} = \overline{g'} = 0 \qquad\qquad \overline{\frac{\partial f'}{\partial s}} = \overline{\frac{\partial^2 f'}{\partial s^2}} = 0 \tag{6.6}$$

and

$$\overline{f'g'} \neq 0 \qquad \text{and} \qquad \overline{\frac{\partial (f'g')}{\partial s}} \neq 0 \tag{6.7}$$

We shall state some rules of operation on mean time-averages herein. If f and g are two dependent variables and if s denotes any one of the independent variables x, y, z, t then

$$\overline{\frac{\partial f}{\partial s}} = \frac{\partial \overline{f}}{\partial s} \; ; \qquad \overline{\int f ds} = \int \overline{f} ds \tag{6.8}$$

The Reynolds decomposition may be expressed as

$$u = \overline{u} + u', \quad v = \overline{v} + v', \quad w = \overline{w} + w', \quad p = \overline{p} + p' \tag{6.9}$$

Plugging in continuity, we get

$$\nabla \cdot \overline{u}_i = 0 \tag{6.10}$$

and

$$\nabla \cdot u'_i = 0 \tag{6.11}$$

Introduction of Reynolds decomposition into the Navier-Stokes equations and subsequent averaging and application of the laws of statistics leads to the appearance of turbulence correlations (turbulent or Reynolds stresses). For example, if we perform the aforesaid exercise on the x momentum equation, we obtain

$$\rho \left\{ \frac{\partial \overline{u}}{\partial t} + \frac{\partial (\overline{u}\,\overline{u})}{\partial x} + \frac{\partial (\overline{u}\,\overline{v})}{\partial y} + \frac{\partial (\overline{u}\,\overline{w})}{\partial z} \right\} = -\frac{\partial \overline{p}}{\partial x}$$

$$+\mu \nabla^2 \overline{u} - \rho \left[\frac{\partial \overline{u'^2}}{\partial x} + \frac{\partial \overline{u'v'}}{\partial y} + \frac{\partial \overline{u'w'}}{\partial z} \right] \tag{6.12}$$

Introducing simplifications arising out of continuity equation we shall obtain

$$\rho \frac{\partial \overline{u}}{\partial t} + \rho \left[\overline{u}\frac{\partial \overline{u}}{\partial x} + \overline{v}\frac{\partial \overline{u}}{\partial y} + \overline{w}\frac{\partial \overline{u}}{\partial z} \right] = -\frac{\partial \overline{p}}{\partial x}$$

$$+\mu \nabla^2 \overline{u} - \rho \left[\frac{\partial \overline{u'^2}}{\partial x} + \frac{\partial \overline{u'v'}}{\partial y} + \frac{\partial \overline{u'w'}}{\partial z} \right] \tag{6.13}$$

Performing a similar treatment on y and z momentum equations, we obtain the y and z momentum equations in the form

$$\rho \frac{\partial \overline{v}}{\partial t} + \rho \left[\overline{u}\frac{\partial \overline{v}}{\partial x} + \overline{v}\frac{\partial \overline{v}}{\partial y} + \overline{w}\frac{\partial \overline{v}}{\partial z} \right] = -\frac{\partial \overline{p}}{\partial y} + \mu \nabla^2 \overline{v}$$

$$-\rho \left[\frac{\partial \overline{u'v'}}{\partial x} + \frac{\partial \overline{v'^2}}{\partial y} + \frac{\partial \overline{v'w'}}{\partial z} \right] \tag{6.14}$$

$$\rho \frac{\partial \overline{w}}{\partial t} + \rho \left[\overline{u}\frac{\partial \overline{w}}{\partial x} + \overline{v}\frac{\partial \overline{w}}{\partial y} + \overline{w}\frac{\partial \overline{w}}{\partial z} \right] = -\frac{\partial \overline{p}}{\partial z} + \mu \nabla^2 \overline{w}$$

$$-\rho \left[\frac{\partial \overline{u'w'}}{\partial x} + \frac{\partial \overline{v'w'}}{\partial y} + \frac{\partial \overline{w'^2}}{\partial z} \right] \tag{6.15}$$

It is to be noted that the terms containing prime symbols were not in the original NS equations.

$$\sigma_T = \begin{bmatrix} \sigma'_{xx} & \tau'_{xy} & \tau'_{xz} \\ \tau'_{xy} & \sigma'_{yy} & \tau_{yz} \\ \tau_{xz} & \tau_{yz} & \sigma_{zz} \end{bmatrix} = -\rho \begin{bmatrix} \overline{u'^2} & \overline{u'v'} & \overline{u'w'} \\ \overline{u'v'} & \overline{v'^2} & \overline{v'w'} \\ \overline{u'w'} & \overline{v'w'} & \overline{w'^2} \end{bmatrix} \tag{6.16}$$

σ_T is the Reynolds stress tensor and written in compact form as $-\rho \overline{u'_i u'_j}$

$$\sigma_{xx} = -p + 2\mu \frac{\partial \overline{u}}{\partial x} - \rho \overline{u'^2} \tag{6.17}$$

$$\tau_{xy} = \mu \left(\frac{\partial \overline{u}}{\partial y} + \frac{\partial \overline{v}}{\partial x} \right) - \rho \overline{u'v'} \tag{6.18}$$

The averaged equations may be written, in a compact form, as

$$\frac{\partial \overline{u}_i}{\partial t} + u_j \frac{\partial \overline{u}_i}{\partial x_j} = -\frac{1}{\rho} \frac{\partial \overline{p}}{\partial x_i} + \frac{\partial}{\partial x_j} \left(\nu \frac{\partial \overline{u}_i}{\partial x_j} - \overline{u'_i u'_j} \right) \tag{6.19}$$

$$\frac{\partial \overline{u}_i}{\partial x_i} = 0 \tag{6.20}$$

We have more unknowns than number of available equations. The modified system of equations cannot be closed within itself unless empirical relations are supplied from experiments to correlate the fluctuating components with the mean motion. This is termed as the closure problem.

In solving the closure problem, the turbulent stresses should be determined by using a turbulence model.

This section discusses only models for turbulent motions. The subgrid-scale models for large-eddy simulations are not included in this discussion.

$$-\overline{u'_i u'_j} = \nu_t \left(\frac{\partial \overline{u}_i}{\partial x_j} + \frac{\partial \overline{u}_j}{\partial x_i} \right) - \frac{2}{3} k \delta_{ij} \tag{6.21}$$

The term ν_t is turbulent (eddy) viscosity. The term involving the Kronecker delta δ_{ij} in Eq. (6.21) is perhaps a somewhat unfamiliar addition to the eddy-viscosity expression. It is necessary to make the expression applicable also to normal stresses (when $i = j$). The first part of Eq. (6.21) involving the velocity gradients would yield the normal stresses.

$$\overline{u'^2} = -2\nu_t \frac{\partial \overline{u}}{\partial x}, \quad \overline{v'^2} = -2\nu_t \frac{\partial \overline{v}}{\partial y}, \quad \overline{w'^2} = -2\nu_t \frac{\partial \overline{w}}{\partial z} \tag{6.22}$$

The sum of the stresses is zero because of the continuity equation. However, all normal stresses are by definition positive quantities, and their sum is twice the kinetic energy k of the fluctuating motion:

$$k = \frac{1}{2} \left(\overline{u'^2} + \overline{v'^2} + \overline{w'^2} \right) \tag{6.23}$$

Inclusion of the second part of the eddy viscosity expression (Eq. (6.21)) assures that the sum of the normal stresses is equal to $2k$. The normal stresses act like pressure forces (i.e., perpendicular to the faces of a control volume). The second part of Eq. (6.21) constitutes a pressure because, like pressure, energy k is a scalar quantity. Therefore, when Eq. (6.21) is used to eliminate $\overline{u'_i u'_j}$ in the momentum equation, this second part can be absorbed by the pressure-gradient term so that in effect the static pressure is replaced as unknown quantity by the pressure $\overline{p} + \frac{2}{3}k$. Therefore the appearance of k in Eq. (6.21) does not necessitate the determination of k, it is the distribution of the eddy viscosity ν_t only that has to be determined. In this context, k can be linked to the intensity of turbulence which is given by

$$I = \sqrt{\frac{1}{3}(\overline{u'^2} + \overline{v'^2} + \overline{w'^2})} \, /U_\infty \tag{6.24}$$

We can also write non-dimensional kinetic energy as

$$\frac{k}{U_\infty^2} = 1.5 \; I^2$$

6.3 REYNOLDS AVERAGED FORM OF ENERGY EQUATION

Let us consider a three-dimensional situation,

$$\frac{\partial T}{\partial t} + u\frac{\partial T}{\partial x} + v\frac{\partial T}{\partial y} + w\frac{\partial T}{\partial z} = \alpha\left[\frac{\partial^2 T}{\partial x^2} + \frac{\partial^2 T}{\partial y^2} + \frac{\partial^2 T}{\partial z^2}\right] \tag{6.25}$$

Using Reynolds decomposition of velocity and temperature

$$u = \bar{u} + u', \quad v = \bar{v} + v', \quad w = \bar{w} + w' \quad T = \bar{T} + T' \tag{6.26}$$

we get

$$\bar{u}\frac{\partial \bar{T}}{\partial x} + \bar{v}\frac{\partial \bar{T}}{\partial y} + \bar{w}\frac{\partial \bar{T}}{\partial z} = \frac{\partial}{\partial x}\left(\alpha\frac{\partial \bar{T}}{\partial x} - \overline{u'T'}\right)$$

$$+\frac{\partial}{\partial y}\left(\alpha\frac{\partial \bar{T}}{\partial y} - \overline{v'T'}\right) + \frac{\partial}{\partial z}\left(\alpha\frac{\partial \bar{T}}{\partial z} - \overline{w'T'}\right) \tag{6.27}$$

The terms $\overline{u'T'}$, $\overline{v'T'}$, and $\overline{w'T'}$ thus, cause additional heat flux in the x, y and z directions respectively, due to turbulent motion. The total heat flux in the three directions will therefore be given by

$$q_x'' = -\rho c_p\left(\alpha\frac{\partial \bar{T}}{\partial x} - \overline{u'T'}\right); \; q_y'' = -\rho c_p\left(\alpha\frac{\partial \bar{T}}{\partial y} - \overline{v'T'}\right);$$

$$q_z'' = -\rho c_p\left(\alpha\frac{\partial \bar{T}}{\partial z} - \overline{w'T'}\right) \tag{6.28}$$

As in the case of turbulent transport of momentum, it is convenient to define an eddy viscosity or turbulent viscosity, to study the turbulent transport of thermal energy, a turbulent thermal diffusivity α_t can be defined.

The total heat flux in x, y and z directions can therefore be given as

$$\frac{q_x''}{\rho c_p} = -(\alpha + \alpha_t)\frac{\partial \bar{T}}{\partial x}, \; \frac{q_y''}{\rho c_p} = -(\alpha + \alpha_t)\frac{\partial \bar{T}}{\partial y}, \; \frac{q_z''}{\rho c_p} = -(\alpha + \alpha_t)\frac{\partial \bar{T}}{\partial z} \tag{6.29}$$

Like eddy viscosity, α_t is not a fluid property but depends on the state of turbulence. In fact the Reynolds analogy between heat and momentum transport suggests

$$\alpha_t = \frac{\nu_t}{\sigma_t} \tag{6.30}$$

The denominator σ_t is called turbulent Prandtl number. Experiments have shown that σ_t varies very little across the flow. Many models make use of σ_t as a constant. For the flow of air a value of 0.9 may be chosen.

Finally the transport equation (6.27) may be written as

$$\overline{u}\frac{\partial \overline{T}}{\partial x} + \overline{v}\frac{\partial \overline{T}}{\partial y} + \overline{w}\frac{\partial \overline{T}}{\partial z} = \frac{\partial}{\partial x}\left[(\alpha + \alpha_t)\frac{\partial \overline{T}}{\partial x}\right]$$

$$+ \frac{\partial}{\partial y}\left[(\alpha + \alpha_t)\frac{\partial \overline{T}}{\partial y}\right] + \frac{\partial}{\partial z}\left[(\alpha + \alpha_t)\frac{\partial \overline{T}}{\partial z}\right] \tag{6.31}$$

6.4 PRANDTL'S MIXING LENGTH

In the case of turbulent flow, velocity, pressure and temperature at a fixed point in space do not remain constant with time but undergo very irregular fluctuations of high frequency. The velocity of the fluid in turbulent flow can be represented by two components

(i) mean motion

(ii) fluctuating component (or eddying motion)

The velocity of fluid is given by

$$u = \overline{u} + u', \quad v = \overline{v} + v', \quad w = \overline{w} + w'$$

Similarly,

$$p = \overline{p} + p', \quad T = \overline{T} + T'$$

However, we must always consider that the fluctuating components do not bring about the bulk displacement of a fluid element. Consider continuity equation in two dimensions: $\frac{\partial u}{\partial x} + \frac{\partial v}{\partial y} = 0$. If we substitute $u = \overline{u} + u'$ and $v = \overline{v} + v'$ and then perform time averaging, we shall obtain

$$\frac{\overline{\partial(\overline{u} + u')}}{\partial x} + \frac{\overline{\partial(\overline{v} + v')}}{\partial y} = 0$$

or

$$\left[\frac{\partial \overline{u}}{\partial x} + \frac{\partial \overline{v}}{\partial y}\right] + \left[\frac{\overline{\partial u'}}{\partial x} + \frac{\overline{\partial v'}}{\partial y}\right] = 0 \tag{6.32}$$

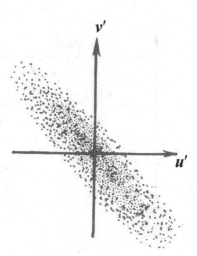

Figure 6.5 Each dot represents a $u'v'$ pair at a given time

But we have seen that

$$\overline{u'} = \overline{v'} = \frac{\partial \overline{u'}}{\partial x} = \frac{\partial \overline{v'}}{\partial y} = 0$$

Invoking

$$\frac{\partial \overline{u'}}{\partial x} = \frac{\partial \overline{v'}}{\partial y} = 0$$

in Eq. (6.32) we get $\frac{\partial \overline{u}}{\partial x} + \frac{\partial \overline{v}}{\partial y} = 0$ which yields

$$\frac{\partial u'}{\partial x} + \frac{\partial v'}{\partial y} = 0$$

we can also write

$$\frac{\partial u'}{\partial x} = -\frac{\partial v'}{\partial y} \tag{6.33}$$

If we consider momentum exchange between two adjacent layers, then on the basis of above equation, it is postulated that if at any instant there is an increase in u' in the x direction it will be followed by an increase in v' in the negative y direction. In other words $\overline{u'v'}$ is nonzero and negative. This is discerned in Fig. 6.5 which shows a cloud of data points (sometimes called a scatter plot). The dots represent the instantaneous values of $u'v'$ pair at different times.

For unit area of the plane PP (Fig. 6.6) the instantaneous turbulent mass transport rate across the plane is $\rho v'$. Associated with this mass transport is a change in the x component of velocity u'. The net momentum flux per unit area, in the x direction, represents the turbulent shear stress at the plane PP (Fig. 6.6) which is $\rho u'v'$. When a turbulent lump movement is in the upward direction ($v' > 0$), it

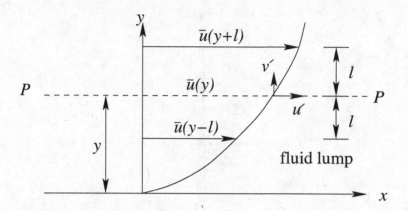

Figure 6.6 Mixing length hypothesis

enters a region of higher $\bar{u}$ and is therefore likely to effect a slowing down fluctuation in u', that is $u' < 0$. A similar argument can be made for $v' < 0$, so that the average turbulent shear stress will be given as

$$\tau_t = -\rho \overline{u'v'} \tag{6.34}$$

Again, let us imagine a turbulent fluid lump which is located a distance l above or below the plane PP. These lumps of fluid move back and forth across the plane and give rise to the eddy or turbulent-shear-stress effect.

At $(y + l)$, the velocity would be

$$\bar{u}(y+l) = \bar{u}\,(y) + l\,\frac{\partial \overline{u}}{\partial y} \tag{6.35}$$

while at $(y - l)$, we can write

$$\bar{u}(y-l) = \bar{u}\,(y) - l\,\frac{\partial \overline{u}}{\partial y} \tag{6.36}$$

Prandtl postulated that the turbulent fluctuation is proportional to the mean of the above two quantities

$$|u'| \approx \frac{1}{2}(|\Delta u_1| + |\Delta u_2|)$$

or

$$u' \approx l\frac{\partial \overline{u}}{\partial y} = C_1\,l\frac{\partial \overline{u}}{\partial y} \tag{6.37}$$

Here l is Prandtl's mixing length. This is analogous to mean free path (average distance a particle travels between collisions) in molecular transport problems. Prandtl also postulated that v' is of the same order of magnitude as u' so that

$$v' = C_2\,l\frac{\partial \overline{u}}{\partial y} \tag{6.38}$$

Now, the turbulent shear stress could be written as

$$\tau_t = -\overline{\rho u' v'} = \rho l^2 \left(\frac{\partial \overline{u}}{\partial y}\right)^2 = \rho \nu_t \frac{\partial \overline{u}}{\partial y} \tag{6.39}$$

The constants C_1 and C_2 can be included in still unknown mixing length, l. The eddy viscosity thus becomes,

$$\nu_t = l^2 \left|\frac{\partial \overline{u}}{\partial y}\right| \tag{6.40}$$

Consider the boundary layer on a flat plate. The shear stress due to laminar flow is given by $\tau_l = \mu \frac{\partial \overline{u}}{\partial y}$. The total shear stress is

$$\tau = \tau_l + \tau_t = \rho \, \nu \, \frac{\partial \overline{u}}{\partial y} + \rho \nu_t \, \frac{\partial \overline{u}}{\partial y} \tag{6.41}$$

or

$$\tau = \rho(\nu + \nu_t) \, \frac{\partial \overline{u}}{\partial y} \tag{6.42}$$

The turbulent viscosity, ν_t in Eq. (6.42) can be determined from Eq. (6.40). However, our problem is still not resolved. How do we determine the value of l, the mixing length? Several correlations, using experimental results for τ_t have been proposed to determine l. In the regime of isotropic turbulence, the most widely used value of mixing length is

$$l = \chi \, y \tag{6.43}$$

where y is the distance from the wall and χ is known as von Karman constant. Experiments have shown that χ is approximately equal to 0.4 (Tennekes and Lumley [1]).

6.5   UNIVERSAL VELOCITY PROFILE (ON FLAT PLATE)

The flow fields that are dominated by high frequency oscillations (small time periods), are isotropic and the mean velocity profiles in such flow fields are universal. In this section we shall discuss the universal velocity profile.

A very thin layer next to the wall behaves like a near wall region of laminar flow. The layer is known as laminar sub-layer and its velocities are such that the viscous forces dominate over the inertia forces (Fig. 6.7). No turbulence exists in it. We know that inertial effects are insignificant in the near wall region and can write

$$\nu \frac{\partial^2 \overline{u}}{\partial y^2} - \frac{\partial \overline{u' v'}}{\partial y} = 0 \tag{6.44}$$

which can be integrated as

$$\nu \frac{\partial \overline{u}}{\partial y} - \overline{u' v'} = \text{constant} \tag{6.45}$$

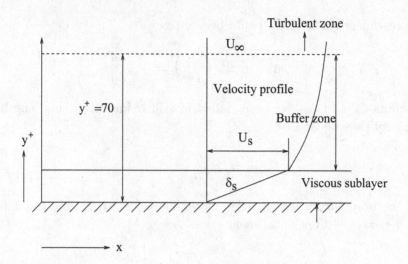

Figure 6.7 Different zones of a turbulent flow past a wall

Again, as we know that the fluctuating components vanish near the wall, the shear stress on the wall is purely viscous and it follows that:

$$\nu \frac{\partial \overline{u}}{\partial y} = \frac{\tau_w}{\rho} \tag{6.46}$$

or

$$\frac{\overline{u} - 0}{y - 0} = \frac{\tau_w}{\rho \, \nu} = \frac{u_\tau^2}{\nu} \qquad \text{where} \qquad u_\tau = \sqrt{\frac{\tau_w}{\rho}} \tag{6.47}$$

The quantity u_τ is known as friction velocity, given by $\tau_w = \rho \, u_\tau^2$. From Eq. (6.47), it is possible to write

$$y \frac{u_\tau}{\nu} = \frac{\overline{u}}{u_\tau} \tag{6.48}$$

Hence a nondimensional coordinate may be defined as $y^+ = y u_\tau / \nu$ and we write down the variation of the nondimensional velocity within the sublayer as

$$\overline{u}^+ = y^+ \tag{6.49}$$

In the turbulent zone, the turbulent shear stress from Prandtl's mixing length model can be written as

$$\tau_t = \rho \, l^2 \left[\frac{\partial \overline{u}}{\partial y} \right]^2 \tag{6.50}$$

Where l is the mixing length $= \chi \, y$ and χ is a von Karman constant we write

$$\frac{\tau_t}{\rho} = \chi^2 y^2 \left[\frac{\partial \overline{u}}{\partial y} \right]^2 \tag{6.51}$$

The shear stress in the turbulent region dictates the velocity U_s at the edge of the sublayer (Fig.6.7). The wall shear stress is estimated as $\tau_w = \mu U_s / \delta_s$.

$$\left[\frac{\partial \bar{u}}{\partial y}\right]^2 = \left[\frac{u_\tau}{\chi y}\right]^2 \qquad [\text{since } \tau_t = \tau_w] \tag{6.52}$$

$$\bar{u} = \frac{u_\tau}{\chi} \ln y + \text{constant} \tag{6.53}$$

or

$$\frac{\bar{u}}{u_\tau} = \frac{1}{\chi} [\ln y - \ln y_0] \qquad \left[y_0 \text{ being very small} = \frac{\beta \nu}{u_\tau}\right] \tag{6.54}$$

or

$$\frac{\bar{u}}{u_\tau} = \frac{1}{\chi} \left[\ln \frac{y u_\tau}{\nu} - \ln \beta\right] \tag{6.55}$$

or

$$\bar{u}^+ = A_1 \ln y^+ + D_1 \qquad \left[\text{put } A_1 = \frac{1}{\chi}\right] \tag{6.56}$$

These constants were determined from experiments. For smooth ducts, A_1 has been observed as $A_1 = 2.5$ and $D_1 = 5.5$ or

$$\bar{u}^+ = 2.5 \ln y^+ + 5.5 \tag{6.57}$$

or $\bar{u}^+ = 2.5 \ln [Ey^+]$ where $E = 9.0$ for smooth walls.

Finally the log-law is defined for the turbulent zone as

$$\bar{u}^+ = \frac{1}{\chi} \ln [Ey^+] \tag{6.58}$$

The coefficient χ is known as the von Karman constant.

The location where log profile and linear profile meet can be calculated as:

$$\underbrace{\bar{u}^+ = \frac{1}{\chi} \ln [Ey^+]}_{\text{log-law}} \tag{6.59}$$

or

$$y^+ = \frac{1}{\chi} \ln [Ey^+] \quad (\text{from Eq. 6.49})$$

or

$$y^+ = 11.63 \tag{6.60}$$

As such, boundary layer thickness is around 1500 y^+.

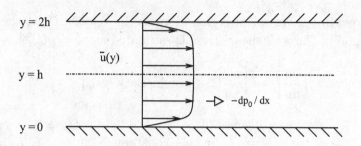

Figure 6.8 Channel flow with pressure gradient

6.6 LAW OF THE WALL AND IMPACT OF PRESSURE GRADIENT

Let us consider a fully developed channel flow as shown in Fig. 6.8.

Assumption: All derivatives with respect to x are zero. However $\frac{dp}{dx}$ is non-zero. Mean velocity field $\bar{u} = [\bar{u}(y), 0, 0]$.

The averaged Navier-Stokes (NS) equations give

$$0 = -\frac{1}{\rho}\frac{\partial p}{\partial x} - \frac{\partial}{\partial y}\left(\overline{u'v'}\right) + \nu\left[\frac{\partial^2 \bar{u}}{\partial y^2}\right] \tag{6.61}$$

$$0 = -\frac{1}{\rho}\frac{\partial p}{\partial y} - \frac{\partial}{\partial y}\left(\overline{v'^2}\right) \tag{6.62}$$

Integrating Eq. (6.62) between 0 and y, with no-slip boundary condition, we get

$$\frac{p}{\rho} + \overline{v'^2} = \frac{p_0}{\rho} \tag{6.63}$$

Thus

$$\frac{\partial p}{\partial x} = \frac{\partial p_0}{\partial x} \tag{6.64}$$

Note also that $p_0 = p_0(x)$.

From Eqs. (6.61) and (6.64)

$$0 = -\frac{1}{\rho}\frac{dp_0}{dx} - \frac{d}{dy}\overline{u'v'} + \nu\left[\frac{d^2\bar{u}}{dy^2}\right] \tag{6.65}$$

Integrating Eq. (6.65) between 0 and y, we get

$$0 = -\frac{y}{\rho}\frac{dp_0}{dx} - \overline{u'v'} + \nu\frac{d\bar{u}}{dy} - \nu\frac{d\bar{u}}{dy}\bigg|_{y=0} \tag{6.66}$$

Set

$$u_\tau^2 = \nu\frac{d\bar{u}}{dy}\bigg|_{y=o}$$

where u_τ is the friction velocity. Equation (6.66) becomes

$$0 = -\frac{y}{\rho}\frac{dp_0}{dx} - \overline{u'v'} + \nu\frac{d\bar{u}}{dy} - u_\tau^2 \tag{6.67}$$

At $y = h$, at the center of the channel, due to symmetry,

$$-\overline{u'v'}|_{y=h} = \nu\frac{d\bar{u}}{dy}\bigg|_{y=h} = 0$$

Thus from Eq. (6.67), we get

$$u_\tau^2 = -\frac{h}{\rho}\frac{dp_0}{dx}$$

Equation (6.67) becomes

$$-\overline{u'v'} + \nu\frac{d\bar{u}}{dy} = u_\tau^2\left(1 - \frac{y}{h}\right) \tag{6.68}$$

Let us substitute

$$u^+ = \frac{\bar{u}}{u_\tau}, \qquad y^+ = \frac{yu_\tau}{\nu}$$

Now Eq. (6.68) reads as

$$-\frac{\overline{u'v'}}{u_\tau^2} + \frac{du^+}{dy^+} = 1 - \frac{y^+}{R_\tau} \tag{6.69}$$

where $R_\tau = u_\tau h/\nu$.

Let y^+ be finite, say of order one, in the limit of $R_\tau \to \infty$. The equation (6.69) becomes

$$-\frac{\overline{u'v'}}{u_\tau^2} + \frac{du^+}{dy^+} = 1 \tag{6.70}$$

Assuming that the wall is smooth and no additional parameters appear in the boundary condition, we expect the solution of Eq. (6.70) to be

$$u^+ = \frac{\bar{u}}{u_\tau} = f(y^+) \tag{6.71}$$

$$-\frac{\overline{u'v'}}{u_\tau^2} = g(y^+)$$

Here, $f(y^+)$ and $g(y^+)$ are the laws of the wall.

(a) Assume $R_\tau \gg 1$, $\overline{u'v'}$ to be negligible as $y^+ \ll R_\tau$. The Eq. (6.69) becomes

$$\frac{du^+}{dy^+} = 1 \tag{6.72}$$

Integrating (6.72) between 0 and y^+ and applying no-slip boundary condition, we obtain

$$u^+ = y^+ \tag{6.73}$$

This law of the wall is valid for $0 \le y^+ < 5$. Such a region is called the viscous sublayer.

(b) Assuming at $\frac{y}{h} \ll 1$, there is a region where the viscous forces are negligible, then from (6.69), we have (for $R_\tau \to \infty$)

$$-\overline{u'v'} = u_\tau^2 \tag{6.74}$$

Consider Prandtl's mixing length hypothesis

$$-\overline{u'v'} = l_m^2 \left(\frac{d\bar{u}}{dy}\right)^2 \tag{6.75}$$

where the mixing length is

$$l_m = \chi y \tag{6.76}$$

The parameter χ is called von-Karman constant.

Here, Karman assumed that l_m should be a function of the distance from the wall in a wall bounded turbulent flow, rather than a constant as taken by Prandtl in the case of free turbulent flows. Substituting Eqs. (6.75) and (6.76) into Eq. (6.74), we obtain

$$\chi^2 y^2 \left(\frac{d\bar{u}}{dy}\right)^2 = u_\tau^2 \tag{6.77}$$

$$\chi y \frac{d\bar{u}}{dy} = u_\tau \tag{6.78}$$

$$\chi \frac{du^+}{dy^+} = \frac{1}{y^+} \tag{6.79}$$

Integrating Eq. (6.79) between 0 and y^+, we get

$$u^+ = \frac{1}{\chi} \ln y^+ + C \tag{6.80}$$

where, $\chi \approx 0.4$ and $C \approx 5.0$.

Equation (6.80) is called the Logarithmic Law of the Wall, which is valid for $30 \le y^+ \le 140$, the inertial sublayer.

6.7 TURBULENT HEAT TRANSFER IN PIPE (SIMPLIFIED ANALYSIS)

Heat flux due to conductivity of a fluid (Fig. 6.9) is:

$$q_l'' = -k \frac{\partial T}{\partial y} \tag{6.81}$$

The mean rate of turbulent heat transfer is given by

$$q_t'' = \rho c_p \overline{v'T'} \qquad \left[c_p \overline{u'T'} < c_p \overline{v'T'}\right] \tag{6.82}$$

The total heat flux is given by

$$q'' = q_l'' + q_t'' = -k \frac{\partial T}{\partial y} + \rho c_p \overline{v'T'} \tag{6.83}$$

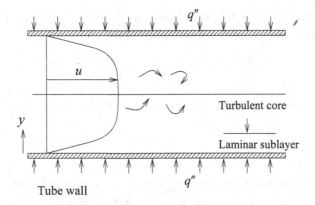

Figure 6.9 Viscous sublayer and turbulent core in a pipe flow

or

$$q'' = -\rho c_p \left[\alpha \frac{\partial \overline{T}}{\partial y} - \overline{v'T'} \right] \qquad (6.84)$$

Turbulent heat exchange is also considered equivalent to an increase in thermal diffusivity of the fluid similar to turbulent momentum exchange

$$\overline{v'T'} = -\alpha_t \frac{\partial \overline{T}}{\partial y} \qquad (6.85)$$

$$q'' = -\rho c_p [\alpha + \alpha_t] \frac{\partial \overline{T}}{\partial y} \qquad (6.86)$$

where α_t is eddy diffusivity.

If q is the total heat transfer, then q may be written as

$$\frac{q}{\rho c_p A} = -(\alpha + \alpha_t) \frac{\partial \overline{T}}{\partial y} \qquad (6.87)$$

In a similar way, shear stress in a turbulent flow could be written as

$$\frac{\tau}{\rho} = (\nu + \nu_t) \frac{\partial \overline{u}}{\partial y} \qquad (6.88)$$

We rely on the Chilton-Colburn analogy at this stage. The analogy says that the heat and momentum are transported as the same rate; that is, $\nu = \alpha$ (Prandtl number, $\frac{\nu}{\alpha} = 1$) and $\nu_t = \alpha_t$ (turbulent Prandtl number, $\sigma_t = 1$). Dividing equation Eq. (6.87) by Eq. (6.88) we get

$$\frac{q}{c_p A \tau} d\overline{u} = -d\overline{T} \qquad (6.89)$$

or

$$(q/c_p \, A \, \tau) \, d\overline{u} = -d \, \overline{T}$$

An additional assumption is invoked here; ratio of heat transfer per unit area to shear stress is constant across the flow field. Thus,

$$\frac{q}{A\tau} = \text{constant} = \frac{q_w}{A_w \tau_w} \tag{6.90}$$

Now Eq. (6.89)can be integrated between $y = y_1$ and $y = y_2$

$$\frac{q_w}{c_p \, A_w \, \tau_w} \int_{y \, = \, y_1}^{y \, = \, y_2} d\bar{u} = \int_{y \, = \, y_1}^{y \, = \, y_2} -d\overline{T}$$

One of the limits of integration may be chosen as y_1 = conditions prevalent at the wall. The other limit $y = y_2$ is chosen in such a way that $\overline{T}(y_2)$ is equal to the bulk mean temperature T_m and $\bar{u}(y_2)$ is approximately the same as the mean velocity u_m

$$\frac{q_w}{c_p \, A_w \, \tau_w} \int_{\bar{u} \, = \, 0}^{\bar{u} \, = \, u_m} d\bar{u} = \int_{\overline{T} \, = \, T_w}^{\overline{T} \, = \, T_m} -d\overline{T} \tag{6.91}$$

or

$$\frac{q_w \, u_m}{A_w \, \tau_w \, c_p} = T_w - T_m \tag{6.92}$$

We know

$$q_w = hA_w \ (T_w - T_m) \tag{6.93}$$

From Fig. 6.10, one can obtain

$$\tau_w = \frac{\Delta p(\pi D^2)}{4\pi DL} = \frac{\Delta p}{4} \frac{D}{L} \tag{6.94}$$

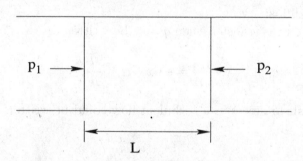

$P_1 \rightarrow$ $\leftarrow P_2$

L

Figure 6.10 Pressure differential in a channel

The pressure drop in a pipe flow can be written as

$$\frac{\Delta p}{\rho} = \frac{f L u_m^2}{2D}$$

Invoking the pressure drop expression in Eq. (6.94) we get

$$\tau_w = \frac{\rho f L u_m^2}{4(2D)} \frac{D}{L} = \frac{f}{8} \, \rho u_m^2 \tag{6.95}$$

Invoking Eqs. (6.93) and (6.95) in Eq. (6.92) we get

$$\frac{8\, h\, A_w\, (T_w - T_m)\, u_m}{A_w\, c_p\, \rho u_m^2\, f} = (T_w - T_m) \tag{6.96}$$

or

$$\frac{h}{\rho c_p u_m} = \frac{f}{8} \tag{6.97}$$

or

$$St = \frac{Nu_D}{Re_D\, Pr} = \frac{f}{8} \tag{6.98}$$

Turbulent friction factor may be assumed as $f = 0.184\ Re_D^{-1/5}$ for $Re \geq 2 \times 10^4$

$$\frac{Nu_D}{Re_D\, Pr} = \frac{0.184\ Re_D^{-1/5}}{8}$$

or

$$Nu_D = 0.023\ Re^{4/5} \tag{6.99}$$

This relation of turbulent heat transfer is highly restrictive because of the $Pr = 1.0$ assumption. Let us return to the Chilton-Colburn analogy for energy and momentum transport. The analogy holds good for $Pr \geq 0.5$ and we can thus write

$$St\, Pr^{2/3} = \frac{C_f}{2} \tag{6.100}$$

Again we know that $C_f = f/4$

$$St\, Pr^{2/3} = \frac{f}{8} \tag{6.101}$$

Now this dependence on Prandtl number works well for turbulent pipe flow. Accordingly Eq. (6.98) can be modified as

$$St\, Pr^{2/3} = \frac{f}{8} \quad \text{with friction factor}\ \ f = 0.184\ Re_D^{-1/5}$$

or

$$\frac{Nu_D}{Re_D\, Pr}\, Pr^{2/3} = \frac{0.184\ Re_D^{-1/5}}{8}$$

or

$$Nu_D = 0.023\ [Re_D]^{4/5}\ [Pr]^{1/3} \tag{6.102}$$

Dittus and Boelter [2] established experimental correlation which is quite close to the above equation. The correlation is:

$$Nu_D = 0.023\ (Re_D)^{4/5}\ (Pr)^n \tag{6.103}$$

$n = 0.4$ for heating, i.e., $T_w > T_m$ and $n = 0.3$ for cooling, i.e., $T_w < T_m$. The result has been confirmed experimentally for the range of conditions: $0.7 \leq Pr \leq$

160, $Re_D \geq 10000$ and $L/D \geq 10$. All fluid properties are evaluated at arithmetic mean of bulk temperature, i.e., $\overline{T}_m = (T_{m,o} + T_{m,i})/2$. For flows with large property variations of the fluids, the correlation due to Sieder and Tate [3] is recommended.

$$Nu_D = 0.027 \, Re_D^{4/5} Pr^{1/3} \left[\frac{\mu}{\mu_w}\right]^{0.14} \tag{6.104}$$

The range of applicability is: $0.7 \leq Pr \leq 16700, Re_D \geq 10000$, and $L/D \geq 10$ where all the properties except μ_w are evaluated at $\overline{T}_m$. Note that, to a good degree of approximation, the correlations may be applied for both the uniform wall temperature and uniform wall heat flux conditions. The μ_w is the property of the fluid, to be evaluated at the wall temperature. The foregoing correlations are restricted to smooth surface conditions. Heat transfer will be enhanced by surface roughness. In fact, heat transfer enhancement is often promoted by using internal ribs, twisted tapes, fins or corrugations.

The above mentioned correlations do not apply to heat transfer in liquid metals ($3 \times 10^{-3} \leq Pr \leq 5 \times 10^{-2}$). For fully developed turbulent flow ($L/D \geq 10$), in smooth circular tubes with uniform wall heat flux, the recommended correlation for liquid metals (due to Skupinski, Tortel and Vautrey [4]) is

$$Nu_D = 4.82 + 0.0185 \, (Pe_D)^{0.827} \, (q_w'' = \text{uniform})$$

Similarly for uniform wall temperature, the correlation (due to Seban and Shimazaki [5]) is

$$Nu_D = 5.0 + 0.025 \, (Pe_D)^{0.8} \, (T_w = \text{uniform})$$

6.8 COMPUTATIONAL APPROACH AND k-ϵ MODEL OF TURBULENCE

The time averaged governing equations for mass, momentum and energy may be written as

$$\frac{\partial \overline{u}_i}{\partial x_i} = 0 \tag{6.105}$$

$$\frac{\partial \overline{u}_i}{\partial t} + \overline{u}_j \frac{\partial \overline{u}_i}{\partial x_j} = -\frac{1}{\rho} \frac{\partial \overline{p}}{\partial x_i} + \frac{\partial}{\partial x_j} \left(\nu \frac{\partial \overline{u}_i}{\partial x_j} - \overline{u_i' u_j'}\right) \tag{6.106}$$

$$\frac{\partial \overline{T}}{\partial t} + \overline{u}_i \frac{\partial \overline{T}}{\partial x_i} = \frac{\partial}{\partial x_i} \left[\alpha \frac{\partial \overline{T}}{\partial x_i} - \overline{u_i' T'}\right] \tag{6.107}$$

with

$$-\overline{u_i' u_j'} = \nu_t \left(\frac{\partial \overline{u}_i}{\partial x_j} + \frac{\partial \overline{u}_j}{\partial x_i}\right) - \frac{2}{3} \, k\delta_{ij} \tag{6.108}$$

$$-\overline{u_i' T'} = \alpha_t \frac{\partial \overline{T}}{\partial x_i} \tag{6.109}$$

and

$$\alpha_t = \nu_t / \sigma_t \tag{6.110}$$

where σ_t is turbulent Prandtl number. The entire system of equations can be closed if σ_t is known and ν_t is determined correctly.

We can construct an expression for turbulent viscosity or eddy viscosity in terms of two attributes of an eddy. The eddy viscosity can be viewed as a product of velocity scale and a length scale. The velocity scale is

$$v = k^{1/2}$$

and the length scale can be shown to be

$$l = k^{3/2}/\epsilon$$

Thus

$$\nu_t = C_\mu \frac{k^2}{\epsilon} \tag{6.111}$$

The quantity C_μ is a model constant to be determined accurately for a specific flow. The kinetic energy k and its dissipation rate ϵ are evaluated at each point in the domain from their governing differential equations. The equation for k can be derived from the Navier-Stokes equations by subtracting the mean equations from the unaveraged equations to obtain an equation for the fluctuating velocity (say u_i'). Taking the scalar product of this equation with the fluctuating velocity (u_j') and averaging yields the equation for the Reynolds stress transport. The transport equation for kinetic energy can be obtained by substituting $j = i$ in the Reynolds stress transport equation. An equation for the dissipation can be derived from the Navier-Stokes equations by an extension of the method deployed to derive the kinetic energy equation.

The governing equation for k is given by

$$\overbrace{\frac{\partial k}{\partial t}}^{(1)} + \overbrace{\overline{u_j}\frac{\partial k}{\partial x_j}}^{(2)} = -\overbrace{\tau_{ij}\frac{\partial \overline{u_i}}{\partial x_j}}^{(3)} - \overbrace{\epsilon}^{(4)}$$
$$\underbrace{-\frac{\partial}{\partial x_j}\left(\frac{1}{2}\overline{u_i'u_i'u_j'} + \frac{1}{\rho}\overline{p'u_j'}\right)}_{(5)} + \underbrace{\nu\nabla^2 k}_{(6)} \tag{6.112}$$

In Eq. (6.112), term-1 is the rate of change of turbulent kinetic energy k, term-2 is the convective transport of k, term-3 is the production by shear, term-4 is the viscous dissipation and term-5 is the third-order turbulent diffusive transport of k. The term-6 is the viscous diffusion of k. The production term represents the transfer of kinetic energy from the mean flow to the turbulent motion. τ_{ij} is the turbulent shear stress.

A scalar dissipation rate ϵ is defined as

$$\epsilon = \frac{1}{2}\epsilon_{ii} = \frac{1}{2}2\nu\overline{\left(\frac{\partial u_i'}{\partial x_j}\frac{\partial u_i'}{\partial x_j}\right)} \tag{6.113}$$

Instead of solving Eq.(6.112), the turbulent kinetic energy, k is obtained by solving the modeled transport equations.

The modeled transport equation for k is

$$\frac{\partial k}{\partial t} + \bar{u}_j \frac{\partial k}{\partial x_j} = \nu_t \left[\frac{\partial \bar{u}_i}{\partial x_j} + \frac{\partial \bar{u}_j}{\partial x_i} \right] \frac{\partial \bar{u}_i}{\partial x_j}$$
$$+ \frac{\partial}{\partial x_j} \left(\frac{\nu_t}{\sigma_k} \frac{\partial k}{\partial x_j} \right) - \epsilon + \nu \nabla^2 k \quad . \tag{6.114}$$

The first term on the right hand side corresponds to production of turbulent kinetic energy. The second term on the right hand side signifies turbulent diffusion of turbulent kinetic energy.

The transport equation for ϵ is also a modeled version of the original transport equation for ϵ. Original transport equation is not being mentioned herein. The modeled equation for ϵ is

$$\frac{\partial \epsilon}{\partial t} + \bar{u}_i \frac{\partial \epsilon}{\partial x_i} = \nu \nabla^2 \epsilon + P_\epsilon + D_\epsilon - \phi_\epsilon \tag{6.115}$$

As to symbols, P_ϵ, D_ϵ and ϕ_ϵ represent the production of dissipation, the turbulent diffusion of dissipation and the turbulent destruction of dissipation respectively. The diffusion term, D_ϵ can be modeled using the gradient transport (similar to the modeled k-equation) hypothesis as

$$D_\epsilon = \frac{\partial}{\partial x_j} \left(\frac{\nu_t}{\sigma_\epsilon} \frac{\partial \epsilon}{\partial x_j} \right) \tag{6.116}$$

where σ_ϵ is an empirical constant.

The destruction of dissipation is assumed to depend only on the length and the time scales. This can be written as

$$\phi_\epsilon = C_{\epsilon 2} \frac{\epsilon^2}{k} \tag{6.117}$$

where $C_{\epsilon 2}$ is an emipirical constant.

The production of dissipation is assumed to be proportional to the production of the turbulent kinetic energy, the length scale and the time scale of turbulence. A simple dimensional analysis for the production term along with the above assumption leads to

$$\frac{\partial \epsilon}{\partial t} + \bar{u}_j \frac{\partial \epsilon}{\partial x_j} = \nu \nabla^2 \epsilon + \frac{\partial}{\partial x_j} \left(\frac{\nu_t}{\sigma_\epsilon} \frac{\partial \epsilon}{\partial x_j} \right)$$
$$+ C_{\epsilon 1} \frac{\epsilon}{k} \nu_t \left(\frac{\partial \bar{u}_i}{\partial x_j} + \frac{\partial \bar{u}_j}{\partial x_i} \right) \frac{\partial \bar{u}_i}{\partial x_j} - C_{\epsilon 2} \frac{\epsilon^2}{k} \tag{6.118}$$

Equation (6.118) is the final expression for the modeled version of the transport equation for the turbulent dissipation rate.

The standard k-ϵ model has five empirical constants $C_\mu, \sigma_k, \sigma_\epsilon, C_{\epsilon 1}$ and $C_{\epsilon 2}$ in its formulation. $C_{\epsilon 2}$ is determined using the experiments on the decay of k behind a grid and a value of $C_{\epsilon 2} = 1.92$ is widely accepted. The experiments on the local-equilibrium-shear-layers suggested that $C_\mu \approx 0.09$. The constants $\sigma_k, \sigma_\epsilon$ and $C_{\epsilon 1}$ were determined and their suggested values are 1.0, 1.3 and 1.44 respectively. It should be noted that the above values for the five constants are not universal and the k-ϵ model may require fine tuning in order to obtain correct results for specific flows. The k-ϵ model has been explained in a greater detail in the pioneering work of Launder and Spalding [6].

In the absence of source terms and under constant property consideration, the energy equation may be written as

$$\frac{D\overline{T}}{Dt} = \frac{\partial}{\partial x}\left[(\alpha + \alpha_t)\frac{\partial \overline{T}}{\partial x}\right] + \frac{\partial}{\partial y}\left[(\alpha + \alpha_t)\frac{\partial \overline{T}}{\partial y}\right] + \frac{\partial}{\partial z}\left[(\alpha + \alpha_t)\frac{\partial \overline{T}}{\partial z}\right] \quad (6.119)$$

where α_t is the turbulent diffusivity and it can be determined from Eq. (6.110). Experiments and analysis recommend the turbulent Prandtl number, σ_t in the range of 0.9. Solution of Eq. (6.119) provides the average temperature distribution in a turbulent flow field.

6.8.1 k-ω Model

Kolmogorov [7] proposed the first two equation model. Kolmogorov chose the kinetic energy of turbulence as one of the parameters, while the other parameter was the dissipation per unit turbulence kinetic energy, ω. Wilcox [8] and Speziale et al. [9] also regard ω as the ratio of ϵ and k. Here we present the model due to Wilcox [8].

The turbulent viscosity is related to k and ω by the expression

$$\mu_t = \rho\frac{k}{\omega} \quad (6.120)$$

The transport equations for turbulent kinetic energy (k) and its dissipation rate per unit turbulence kinetic energy (ω) are

$$\rho\frac{\partial k}{\partial t} + \rho\overline{u}_i\frac{\partial k}{\partial x_i} = \frac{\partial}{\partial x_i}\left(\mu + \frac{\mu_t}{\sigma_k}\frac{\partial k}{\partial x_i}\right) + \mu_t\left(\frac{\partial \overline{u}_i}{\partial x_j} + \frac{\partial \overline{u}_j}{\partial x_i}\right)\frac{\partial \overline{u}_i}{\partial x_j} - \rho\,k\,\omega\,\beta^* \quad (6.121)$$

$$\rho\frac{\partial \omega}{\partial t} + \rho\overline{u}_i\frac{\partial \omega}{\partial x_i} = \frac{\partial}{\partial x_i}\left(\mu + \frac{\mu_t}{\sigma_\omega}\frac{\partial \omega}{\partial x_i}\right) + \alpha\frac{\omega}{k}\mu_t\left(\frac{\partial \overline{u}_i}{\partial x_j} + \frac{\partial \overline{u}_j}{\partial x_i}\right)\frac{\partial \overline{u}_i}{\partial x_j} - \beta\,\rho\,\omega^2 \quad (6.122)$$

The coefficients have the following empirically derived values

$$\alpha = 5/9, \beta = 3/40, \beta^* = 9/100, \sigma_k = 2.0, \sigma_\omega = 2.0$$

In the k-ϵ family of models the non-local property of turbulence is accounted for usually by one turbulent length scale that is dictated by a model equation which is derived on the basis of closure assumptions for most parts of the governing processes. Such models need several empirical coefficients and for obvious reasons such coefficients cannot be universal constants. Despite these limitations, many times, k-ϵ family of eddy viscosity models have produced fairly acceptable results in predicting flows which have predominantly small scale turbulence structures and which can be considered to be interpolates of basic experiments for which the coefficients have been derived.

Deb, Biswas and Mitra [10] have computed three-dimensional turbulent flows with longitudinal vortices embedded in the boundary layer on a channel wall. Although the behaviour of the k-ϵ model is not known for such highly vortical flows, comparison between the measurements due to Pauley and Eaton [11, 12] and the computed results shows that the interaction of the longitudinal vortices with boundary layer within a turbulent channel flow is captured reasonably well by the numerical simulation.

6.8.2 Special Features of Near Wall Flow

In the experiments on turbulent flows, the presence of solid walls is observed to have a damping effect on the transport of turbulence. In numerical calculation of flows, the velocity components and the turbulence energy may be set to zero at wall through the no-slip condition. However, ambiguity exists about the values of the turbulence dissipation parameter ϵ or ω at the solid boundary. Across a turbulent boundary layer, the flow has to undergo a transition from fully turbulent to completely laminar within the thin viscosity-dominated sublayer adjacent to the solid surface. In this laminar and transitional layer, the molecular viscosity has a direct damping effect on the turbulence. This phenomenon is called low Reynolds number turbulence and the transition from high to low Reynolds number regions is determined by the local turbulence Reynolds number, $R_t = \rho k^2/(\mu\epsilon)$ where k is the turbulence kinetic energy and ϵ is its dissipation. Two significant physical effects of the presence of a wall are the following:

(i) Molecular viscosity diffuses vorticity and damps turbulence. Near a solid wall, the viscous diffusion terms which are usually negligible compared to other terms in the regions away from wall, become important.

(ii) Significant reduction of velocity fluctuation normal to the wall is brought about by the pressure reflection mechanism. The isotropic eddy viscosity based models, such as the k-ϵ or k-ω are unable to separate this effect from the viscosity effect.

6.8.3 Near Wall Treatment in Transport Equation based Models

Usually the wall function approach is used to handle the zone in the vicinity of solid wall boundaries in the framework of two equation models (k-ϵ, or k-ω) of turbulence.

Wall Function Approach

This method has been employed successfully in many applications (Launder and Spalding [6]). The method uses Logarithmic Law of the Wall, described earlier, as the constitutive relation between velocity and the wall shear stress. In terms of the velocity at the grid point closest to the wall surface, one assumes the Law of the Wall to hold. Such a relation between wall and the fully turbulent layer away from the walls allows prescription of the boundary conditions for velocity and turbulence quantities at a point placed outside the velocity affected near wall layer.

Based on the equilibrium consideration (production of k = dissipation of k), k and ϵ or k and ω at the near wall node (P) are prescribed as follows:

$$u_\tau = C_\mu^{1/4} k_P^{1/2}, \qquad \epsilon_P = \frac{u_\tau^3}{\chi \, y_P} \quad \text{and} \quad \omega_P = \frac{k^{1/2}}{C_\mu^{1/4} \chi \, y_P} \qquad (6.123)$$

where C_μ is a closure coefficient and χ is von Karman constant. However, the friction velocity $u_\tau = \left(\sqrt{\tau_w/\rho} \right)$ is not known a priori and it is calculated by iterative type solution algorithm where $u_\tau, k_P, \overline{u}_P$ and ϵ_P or ω_P at the first near wall point are coupled through the relevant equations. However, the logarithmic law is not strictly valid for flows where strong pressure gradients are involved and the standard wall function methods are not suitable. For shear layers with strong secondary flows (curved ducts) or for transitional boundary layer flows wall function approach requires modification.

Wall function approach is applicable for turbulent thermal energy equation too. The universal temperature profile is defined over the fully developed turbulent flow regime. The detailed technique pertaining to implementation of universal temperature profile is available elsewhere [10, 12]. The universal temperature profile is given by

$$\overline{T}^+ = \frac{1}{\chi_C} ln[E^* y^+] \qquad (6.124)$$

The value of χ_C is 0.46. The quantity E^* is similar to E in Eq. (6.58).

6.9   RNG k-ϵ MODEL AND KATO-LAUNDER MODEL

In this section we shall discuss two variants of k-ϵ turbulence models. The RNG based k-ϵ turbulence model follows the two equation turbulence modeling framework and has been derived from the original governing equations for fluid flow using a mathematical technique called the Renormalization Group (RNG) method due to Yakhot and Orszag [13]. The RNG method is applicable to scale invariant phenomena lacking externally imposed length and time scales. For turbulence, this signifies that the method can describe the small scales which should be statistically independent of the external initial conditions and dynamical forces that create them through different instability phenomena. The RNG method gives a theory of the Kolmogorov equilibrium range of turbulence, comprising the inertial range of small-scale eddies whose energy spectrum follows the Kolmogorov law $E(k) \sim k^{-5/3}$. The scales of effective excitation in turbulence, range from the low wave number $k_0 = 2\pi/L$ (large scale eddies) to high wave number viscous cutoff Λ (corresponding to smallest energy containing eddies).

The RNG method removes a narrow band of modes near Λ by representing these modes in term of lower modes in the interval $k_0 < k < \Lambda e^{-l}(l << 1)$. When this narrow band of modes is removed, the resulting equations of motion for the remaining modes represent a modified system of Navier-Stokes equations. The equations are dictated by a modified viscosity. The first band of modes is removed

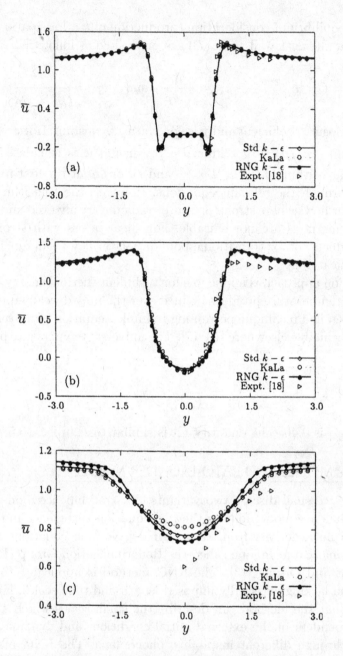

Figure 6.11   Time-averaged streamwise velocity profiles at: (a) $x = 0$, (b) $x = 1.0$ and (c) $x = 5$

from the dynamics and the process of removal of degrees of freedom is repeated (also see Choudhury [14]). In this way the RNG method enables computation of Navier-Stokes equations on relatively coarser grids at high Reynolds numbers.

In standard k-ϵ model, high generation rates are predicated in the regions of high strain rate or streamline curvature (Launder and Spalding, [6]). This excess of turbulence energy creates high levels of turbulent viscosity. The high turbulent viscosity overpredicts mixing and delay separation. The Kato-Launder (KaLa) modified model [15] remedies this problem.

Two different high Reynolds number versions of the two-equations model that have been used in the present text are (i) RNG k-ϵ and (ii) Kato-Launder models. Both models relate the turbulent viscosity ν_t to the turbulent kinetic energy k and its rate of dissipation ϵ. In the standard k-ϵ model, the transport equations for k and ϵ are

$$\frac{\partial k}{\partial t} + \frac{\partial [\overline{u}_i k]}{\partial x_i} = \frac{\partial}{\partial x_i} \left[\frac{\nu_t}{\sigma_k} \frac{\partial k}{\partial x_i} \right] + P_k - \epsilon \tag{6.125}$$

$$\frac{\partial \epsilon}{\partial t} + \frac{\partial [\overline{u}_i \epsilon]}{\partial x_i} = \frac{\partial}{\partial x_i} \left[\frac{\nu_t}{\sigma_\epsilon} \frac{\partial \epsilon}{\partial x_i} \right] + C_{\epsilon 1} P_k \frac{\epsilon}{k} - C_{\epsilon 2} \frac{\epsilon^2}{k} \tag{6.126}$$

where the production term is

$$P_k = C_\mu \epsilon S^2, \quad S = \frac{k}{\epsilon} \sqrt{\frac{1}{2} \left[\frac{\partial \overline{u}_i}{\partial x_j} + \frac{\partial \overline{u}_j}{\partial x_i} \right]^2} \tag{6.127}$$

The KaLa model is similar to the standard k-ϵ, except that the turbulence production term in Eq. (6.127) is replaced by

$$P_k = C_\mu \epsilon S \Omega, \quad \Omega = \frac{k}{\epsilon} \sqrt{\frac{1}{2} \left[\frac{\partial (\overline{u}_i)}{\partial x_j} - \frac{\partial (\overline{u}_j)}{\partial x_i} \right]^2} \tag{6.128}$$

The quantity Ω is related to the average rotation of a fluid element. In the simple shear flow context, S and Ω are equal. In stagnation flows, $\Omega = 0$ and $S > 0$. This leads to the desired reduction of the production of kinetic energy near the forward stagnation point of the bluff objects. This has an important effect of lowering eddy viscosity in the boundary-layers and permits vortices to be shed from the rear side. In the RNG k-ϵ model, P_k is given by Eq. (6.127) and Eq. (6.128) is augmented on the right-hand side by an extra strain-rate term R given by

$$R = -\frac{C_\mu \eta^3 (1 - \eta/\eta_0) \epsilon^2}{(1 + \beta_0 \eta^3) k} \tag{6.129}$$

where the quantity η as given by

$$\eta = \frac{k}{\epsilon} \left[\left(\frac{\partial \overline{u}_i}{\partial x_j} + \frac{\partial \overline{u}_j}{\partial x_i} \right) \frac{\partial u_i}{\partial x_j} \right]^{1/2} \tag{6.130}$$

The eddy viscosity ν_t for the standard k-ϵ and KaLa models is determined from the expression

$$\nu_t = C_\mu \frac{k^2}{\epsilon} \tag{6.131}$$

For the RNG k-ϵ, the eddy viscosity expression is

$$\nu_t = \nu \left[1 + \left(\frac{C_\mu}{\nu} \right)^{1/2} \frac{k}{\epsilon^{1/2}} \right]^2 \tag{6.132}$$

The parameters for each of the above models appearing in Eqs. (6.125) – (6.132) are given in Table 6.1.

Table 6.1
Model parameters

Models	C_μ	$C_{\epsilon 1}$	$C_{\epsilon 2}$	σ_k	σ_ϵ	β_0	η_0
Standard k-ϵ and $KaLa$	0.09	1.44	1.92	1.0	1.3	-	-
RNG k-ϵ	0.0845	1.42	1.68	0.7179	0.7179	0.012	4.38

The RNG theory and its application to turbulence are described by Yakhot and Orszag [13]. The scale elimination procedure in the RNG theory results in a differential equation for turbulent viscosity, which is integrated to obtain an accurate description of how the effective turbulent transport varies with the effective Reynolds number and near-wall flows. In the high Reynolds number limit, the expression of turbulent viscosity is the same as in the standard k-ϵ model.

The RNG model is reliable for a wider class of flows than the standard k-ϵ model. However, the accuracy for predicting the turbulent flows using the RNG model is reported as poorer than those for other k-ϵ models for vortex shedding behind the bluff objects, such as square cylinders and circular tubes in a heat exchanger. In these cases, the separated flows are not well predicted as in chevron or corrugated plate geometry. Such vortex shedding has low frequency modulations. Saha et al. [16] compared three turbulence models to capture the essence of time-averaged flow quantities in a vortex shedding dominated flow field through the turbulence models in two dimensions. They used the Launder and Spalding [6] standard k-ϵ model, the Kato-Launder k-ϵ model [15], and the RNG k-ϵ model of Yakhot et al. [17]. In terms of the parameters such as Strouhal number, lift and drag coefficients, the predictions due to the Kato-Launder and the standard k-ϵ models were close to each other. Figure 6.11 reveals good comparison of velocity profiles at the downstream of the cylinder due to all three models. The predictions are reasonably close to experiments of Lyn et al. [18].

6.10 SUMMARY

Turbulence is a difficult subject and we probably cannot expect a universal solution to problems involving turbulent flows. Although we have numerous models, each one has its own area of applicability.

All the variants of eddy viscosity models, contain a diffusive term involving ν_t. For high Reynolds number turbulence, the Reynolds stress terms in the mean-flow equation dominate over the genuine viscous terms, that is $\nu_t >> \nu$ except for thin viscous layers near walls. The turbulent Reynolds number is not high enough for validity of the k-ϵ model in the viscous layers and it must be modified if a realistic description of those layers is envisaged. Low-Reynolds number models attempt to reproduce the experimentally observed viscous layer scaling laws, based on ν and the turbulent friction velocity u_τ. However, one should not expect them to perform well near separation. The standard values of model constants were given earlier, but better results can usually be obtained by tuning the parameters to particular flows. At the least, these should be tuned with the same class of flows. Even though the basic model provides a rather crude representation of the physics of turbulence, the flexibility provided by the parameter can lead to a somewhat better situation. Once the model is fine tuned for one member of a class of similar flows, it is expected to predict the other flows in that class reasonably well. This is what is needed in many design applications, where the basic form of the flow geometry can be set through the mean flow calculations.

REFERENCES

1. H. Tennekes and J. L. Lumley, A First Course in Turbulence, MIT Press, 1972.

2. P. W. Dittus and L. M. K. Boelter, Heat Transfer in Automobile Radiators of the Turbulent Type, Univ. Calif. Publ. Eng., Vol. 2, pp. 443-461, 1930 (reprinted in Int. Commun. Heat Mass Transfer Vol. 12, pp. 3-22, 1985).

3. E. N. Sieder and C. E. Tate, Heat Transfer and Pressure Drop of Liquids in Tubes, Ind. Eng. Chem., Vol. 28, pp. 1429-1436, 1936.

4. E. S. Skupinski, J. Tortel, and L. Vautrey, Determination des Coefficients de Convection d'un Alliage Sodium-Potassium dans un tube Circulaire, Int. J. Heat Mass Transfer, Vol. 8, pp. 937, 1965.

5. R. A. Seban and T. T. Shimazaki, Heat Transfer to a Fluid Flowing Turbulently in a Smooth Pipe with Wall at Constant Temperature, Trans. ASME, Vol. 73, pp. 803-809, 1951.

6. B. E. Launder, and D. B. Spalding, The Numerical Computation of Turbulent Flows, Comput. Meth. Appl. Mech. Eng., Vol. 3, pp. 269-289, 1974.

7. A. N. Kolmogorov, The Equations for Turbulent Motion in an Incompressible Fluid, IZU Acad Sci., USSR Phys., Vol. 6, pp. 56-58, 1942.

8. D. C. Wilcox, Reassessment of the Space Determining Equation of Advanced Turbulence Models. AIAA J., Vol. 26, pp. 1299-1310, 1988.

9. C. G. Speziale, R. Abid and E. C. Anderson, A Critical Evolution of Two Equation Models for Near Wall Turbulence, AIAA Paper 90-1481, Seattle WA, 1990.

10. P. Deb, G. Biswas, and N. K. Mitra, Heat Transfer and Flow Structure in Laminar and Turbulent Flows in a Rectangular Channel with Longitudinal Vortices, Int. J. Heat Mass Transfer, Vol. 38, pp. 2427-2444, 1995.

11. W. R. Pauley and J. K. Eaton, Experimental Study of the Development of Longitudinal Vortex Pairs Embedded in a Turbulent Boundary Layer, AIAA J., Vol. 26(7), pp. 816-823, 1988.

12. G. Biswas and V. Eswaran, Turbulent Flows: Fundamentals, Experiments and Modeling, Narosa Publishing House, New Delhi, 2002.

13. V. Yakhot and S. A. Orszag, Renormalization Group Analysis of Turbulence: 1. Basic Theory, J. Sci. Comput., Vol. 1, pp. 1-51, 1986.

14. D. Choudhury, Introduction to the Renormalization Group Method and Turbulence Modeling, ANSYS-Fluent, TM 107, 1993.

15. M. Kato, and B. E. Launder, The Modelling of Turbulent Flow around Stationary and Vibrating Square Cylinders, Proc. 9th Symposium on Turbulent Shear Flows, Kyoto, Japan, 10-4-1 to 10-4-6, 1993.

16. A. K. Saha, G. Biswas., and K. Muralidhar, Numerical Study of Turbulent Unsteady Wake Behind a Partially Enclosed Cylinder Using RANS. Comput., Meth. Appl. Mech. Eng., Vol. 178, pp. 323-341, 1999.

17. V. Yakhot, S. A. Orszag, S. Thangam, T. B. Gatski, and C. G. Speziale, Development of Turbulence Models for Shear Flows by a Double Expansion technique, Phys. Fluids, Vol. A4, pp. 1510-1520, 1992.

18. D. A. Lyn, S. Einav, W. Rodi and J. H. Park, A Laser-Doppler Velocimetry Study of Ensemble-Averaged Characteristics of Turbulent Near Wake of a Square Cylinder, J. Fluid Mech., Vol. 304, pp. 285-319, 1995.

EXERCISES

1. Two-dimensional instantaneous Navier-Stokes Equations are given by

$$\rho\left(\frac{\partial u}{\partial t} + u\frac{\partial u}{\partial x} + v\frac{\partial u}{\partial y}\right) = -\frac{\partial p}{\partial x} + \mu\left(\frac{\partial^2 u}{\partial x^2} + \frac{\partial^2 u}{\partial y^2}\right)$$

$$\rho\left(\frac{\partial v}{\partial t} + u\frac{\partial v}{\partial x} + v\frac{\partial v}{\partial y}\right) = -\frac{\partial p}{\partial y} + \mu\left(\frac{\partial^2 v}{\partial x^2} + \frac{\partial^2 v}{\partial y^2}\right)$$

Apply Reynolds decomposition related to the average and fluctuating components for the velocity as well as the pressure. Derive the time averaged equation for the turbulent momentum transport in a channel. Also show the Reynolds stress tensor.

2. Theodore von Karman suggested the mixing length to be

$$l = \chi \left| \frac{d\bar{u}/dy}{d^2\bar{u}/dy^2} \right|$$

where χ is von Karman constant.

Show that the turbulent velocity profile on a flat plate beyond the viscous sublayer is given by $\frac{\bar{u}}{u_\tau} = A_1 \ln y^+ + D_1$, where A_1 and D_1 are arbitrary constants.

3. Two-dimensional instantaneous temperature field in flow field is given by

$$\frac{\partial T}{\partial t} + u\frac{\partial T}{\partial x} + v\frac{\partial T}{\partial y} = \alpha \left(\frac{\partial^2 T}{\partial x^2} + \frac{\partial^2 T}{\partial y^2} \right)$$

Apply Reynolds decomposition related to the average and fluctuating components for the velocity as well as the scalar field. Derive the time averaged equation for the turbulent energy equation in a channel. Also derive the non-dimensional form of the equation. The Reynolds and the Prandtl numbers are given by

$$Re = \frac{U_{ref}H}{\nu} \quad \text{and} \quad Pr = \frac{\nu}{\alpha}$$

4. Consider turbulent flow over a flat plate. The turbulent boundary layer equation is given by (all the symbols have their usual meaning)

$$\bar{u}\frac{\partial \bar{u}}{\partial x} + \bar{v}\frac{\partial \bar{u}}{\partial y} = \frac{\partial}{\partial y}\left(\nu\frac{\partial \bar{u}}{\partial y} \right) - \frac{\partial}{\partial y}(\overline{u'v'})$$

Near the wall, within the viscous sublayer, the inertial effects are insignificant. Also we know that the fluctuating components vanish near the wall. Find out a wall coordinate and indicate the velocity variations with the wall coordinate. Consider Prandtl's mixing length model outside the sublayer, i.e., in the turbulent flow zone and show that the velocity variation within the turbulent zone follows a log-law. Identify the location (the value in $y+$ coordinate) where the log-law profile and the viscous-zone linear profile meet.

5. Consider a fully developed turbulent pipe flow. The total heat flux to the flowing fluid may be considered as the summation of laminar heat flux and the turbulent heat flux. The shear stress at any layer is also thought to be addition of laminar shear stress and the turbulent shear stress. Applying Chilton–Colburn analogy, derive an expression for Nusselt number based on the duct diameter. Propose a technique to develop a more versatile expression that includes Prandtl number.

## CHAPTER 7

Free Convection

7.1 INTRODUCTION

Now we shall consider situations for which there is no forced velocity, yet convection currents exist within the fluid. They originate when a body force acts on a fluid in which there are density gradients. The net effect is a buoyancy force, which induces fluid velocity. The density gradient is due to a temperature gradient, and the body force is due to the gravitational field. The seminal contributions of Ostrach [1] and Gebhart, Jaluria et al. [2] provide excellent insight into natural convective flows and its importance in several areas of energy transfer.

7.2 FREE CONVECTION OVER VERTICAL FLAT PLATE

To begin with, the focus is on a classical example of boundary layer development on a heated vertical plate (Fig. 7.1). The fluid close to plate is less dense than fluid that is further removed. Buoyancy forces therefore induce a free convection boundary layer in which the heated fluid rises vertically, entraining fluid from quiescent region. The resulting velocity distribution is unlike that associated with forced convection boundary layers. In particular, the velocity is zero as $y \to \infty$, as well as at $y = 0$. With one exception, assume the fluid to be incompressible. The exception involves accounting for the effect of density variation due to temperature causing buoyancy force (the Boussinesq approximation). The appropriate form of the x-momentum equation is then

$$\rho \left(u \frac{\partial u}{\partial x} + v \frac{\partial u}{\partial y} \right) = -\frac{\partial p}{\partial x} - \rho g + \mu \frac{\partial^2 u}{\partial y^2} \tag{7.1}$$

Outside the boundary layer, we can write

$$\frac{\partial p}{\partial x} = -\rho_\infty g \tag{7.2}$$

Substituting (7.2) into (7.1), we obtain the following equation

$$\rho \left(u \frac{\partial u}{\partial x} + v \frac{\partial u}{\partial y} \right) = (\rho_\infty - \rho)g + \mu \frac{\partial^2 u}{\partial y^2} \tag{7.3}$$

Introducing the volume coefficient of expansion $\beta = \frac{1}{\forall}\left(\frac{\partial \forall}{\partial T}\right)_p$, $\forall \equiv$ volume, we can write

$$\beta = \frac{1}{\forall_\infty}\left(\frac{\forall - \forall_\infty}{T - T_\infty}\right) = \frac{\rho_\infty}{1}\left(\frac{\rho_\infty - \rho}{\rho\rho_\infty}\right)\frac{1}{T - T_\infty}$$

or

$$\beta = \frac{(\rho_\infty - \rho)}{\rho(T - T_\infty)} = -\frac{1}{\rho}\left(\frac{\partial \rho}{\partial T}\right)_p \tag{7.4}$$

From (7.3) and (7.4) we can write

$$\rho\left(u\frac{\partial u}{\partial x} + v\frac{\partial u}{\partial y}\right) = \beta g\rho(T - T_\infty) + \mu\frac{\partial^2 u}{\partial y^2} \tag{7.5}$$

Similar to forced flow boundary layer, y momentum equation takes the form, $\frac{\partial p}{\partial y} = 0$. The set of equations governing natural convection over a vertically heated plate becomes

$$\frac{\partial u}{\partial x} + \frac{\partial v}{\partial y} = 0 \tag{7.6}$$

$$\left(u\frac{\partial u}{\partial x} + v\frac{\partial u}{\partial y}\right) = g\beta(T - T_\infty) + \nu\frac{\partial^2 u}{\partial y^2} \tag{7.7}$$

$$u\frac{\partial T}{\partial x} + v\frac{\partial T}{\partial y} = \alpha\frac{\partial^2 T}{\partial y^2} \tag{7.8}$$

with $\beta = -\frac{1}{\rho}\left(\frac{\partial \rho}{\partial T}\right)_p = \frac{1}{\rho}\left(\frac{p}{RT^2}\right) = \frac{1}{T}$ (this might be needed for calculating β for perfect gases).

Boundary conditions at $y = 0 : u = 0, v = 0$; at $y = \delta : u = 0, \frac{\partial u}{\partial y} = 0$; at $y = 0 : T = T_w$; at $y = \delta : T = T_\infty$ and $\frac{\partial T}{\partial y} = 0$. Integration of Eq. (7.7) within the boundary layer yields

$$\int_o^\delta \left[u\frac{\partial u}{\partial x} + v\frac{\partial u}{\partial y}\right]dy = \int_o^\delta g\beta(T - T_\infty)dy + \int_o^\delta \nu\frac{\partial^2 u}{\partial y^2}dy$$

Using continuity equation

$$\int_o^\delta \left[\frac{\partial u^2}{\partial x} + \frac{\partial}{\partial y}(uv)\right]dy = g\beta\int_o^\delta (T - T_\infty)dy - \nu\frac{\partial u}{\partial y}\bigg|_{y=0}$$

or

$$\frac{d}{dx}\int_o^\delta u^2 dy + [uv]_o^\delta = g\beta\int_o^\delta (T - T_\infty)dy - \nu\frac{\partial u}{\partial y}\bigg|_{y=0}$$

or

$$\frac{d}{dx}\int_o^\delta u^2 dy = g\beta\int_o^\delta (T - T_\infty)dy - \nu\frac{\partial u}{\partial y}\bigg|_{y=0} \tag{7.9}$$

Integration of Eq. (7.8) yields

$$\int_0^\delta \left(u\frac{\partial T}{\partial x} + v\frac{\partial T}{\partial y} \right) dy = \int_0^\delta \alpha \frac{\partial^2 T}{\partial y^2} dy$$

Integrating the second term by parts and invoking from Eq. (7.6), we get

$$\int_0^\delta \left(u\frac{\partial T}{\partial x} \right) dy + \int_0^\delta T\frac{\partial u}{\partial x} dy - T_\infty \int_0^\delta \frac{\partial u}{\partial x} dy = -\alpha \left. \frac{\partial T}{\partial y} \right|_{y=0}$$

$$\int_0^\delta \frac{\partial}{\partial x} \{u(T - T_\infty)\} \, dy = -\alpha \left. \frac{\partial T}{\partial y} \right|_{y=0}$$

$$\frac{d}{dx} \int_0^\delta u(T - T_\infty) dy = -\alpha \left. \frac{\partial T}{\partial y} \right|_{y=0} \tag{7.10}$$

Defining $\theta = T - T_\infty$ and substituting this in Eq. (7.10), yields

$$\frac{d}{dx} \int_0^\delta u\theta dy = -\alpha \left. \frac{\partial \theta}{\partial y} \right|_{y=0} \tag{7.11}$$

Defining $\theta_m = T_w - T_\infty$, temperature distribution may be assumed as

$$\frac{\theta}{\theta_m} = \frac{T - T_\infty}{T_w - T_\infty} = \left(1 - \frac{y}{\delta} \right)^2 \tag{7.12}$$

This satisfies at $y = 0, T = T_w$ i.e., at $y = 0, \theta = \theta_m$; at $y = \delta, T = T_\infty$ i.e., at $y = \delta, \theta = 0$; at $y = \delta, \frac{\partial T}{\partial y} = 0$ i.e., at $y = \delta, \frac{\partial}{\partial y}\left[(T_w - T_\infty)\left(1 - \frac{y}{\delta}\right)^2 \right] = 0$.

The nature of velocity profile shown in Fig. 7.1 requires velocity in the boundary layer to be of the form

$$u = a + by + cy^2 + dy^3 \tag{7.13}$$

The boundary conditions are at $y = 0, u = 0$; at $y = \delta, u = 0$; at $y = 0, \nu\frac{\partial^2 u}{\partial y^2} = -\beta g(T_w - T_\infty)$; at $y = \delta, \frac{\partial u}{\partial y} = 0$. These boundary conditions will finally produce

$$a = 0, \quad b = \frac{\delta\beta g}{4\nu}(T_w - T_\infty), \quad c = -\frac{\beta g}{2\nu}(T_w - T_\infty), \quad d = \frac{\beta g}{4\nu\delta}(T_w - T_\infty)$$

Substituting these values Eq. (7.13) we get

$$u = \frac{g\beta\delta^2(T_w - T_\infty)}{4\nu}\frac{y}{\delta}\left(1 - \frac{y}{\delta} \right)^2 \tag{7.14}$$

The terms $(T_w - T_\infty)$, δ^2, $g\beta/4\nu$ can together form the characteristic velocity, u_x so that we can write

$$\frac{u}{u_x} = \frac{y}{\delta}\left(1 - \frac{y}{\delta} \right)^2 \tag{7.15}$$

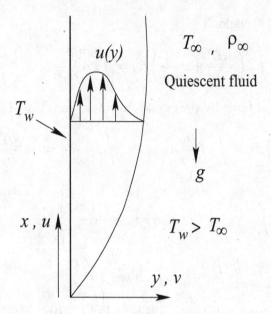

Figure 7.1 Boundary layer development on a heated vertical plate

The maximum velocity and its position (distance in y direction) at any x can be obtained from Eq. (7.15)

$$\frac{d}{dy}(u) = \frac{d}{dy}\left[u_x \left\{ \frac{y}{\delta}\left(1 - \frac{y}{\delta}\right)^2 \right\} \right] = 0 \tag{7.16}$$

$$\frac{d}{dy}\left[\frac{y}{\delta}\left(1 - \frac{2y}{\delta} + \frac{y^2}{\delta^2}\right) \right] = 0$$

or $\dfrac{1}{\delta} - \dfrac{4y}{\delta^2} + \dfrac{3y^2}{\delta^3} = 0$

or $\delta^2 - 4\delta y + 3y^2 = 0$

$$(\delta - y)\,(\delta - 3y) = 0 \tag{7.17}$$

At $y = \delta$ velocity is zero.

Therefore at $y = \frac{\delta}{3}$, velocity is maximum and

$$\frac{u_{max}}{u_x} = \frac{1}{3}\left[1 - \frac{1}{3}\right]^2$$

or, $u_{max} = \dfrac{4}{27} u_x$

Energy equation, Eq. (7.11), can be now written as

$$\frac{d}{dx}\int_o^\delta \theta_m \left(1 - \frac{y}{\delta}\right)^2 u_x \frac{y}{\delta}\left(1 - \frac{y}{\delta}\right)^2 dy = - \alpha\,\theta_m \frac{\partial}{\partial y}\left[\left(1 - \frac{y}{\delta}\right)^2\right]\Bigg|_{y=0}$$

or

$$\frac{1}{30}\frac{d}{dx}(u_x\,\delta)\,\theta_m = \frac{2\alpha\,\theta_m}{\delta} \tag{7.18}$$

The momentum equation, Eq. (7.9), may be written as

$$\frac{d}{dx} \int_o^\delta \left[u_x \frac{y}{\delta} \left(1 - \frac{y}{\delta} \right)^2 \right]^2 dy = g\beta \int_o^\delta \theta_m \left(1 - \frac{y}{\delta} \right)^2 dy - \nu u_x \frac{d}{dy} \left[\frac{y}{\delta} \left(1 - \frac{y}{\delta} \right)^2 \right]_{y=0}$$

or

$$\frac{1}{105} \frac{d}{dx} \left(u_x^2 \, \delta \right) = \frac{g \, \beta \, \theta_m \, \delta}{3} - \frac{\nu u_x}{\delta} \tag{7.19}$$

In Eqs. (7.18) and (7.19), u_x and δ are dependant variables and x is the independant variable. To solve the above two equations, u_x and δ are considered as functions of x.

Assume u_x and δ vary according to the following functions

$$u_x = C_1 x^m \qquad \text{and} \qquad \delta = C_2 x^n$$

Substitution in Eq. (7.18) will produce

$$\frac{m+n}{30} C_1 C_2 (x)^{m+n-1} = \frac{2\alpha}{C_2} x^{-n} \tag{7.20}$$

Substitution in Eq. (7.19) will produce

$$\frac{2m+n}{105} C_1^2 C_2 (x)^{2m+n-1} = \frac{1}{3} g \, \beta \, \theta_m \, C_2 x^n - \frac{\nu C_1 x^{m-n}}{C_2} \tag{7.21}$$

The equations have to be treated in the following way. First let us consider Eq. (7.20). If a monomial nonlinear expression is dependent on a variable (say x) that is equivalent to an expression independent of x, then for universal validity of the equation as a function of x, the power of the nonlinear expression must vanish. If we divide the entire Eq. (7.20) by its right-hand side, then for the universal validity of the expression, we can say

$$m + 2n - 1 = 0$$

Now, let us look at Eq. (7.21). Any number of terms with non-zero degree can combine to result into a single term, provided each of the terms has same degree. The degree of the resulting single term will be equal to that of other combining terms. This will lead to

$$2m + n - 1 = n \qquad \text{and} \qquad 2m + n - 1 = m - n$$

or

$$2m - 1 = 0 \qquad \text{and} \qquad m + 2n - 1 = 0$$

The conclusion derived from Eq. (7.20) and one of the two conclusions derived from Eq. (7.21) is the same. Now by solving these, we get

$$m = 1/2 \qquad \text{and} \qquad n = 1/4$$

Substituting the values of m and n in Eq. (7.20) and Eq. (7.21), we get

$$C_1 C_2^2 = 80\alpha \tag{7.22}$$

$$\frac{C_1^2 C_2}{84} = \frac{1}{3} g\,\beta\,\theta_m C_2 - \frac{C_1}{C_2}\,\nu \tag{7.23}$$

From Eq. (7.22) we get

$$C_1 = \frac{80\alpha}{C_2^2}$$

Substitution in Eq. (7.23) produces

$$\left(\frac{80\alpha}{C_2^2}\right)^2 \frac{C_2}{84} = \left(\frac{1}{3}g\beta\theta_m\right) C_2 - \frac{80\alpha}{C_2^2}\frac{\nu}{C_2}$$

or

$$76\alpha^2 = \left(\frac{1}{3}g\beta\theta_m\right) C_2^4 - 80\alpha\nu$$

or

$$\left(\frac{1}{3}g\beta\theta_m\right) C_2^4 = 76\alpha^2 + 80\alpha\nu$$

or

$$C_2 = \alpha^{1/4}\left(\frac{76}{80} + \frac{\nu}{\alpha}\right)^{1/4} (80\alpha)^{1/4} \left(\frac{1}{3}\frac{g\beta\theta_m}{\nu^2}\right)^{-1/4} \nu^{-1/2}$$

or

$$C_2 = 3.93 \left(\frac{\nu}{\alpha}\right)^{-1/2} \left(\frac{\nu}{\alpha} + \frac{20}{21}\right)^{1/4} \left(\frac{g\beta\theta_m}{\nu^2}\right)^{-1/4}$$

The back subsitution in the expression for C_1 produces

$$C_1 = \frac{80\alpha}{C_2^2} = 5.17\,\nu \left(\frac{\nu}{\alpha} + \frac{20}{21}\right)^{-1/2} \left(\frac{g\beta\theta_m}{\nu^2}\right)^{1/2}$$

Therefore, the values of C_1 and C_2 are given as

$$C_1 = 5.17\,\nu \left[\frac{\nu}{\alpha} + \frac{20}{21}\right]^{-1/2} \left(\frac{g\,\beta\,\theta_m}{\nu^2}\right)^{1/2} \tag{7.24}$$

$$\text{and,}\quad C_2 = 3.93 \left(\frac{\nu}{\alpha}\right)^{-1/2} \left[\frac{\nu}{\alpha} + \frac{20}{21}\right]^{1/4} \left(\frac{g\,\beta\,\theta_m}{\nu^2}\right)^{-1/4} \tag{7.25}$$

Substituting C_1 and C_2 for evaluating u_x and δ and substituting Pr for ν/α we get

$$u_x = (5.17)\,\nu \left(Pr + \frac{20}{21}\right)^{-1/2} \left(\frac{g\,\beta\,\theta_m}{\nu^2}\right)^{1/2} x^{1/2}$$

$$\delta = 3.93\,(Pr)^{-1/2} \left(Pr + \frac{20}{21}\right)^{1/4} \left(\frac{g\,\beta\,\theta_m}{\nu^2}\right)^{-1/4} x^{1/4}$$

or

$$\frac{\delta}{x} = 3.93 \, (Pr)^{-1/2} \left(Pr + \frac{20}{21}\right)^{1/4} \left(\frac{g \, \beta \, \theta_m}{\nu^2}\right)^{-1/4} x^{-3/4}$$

or

$$\frac{\delta}{x} = 3.93 \, (Pr)^{-1/2} \left(Pr + \frac{20}{21}\right)^{1/4} \left(\frac{g \, \beta \, \theta_m x^3}{\nu^2}\right)^{-1/4}$$

Introducing $Gr_x = $ Grashof number $\equiv \frac{g \, \beta \, \theta_m x^3}{\nu^2}$ we get

$$\frac{\delta}{x} = 3.93(Pr)^{-1/2} \left(Pr + \frac{20}{21}\right)^{1/4} Gr_x^{-1/4}$$

or

$$\frac{\delta}{x} = 3.93 \left[\frac{Pr + \frac{20}{21}}{Pr^2 \, Gr_x}\right]^{1/4} \tag{7.26}$$

Equation (7.26) gives the variation of boundary layer thickness along the height of the plate. Here thermal boundary layer thickness and hydrodynamic boundary layer thickness are assumed to be same.

7.2.1 Heat transfer coefficient

The heat flux from the wall may be expressed as

$$q'' = -k \left.\frac{\partial T}{\partial y}\right|_{y=0} = h_x (T_w - T_\infty)$$

The local heat transfer coefficient at any x- location is

$$h_x = \frac{-k(\partial T/\partial y)_{y=0}}{T_w - T_\infty}$$

or

$$h_x = -\frac{k}{(T_w - T_\infty)} \frac{\partial}{\partial y} \left[T_\infty + (T_w - T_\infty)\left(1 - \frac{y}{\delta}\right)^2\right]_{y=0} = \frac{2k}{\delta}$$

or

$$\frac{h_x x}{k} = \frac{2x}{\delta} = \frac{2}{3.93} \left[\frac{Pr^2 \, Gr_x}{Pr + \frac{20}{21}}\right]^{1/4}$$

The local Nusselt number is given by

$$Nu_x = 0.508 \left[\frac{Pr^2 Gr_x}{(Pr + \frac{20}{21})}\right]^{1/4} \tag{7.27}$$

The average Nusselt number over the plate is given by

$$\overline{Nu}_L = \frac{1}{L} \int_0^L Nu_x dx = \frac{4}{7} Nu_L$$

where L is the length of the plate.

Note: In a field where free convection and forced convection both are present, free convection is negligible if $Gr/Re^2 \ll 1$, forced convection is negligible if $Gr/Re^2 \gg 1$. For a situation with combined free and forced convection, the condition is, $Gr/Re^2 \sim 1$.

We can also write

$$\frac{g\beta\theta_m L^3}{\nu^2} \sim \frac{U^2 L^2}{\nu^2}$$

An equivalent free convection velocity may be defined as

$$u \sim \sqrt{g\beta\theta_m L}$$

The above Eq. (7.27) is used for laminar boundary layers. It is customary to correlate the transition to turbulence if the Rayleigh number crosses a certain limit.

$$Ra_{x,c} = Gr_{x,c} \; Pr = \frac{g\beta(T_w - T_\infty)x^3}{\nu^2} \; \frac{\nu}{\alpha}$$

Flow becomes turbulent if $Ra_{x,c} \geq 10^9$

Some important correlations:

- Vertical plate: $\overline{Nu}_L = 0.68 + \dfrac{0.67 Ra_L^{1/4}}{\left[1 + (0.492/Pr)^{9/16}\right]^{4/9}}$, $10^{-1} < Ra_L < 10^9$.

- Horizontal plate: $L = \frac{A_s}{P} = \frac{\text{Surface area}}{\text{Perimeter}}$. If upper side of the plate is heated, then $\overline{Nu}_L = 0.54 Ra_L^{1/4}$, $10^4 \leq Ra_L \leq 10^7$.

- Long horizontal cylinder (due to Churchill and Chu [3]) $\overline{Nu}_D = \left[0.60 + \dfrac{0.387(Ra_D)^{1/6}}{\left[1 + (0.559/Pr)^{9/16}\right]^{8/27}}\right]^2$ $10^{-5} \leq Ra_D \leq 10^{12}$.

7.3 VERTICAL CYLINDER

For the fluids, having Prandtl numbers of 0.7 and higher, a vertical cylinder may be treated as a vertical flat plate [2] when

$$\frac{D}{L} \geq \frac{35}{(Gr_L)^{1/4}}$$

Basically, if the thermal boundary layer is significantly smaller than the radius of the cylinder, the vertical flat plate and vertical cylinder have considerable geometrical similarity.

Figure 7.2 shows a plot of the ratio of the average Nusselt number for a vertical cylinder to that of a vertical plate as a function of the parameter $\xi = \frac{2\sqrt{2}}{(Gr_L)^{1/4}} \cdot \left(\frac{L}{R}\right)$. Here, R is the radius of the cylinder and L is the length.

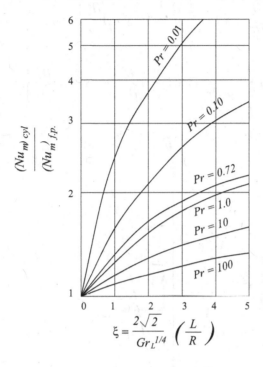

Figure 7.2 The ratio of the Nusselt number for a vertical plate to that of a vertical cylinder (courtesy of Ozisik [4])

7.4 PLUMES

A plume is the buoyancy induced flow resulting in a fluid, when energy is supplied continuously at just one location in the fluid. In practice free plumes, like free jets are generally turbulent. An axisymmetric plume is found to be laminar only if the Rayleigh number based on its heat source(q'') and height of the plume(H) is less than 10^{10}. Rayleigh number is given by

$$Ra_H = \frac{g\beta H^4 q''}{\alpha\nu k}$$

7.4.1 Integral Analysis of Steady Plume

The governing equations for the flow in a plume are the continuity equation, the Navier-Stokes equations, and the energy equation. Assuming the mean flows in a plume to be steady and axisymmetric (Fig. 7.3) the Reynolds averaged form of these equations are

$$\frac{1}{r}\frac{\partial}{\partial r}(r\bar{u}) + \frac{\partial \bar{v}}{\partial y} = 0 \tag{7.28}$$

$$\frac{\partial}{\partial y}(\bar{v}^2) + \frac{1}{r}\frac{\partial}{\partial r}(r\bar{u}\bar{v}) = \frac{1}{r}\frac{\partial}{\partial r}\left[r(\nu + \nu_t)\frac{\partial \bar{v}}{\partial r}\right] + g\beta(\bar{T} - T_\infty) \tag{7.29}$$

$$\frac{\partial}{\partial y}(\bar{v}\bar{T}) + \frac{1}{r}\frac{\partial}{\partial r}(r\bar{u}\bar{T}) = \frac{1}{r}\frac{\partial}{\partial r}\left[r(\alpha + \alpha_t)\frac{\partial \bar{T}}{\partial r}\right] \tag{7.30}$$

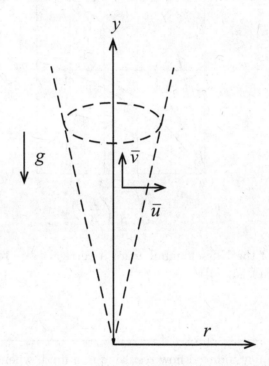

Figure 7.3 Model flow domain (dotted lines show time averaged flow region)

Integration of Eq. (7.28) from $r = 0$ to $r \to \infty$ gives

$$(r\bar{u})_\infty + \frac{d}{dy}\int_0^\infty \bar{v}r\,dr = 0 \tag{7.31}$$

Similarly integration of the momentum equation, Eq. (7.29), and the energy equation, Eq. (7.30), subject to the boundary conditions that at $r \to \infty$, $\bar{v} = 0$, $\bar{T} = T_\infty$ and conditions of symmetry at the centerline, yields

$$\frac{d}{dy}\int_0^\infty \bar{v}^2 r\,dr = g\beta\int_0^\infty (\bar{T} - T_\infty)r\,dr \tag{7.32}$$

$$\frac{d}{dy} \int_0^\infty \bar{v}(\bar{T} - T_\infty) r \, dr = 0$$

$$\text{or} \quad \int_0^\infty \bar{v}(\bar{T} - T_\infty) r \, dr = \frac{q}{2\pi \rho c_p} \tag{7.33}$$

where q is the rate of heat release (the strength of the heat source) at the origin of the plume.

For proceeding further with the analysis, it is essential to make assumptions for the velocity and temperature profiles. The experimental observations show a Gaussian profile for both velocity and temperature. Thus the following profiles are assumed.

$$\bar{v} = \bar{v}_c \, exp\left[-\left(\frac{r}{b}\right)^2\right] \tag{7.34}$$

$$\bar{T} - T_\infty = (\bar{T}_c - T_\infty) \, exp\left[-\left(\frac{r}{b_T}\right)^2\right] \tag{7.35}$$

where b and b_T are characteristic radial dimensions proportional to the plume thickness and the ratio of b and b_T is of the order of unity. $\bar{v}_c$ and $\bar{T}_c$ are the centerline velocity and temperatures respectively. To solve the resultant equations for all the unknowns, an additional assumption is required for the entrainment term $(r\bar{u})_\infty$. Here, we can make use of another experimental observation that the radius of the plume increases linearly with its height or

$$b \sim y \tag{7.36}$$

Using this assumption in the continuity equation leads to the result

$$\bar{v}_c b \sim (r\bar{u})_\infty$$

$$\text{or} \quad (r\bar{u})_\infty = -\hat{\alpha} b \bar{v}_c \tag{7.37}$$

where $\hat{\alpha}$ is the entrainment coefficient to be determined experimentally and has an approximate value of 0.12. More insight into turbulent-plume problem is available in Bejan [5].

Using the profiles (7.34) and (7.35) and the relation (7.37) leads to the following solution of Eqs. (7.31)-(7.33).

$$b = \frac{6}{5}\hat{\alpha}y \tag{7.38}$$

$$\bar{v}_c = \left[\frac{25}{24\pi\hat{\alpha}^2} \frac{qg\beta}{\rho c_p y}\left(1 + \frac{b_T^2}{b^2}\right)\right]^{1/3} \tag{7.39}$$

$$\bar{T}_c - T_\infty = 0.685\left(1 + \frac{b^2}{b_T^2}\right)\left(1 + \frac{b_T^2}{b^2}\right)^{-1/3}\left(\frac{q}{\pi\rho c_p}\right)^{2/3}\hat{\alpha}^{-4/3}y^{-5/3}(g\beta)^{-1/3} \tag{7.40}$$

7.5 MIXED CONVECTION IN RECTANGULAR CHANNEL

Let us consider flow of air or gases through a horizontal channel as shown in Fig. 7.4. Two important non-dimensional groups may be introduced here. One is Archimedes number, A_r and the other is Froude number, F_r. The reference temperature is T_∞ and the reference pressure is P_∞.

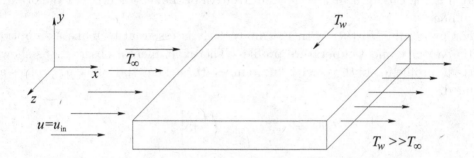

Figure 7.4 Interaction of buoyancy induced flow and forced flow through horizontal channel

The Froude number is given by, $F_r = \frac{V}{\sqrt{g\ell}}$ and Archimedes number is given by $A_r = \frac{\beta(T_w - T_\infty)}{F_r^2} = \beta(T_w - T_\infty)g\ell/V^2$. In the above mentioned definitions, V is the velocity and l is the length scale. The Archimedes number is linked to Grashof number in the following way

$$Gr = Ar \times Re^2 = \frac{\beta(T_w - T_\infty)g\ell}{V^2}\frac{V^2\ell^2}{\nu^2}$$

$$Gr = \frac{g\beta(T_w - T_\infty)\ell^3}{\nu^2}$$

Figure 7.4 shows combined buoyancy induced flow and forced convection flow through a horizontal channel. The following are the governing equations

Continuity:

$$\nabla.\mathbf{V} = 0$$

x mom:

$$\frac{\partial u}{\partial t} + (\mathbf{V}.\nabla)u = -\frac{1}{\rho}\frac{\partial p}{\partial x} + \nu\nabla^2 u$$

y mom:

$$\frac{\partial v}{\partial t} + (\mathbf{V}.\nabla)v = -\frac{1}{\rho}\frac{\partial p}{\partial y} + \nu\nabla^2 v + g\beta(T - T_\infty)$$

z mom:

$$\frac{\partial w}{\partial t} + (\mathbf{V}.\nabla)w = -\frac{1}{\rho}\frac{\partial p}{\partial z} + \nu\nabla^2 w$$

Energy:

$$\frac{\partial T}{\partial t} + (\mathbf{V}.\nabla)T = \alpha[\nabla^2 T]$$

After non-dimensionalization,

$$u^* = u/u_{in}$$
$$v^* = v/u_{in}$$
$$w^* = \frac{w}{u_{in}}$$
$$x^* = \frac{x}{L}$$
$$y^* = y/L$$
$$z^* = z/L$$
$$p^* = p/\rho u_{in}^2$$
$$\theta = (T - T_\infty)/(T_w - T_\infty)$$
$$t^* = t/(L/u_{in})$$
$$Re = \frac{u_{in}L}{\nu}$$
$$Gr = g\beta(T_w - T_\infty)L^3/\nu^2$$
$$Pr = \nu/\alpha$$

we get

$$\frac{\partial u^*}{\partial x*} + \frac{\partial v^*}{\partial y^*} + \frac{\partial w^*}{\partial z^*} = 0 \tag{7.41}$$

$$\frac{Du^*}{Dt^*} = -\frac{\partial p^*}{\partial x^*} + \frac{1}{Re}\nabla^2 u^* \tag{7.42}$$

$$\frac{Dv^*}{Dt^*} = -\frac{\partial p^*}{\partial y^*} + \frac{1}{Re}\nabla^2 v^* + \frac{Gr}{Re^2}\theta \tag{7.43}$$

$$\frac{Dw^*}{Dt^*} = -\frac{\partial p^*}{\partial z^*} + \frac{1}{Re}\nabla^2 w^* \tag{7.44}$$

$$\frac{D\theta}{Dt} = \frac{1}{Re \,.\, Pr}\nabla^2\theta \tag{7.45}$$

Gr/Re^2 is called Richardson number $= R_i$.

The problem can be solved numerically using coupled (Navier-Stokes and energy equations) solution technique applied to Eqs. (7.41) to (7.45). Many useful applications of a similar flow configuration have been discussed by Incropera et al. [6].

7.6 MIXED CONVECTION INCLUDING BUOYANCY AIDED AND BUOYANCY OPPOSED FLOWS

Let us consider the temperature conditions as depicted in Fig. 7.5

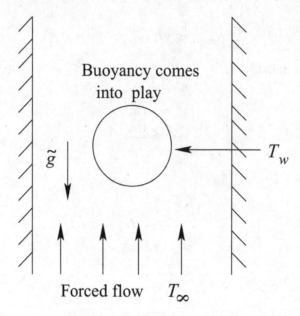

Figure 7.5 Mixed convection over a cylinder through parallel plate

$$T_w > T_\infty, R_i + ve$$
$$T_w < T_\infty, R_i - ve$$

In the mixed convection problem shown in Fig. 7.5 vortex shedding is found to stop at a critical Richardson number. The forced flow Reynolds number may be considered as 100. The following explanations are interesting (Singh et al. [7]).

The physical significance of negative Ri is culmination of a buoyancy force in a direction opposite to the flow direction. In the viscous region aft of the cylinder, the inertia force is opposed by the force due to an adverse pressure gradient, the negative buoyancy force, and the viscous force. Depending on the magnitude of Ri, the point of separation moves toward the leading edge, causing an early separation. This also broadens the wake. Vortex shedding becomes prominent in the entire regime of negative Ri. The shedding phenomenon continues to characterize the flow even at $Ri = 0$ and 0.1 (Fig. 7.6). For $Ri = 0.15$ (streamlines are not shown here) and above, the flow structure becomes different. The aiding buoyancy changes the dynamics altogether. Near the vicinity of the cylinder, even aft of the cylinder, the buoyancy force is added up with the inertia force in the face of combined pressure force and viscous force. The phenomenon results in a separation delay, and vortex shedding is also stopped. This observation is in agreement with earlier study by Chang and Sa [8]. Figure 7.6 shows isotherms for various Ri. The buoyancy opposed and buoyancy aiding cases are discerned in this figure. It can readily be

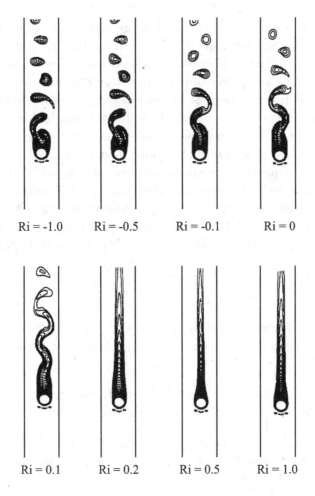

Ri = -1.0   Ri = -0.5   Ri = -0.1   Ri = 0

Ri = 0.1   Ri = 0.2   Ri = 0.5   Ri = 1.0

Figure 7.6 Isotherms for different Richardson numbers at $Re = 100$ and $D/H = 0.25$.

appreciated that the waviness of the isotherms downstream of the cylinder is greater for the buoyancy-opposed convection than for buoyancy-aiding convection. Beyond $Ri = 0.2$, the plume spread becomes narrow, and waviness disappears.

Concluding Remarks

Free convection has wide range of applications in cooling of electronic devices and computers. Some heat-generating electronic components are cooled using a fan or a pump too. Therefore, a decision has to emerge whether only free convection or mixed convection cooling is to be used for cooling of electronic components. The choice depends on the maximum allowable operating temperature of the equipment. The convection heat transfer rate from a surface at temperature T_w in a medium at T_∞ is given by

$$Q_{conv} = hA_w(T_w - T_\infty)$$

where h is the heat transfer coefficient and A_w is the surface area. Therefore, the surface temperature will be higher if h is low for the equipment. Free convection heat transfer coefficient is usually low.

Free convection or natural convection is a preferred mode of heat transfer. Usually blowers or pumps are not needed for free convection and as a consequence the problems, such as vibration, noise, etc. can be avoided. Free convection is usually appropriate for cooling low-power devices. It is very effective for the extended surfaces too. For the high-power devices, blowers or pumps are needed to maintain the operating temperature below the allowable limit. For even higher power devices, in order to maintain the surface temperature below the prescribed limit, phase-change cooling is utilized. Heat pipes are such cooling devices.

The rates of heat transfer in free or natural convection can be made much higher by operating the devices in the near-critical or supercritical regions of fluids. In such regimes, small changes in temperature or pressure yield large changes in the fluid thermophysical properties of the fluid. Significant change in the thermophysical properties of the fluid can bring about appreciable changes in the rate of convective heat transfer.

REFERENCES

1. S. Ostrach, Natural Convection in Enclosers, in Advances in Heat Transfer, Eds. J. P. Hertnett and Thomas F. Irvine, Academic Press, Vol. 8, pp. 161-227, 1972.

2. B. Gebhart, Y. Jaluria, R. L. Mahajan and B. Sammakia, Buoyancy Induced Flows and Transport, Hemisphere Publishing, Washington, 1988.

3. S. W. Churchill and H. H. S. Chu, Correlating Equation for Laminar and Turbulent Free Convection for a Vertical Plate, Int. J. Heat Mass Transfer, Vol. 18, pp. 1323, 1975.

4. M. N. Ozisik, Heat Transfer: A Basic Approach, McGraw Hill, 1985.

5. A. Bejan, Convective Heat Transfer, John Wiley & Sons, 2004.

6. F. P. Incropera, A. L. Knox and J. R. Maughan, Mixed Convection Flow and Heat Transfer in the Entry Region of a Horizontal Rectangular Duct, J. Heat Transfer, Vol. 109, pp. 434-439, 1987.

7. S. Singh, G. Biswas and A. Mukhopadhyay, Effect of Thermal Buoyancy on the Flow through a Vertical Channel with a Built-in-Circular Cylinder, Num. Heat Transfer, Part A, Vol. 34, pp. 769-789, 1998.

8. K. S. Chang and J. Y. Sa, The Effect of Buoyancy on Vortex Shedding in the Near Wake of a Circular Cylinder, J. Fluid Mech., Vol. 220, pp. 253-266, 1990.

EXERCISES

1. (a) What is Boussinesq approximation for the free convective flows?

(b) Write an expression for the free convective velocity, if the difference between the maximum and the minimum temperatures in the field is known.

(c) Invoke Boussinesq approximation in the governing equations for a free convective flow over a vertical heated flat surface.

(d) Find out the integral form of the momentum and energy equations.

(e) Indicate the velocity profile that is needed to solve the integral form of the equations.

(f) Find the distance from the solid surface where you expect the maximum velocity.

(g) Express Grashof number in terms of Archimedes number and Reynolds number.

2. Consider a natural convective flow over a heated vertical flat plate. The wall temperature is T_w and the quiescent fluid temperature is T_∞. The vertical direction coordinate is x and the normal direction is y. The x-velocity component is u and the y-velocity is given by v. The governing momentum equation is given by

$$\rho\left(u\frac{\partial u}{\partial x} + v\frac{\partial u}{\partial y}\right) = -\frac{\partial p}{\partial x} - \rho g + \mu\frac{\partial^2 u}{\partial y^2}$$

The reference temperature is T_∞ and the reference density is ρ_∞. Express the buoyancy term as a function of volume coefficient of expansion. Using the concept of a reference velocity, write the non-dimensional form of the resulting equation. Show that the non-dimensional coefficient arising out of first two terms of the right-hand side can be expressed in terms of Grashof number and Reynolds number.

3. Consider the dimensional form of equation that results from the governing equation given in Problem 2 by invoking the expression related to the volume coefficient of expansion. For both the momentum and the energy equations, a similarity solution can be obtained using the following similarity variables.

$$\frac{u}{u_x} \quad \text{and} \quad \eta = \frac{y}{x}\left(\frac{Gr_x}{4}\right)^{1/4} \qquad \text{where} \quad u_x = \sqrt{4g\beta(T_w - T_\infty)x}$$

Derive the expressions for appropriate stream function and the u-velocity.

4. Formulate a three-dimensional mixed convection problem for the flow of air through a rectangular heated duct and show that y-momentum equation is connected with temperature term through Richardson number. The gravity acts in the negative y-direction.

5. Start with a finite control volume approach (Fig. 7.7) and find out an expression for modified y- momentum equation for a mixed convection flow.

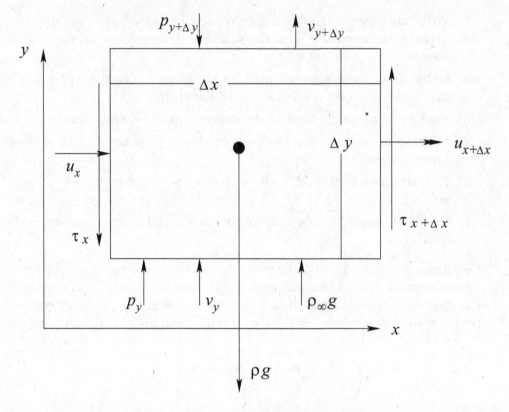

Figure 7.7 A control volume in a mixed convection flow field

6. Consider a natural convective flow over a heated vertical flat plate. The wall temperature is T_w and the quiescent fluid temperature is T_∞. The vertical direction coordinate is x and the normal direction is y. The x-velocity component is u and the y-velocity is given by v. The governing momentum equation is given by

$$\rho\left(u\frac{\partial u}{\partial x} + v\frac{\partial u}{\partial y}\right) = -\frac{\partial p}{\partial x} - \rho g + \mu\frac{\partial^2 u}{\partial y^2}$$

The reference temperature is T_∞ and the reference density is ρ_∞. Express the buoyancy term as a function of volume coefficient of expansion β. The energy equation is given by

$$u\frac{\partial T}{\partial x} + v\frac{\partial T}{\partial y} = \alpha\frac{\partial^2 T}{\partial y^2}$$

Integrate the momentum equation and the energy equation within the boundary layer. Show that the momentum integral equation is given by

$$\frac{d}{dx}\int_o^\delta u^2\,dy = g\beta\int_o^\delta (T - T_\infty)\,dy - \nu\left.\frac{\partial u}{\partial y}\right|_{y=0}$$

and the energy integral equation is given by

$$\frac{d}{dx} \int_o^\delta u\theta dy = -\alpha \left. \frac{\partial \theta}{\partial y} \right|_{y=0} \qquad \text{where} \quad \theta = (T - T_\infty)$$

The boundary conditions are :

at $y = 0 : u = 0, v = 0, \nu \frac{\partial^2 u}{\partial y^2} = -g\beta(T_w - T_\infty)$; at $y = \delta : u = 0, \frac{\partial u}{\partial y} = 0$ and

at $y = 0 : T = T_w$; at $y = \delta : T = T_\infty$ and $\frac{\partial T}{\partial y} = 0$

The temperature profile that satisfies the above mentioned temperature boundary conditions is given by

$$\frac{\theta}{\theta_m} = \left(1 - \frac{y}{\delta}\right)^2 \qquad \text{where} \quad \theta_m = T_w - T_\infty$$

Show that the velocity profile is given by

$$\frac{u}{u_x} = \frac{y}{\delta} \left(1 - \frac{y}{\delta}\right)^2 \qquad \text{where} \quad u_x' = \frac{g\beta\delta^2(T_w - T_\infty)}{4\nu}$$

## CHAPTER 8

Introduction to Boiling

8.1 POOL BOILING

Heat is transferred from solid surfaces to liquid pools with phase change occurring on the surface. For liquid at its saturation temperature T_{sat}, and surface at temperature T_w, $q_w = h(T_w - T_{sat}) = h\Delta T$. Heat transfer is strongly influenced by the formation, growth and departure of vapor bubbles.

8.1.1 Boiling Regimes

Nukiyama [1] identified different regimes of pool boiling by using an apparatus similar to that shown in the Fig. 8.1.

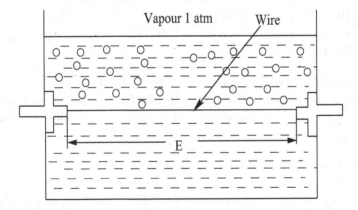

Figure 8.1 Nukiyama's power-controlled heating apparatus for demonstrating the boiling curve

1. **Free Convection:** This phenomenon occurs at low temperatures (excess temperature $\leq 5°C$). Heat transfer from the heated surface to the liquid in its vicinity causes the liquid to be superheated. The superheated liquid rises to the free liquid surface by free or natural convection where vapor is produced by evaporation.

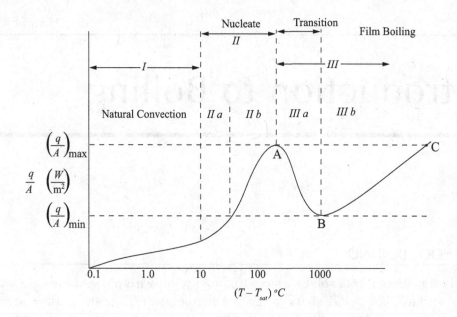

Figure 8.2 Typical boiling curve for water at one atmosphere pressure with surface heat flux $\left(\frac{q}{A}\right)$ as a function of excess temperature $\Delta T = T_w - T_{sat}$

2. **Nucleate Boiling:** It commences with the increased difference between $(T_w - T_{sat})$ (see Fig. 8.2). IIa: The bubbles formed are very few in number. These bubbles grow in size, separate from the heated surface and rise to the free surface. IIb: The rate of bubble formation increases and number of locations where they are formed increases too. Bubbles begin to merge both laterally and in the vertical direction.

3. **Transition Boiling:** The regime corresponding to $(T_w - T_{sat}) \approx 30° - 150°C$ is known as transition boiling IIIa. The bubble formation is rapid and vapor film begins to form on the surface. The surface condition oscillates between film and nucleate boiling.

4. **Film Boiling:** Rapid bubble formation rate causes them to coalesce and cover the surface $T_w \simeq 250°C$, with a vapor film IIIb. The vapor film interface is unstable and involves cyclic release of bubbles from the interface. At high surface temperature difference (1000°C), the radiative transfer across the vapor film becomes significant.

In the region of free convection, the heat flux is proportional to $(T_w - T_{sat})^n$ where n is slightly greater than unity (≈ 1.3). When the transition from natural convection to nucleate boiling occurs, heat flux starts to increase more rapidly with the temperature difference, the value of n increasing to about 3. At the end of the region pertaining to nucleate boiling the boiling curve reaches a peak (point A). In transition boiling wall heat flux decreases because the thermal resistance to heat flow increases with the coverage of most of the surface with vapor film. As more and more of the surface is covered with vapor, heat flux decreases whereas surface temperature increases. At the end of this regime, heat flux passes through a minimum (point B). It starts increasing again with $(T_w - T_{sat})$ only when stable film boiling begins and radiation becomes important.

It is of interest to know how the temperature of the heating surface changes as the heat flux is steadily increased from zero. Up to point A, nucleate boiling occurs and the temperature of the heating surface is obtained by reading the value $(T_w - T_{sat})$ from the boiling curve and adding it to the value of T_{sat}. If the heat flux is increased a little beyond the value at A, the temperature of the surface shoots up and the process goes through AB-BC very rapidly before settling at point C. For many surfaces, the temperature at C is high enough to cause the material to melt. Thus in many practical situations, it is undesirable to cross this value. The value of heat flux at A is therefore of significance in engineering and is called the burn out, *critical or peak heat flux.*

8.2   NUCLEATION AND BUBBLE GROWTH

Nucleation is a process in which finite size clusters of molecules encompassing properties of the second phase appear in the host liquid. This could be the initiation of bubble formation during boiling or cavitation and crystal formation during solidification. The nucleation process is generally subdivided into two categories: homogeneous nucleation and heterogeneous nucleation. The homogeneous nucleation refers to the first appearance of a vapor bubble in the liquid pool and far away from the bounding walls of the container. The heterogeneous nucleation is a process in which bubbles form at pits, scratches and grooves on a heated surface submerged in a pool of liquid. Before going on to the detailed description and analyses of these processes, let us first look at the conditions for mechanical and thermal equilibrium of a bubble with the host liquid.

The condition for mechanical equilibrium simply states that the pressure in a perfectly spherical bubble of radius, R, must exceed the ambient by $2\sigma/R$, where σ is the interfacial tension.

$$p_b - p_l = \frac{2\sigma}{R} \tag{8.1}$$

In a gravitational field, the bubble being lighter than the surrounding liquid will buoy away. However here we are considering a static equilibrium such that the bubble is relatively small in size and is held in place. Now if the bubble is in thermal equilibrium with the liquid, the temperature of the bubble and the liquid should be the same, i.e.,

$$T_b = T_l \tag{8.2}$$

Since from thermodynamic equilibrium consideration, vapor must be at the saturation temperature corresponding to pressure in the bubble and we know from the equation of state that saturation temperature increases with pressure, the bubble or the liquid temperature should be higher than the saturation temperature corresponding to the pressure in the liquid. In other words, the liquid must be superheated. The degree of superheat can be determined if we combine the mechanical equilibrium condition with the Clausius-Clapeyron equation; which relates the differential change in saturation temperature with pressure as

$$\frac{\Delta p}{\Delta T} = \frac{h_{fg}}{v_{fg} \left(T_{sat} + \frac{\Delta T}{2}\right)} \tag{8.3}$$

In Eq. (8.3), T_{sat} is the absolute saturation temperature corresponding to the system pressure and v_{fg} is the difference in specific volumes of vapor and liquid at the system pressure. For $\Delta T \ll T_{sat}$, Eq. (8.3) can be approximately written as

$$\frac{\Delta p}{\Delta T} = \frac{h_{fg}}{v_{fg} T_{sat}} \tag{8.4}$$

Substituting for Δp in Eq. (8.4), the liquid superheat is obtained as

$$(T_l - T_{sat}) \equiv (T_b - T_{sat}) = \frac{2\sigma v_{fg} T_{sat}}{R h_{fg}} \tag{8.5}$$

At low pressures the specific volume of vapor is much larger than that of the liquid such that v_{fg} in Eq. (8.5), can be replaced by $1/\rho_v$ without making any appreciable error. With this change Eq. (8.5) becomes

$$(T_l - T_{sat}) = \frac{2\sigma T_{sat}}{R \rho_v h_{fg}} \tag{8.6}$$

Example 1: Calculate the required water superheat if a stream bubble of radius 10^{-3} cm is to coexist with the liquid at a system pressure of one atmosphere.

$$T_l - T_{sat} = \frac{2 \times 58.9 \times 10^{-3} \times 373}{10^{-5} \times 0.5977 \times 2.257 \times 10^6} = 3.26 K$$

Thus we see that for a relatively small bubble, the required liquid superheat is not very large. The superheat will tend to decrease as the system pressure is increased. Another observation that can be made about the requirement of thermal equilibrium is that a bubble in the absence of any mass transfer will have to shrink upon introduction into a pool of liquid if the liquid superheat is more than the superheat corresponding to the initial bubble radius. Similarly if the superheat is smaller than that corresponding to the temperature in the bubble, the bubble will have to expand. However, mass transfer at the interface will cause the bubble to expand or shrink depending on the bulk temperature.

8.2.1 Homogeneous Nucleation

The classical theory of homogeneous bubble nucleation is a **mixture of macroscopic and molecular concepts**. The process of creation of a bubble in a superheated liquid is initiated by a cluster of activated molecules. These molecules have energies that are considerably in excess of the average. Generally a cluster is constantly growing or diminishing in size as additional activated molecules join or break away from it. When a cluster attains a size such that its excess availability[1] is maximum it takes on the form of a bubble. If the activated cluster has a density which is different than the surrounding liquid, surface tension can be presumed to exist at the interface between the two. As described by Cole [2] for a certain volume of liquid containing an activated cluster of radius r, the availability can be written as

$$A = (n_t - n_a) g_l + n_a g_a + 4\pi r^2 \sigma \tag{8.7}$$

Here n_t is the total number of molecules, n_a is the number of activated molecules in the cluster and g_l and g_a are the specific Gibbs functions of the nonactivated and activated molecules respectively. If ρ_a is the fluid density in the activated cluster, M is the molecular weight and N_A is the Avogadro's number, an expression for the number of activated molecules in a cluster of radius, r, can be written as

$$n_a = \frac{4}{3}\pi r^3 \rho_a \frac{N_A}{M} \tag{8.8}$$

Substitution of (8.8) in (8.7) yields

$$A = n_t g_l + \frac{4}{3}\pi r^3 \rho_a \frac{N_A}{M}(g_a - g_l) + 4\pi r^2 \sigma \tag{8.9}$$

Since in the absence of a cluster of activated molecules, the availability of the system will be $n_t g_l$, the excess availability in the presence of an activated cluster can be written from Eq. (8.9) as

$$\Delta A = \frac{4}{3}\pi r^3 \rho_a \frac{N_A}{M}(g_a - g_l) + 4\pi r^2 \sigma \tag{8.10}$$

At the metastable equilibrium, the excess availability of the system with respect to the cluster radius will be maximum. The equilibrium or critical cluster radius, r_c, can be obtained by differentiating Eq. (8.10) with respect of r and setting it equal to zero. If we do so, we obtain

$$\frac{d(\Delta A)}{dr} = \frac{4\pi r_c^2 \rho_a N_A}{M}(g_a - g_l) + 8\pi r_c \sigma = 0$$

or

$$r_c = \frac{2\sigma M}{\rho_a N_A (g_l - g_a)} \tag{8.11}$$

Substituting for $(g_a - g_l)$ from Eq. (8.11) into Eq. (8.10), the excess availability is obtained as

$$\Delta A = 4\pi r^2 \sigma \left(1 - \frac{2}{3}\frac{r}{r_c}\right) \tag{8.12}$$

Figure 8.3 shows the dimensionless excess availability as a function of dimensionless cluster radius r/r_c. The maximum value of the excess availability is

$$\Delta A_{max} = \frac{4\pi r_c^2 \sigma}{3} \tag{8.13}$$

---

[1]Availability is an intrinsic property of a system and represents its potential to do work.

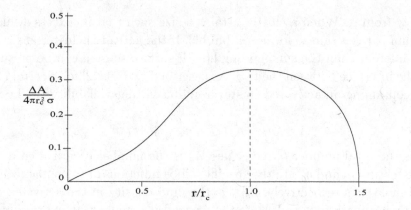

Figure 8.3 Excess availability as a function of activated cluster radius

Neglecting activation energy for diffusion of molecules in the liquid, the rate at which activated clusters appear per unit volume of the liquid per unit time can be written as

$$J = f e^{-(\Delta A_{max}/kT_l)} \tag{8.14}$$

where f is the frequency of molecular collisions per unit volume and k is Boltzmann constant. In writing Eq. (8.14) it is assumed that nucleation will occur when the excess availability of the cluster is several times that of the kinetic energy of a molecule which is characterized by kT_l. As such the ratio $\Delta A_{max}/kT_l$ represents a dimensionless number rather than number of molecules. From kinetic theory, the frequency of molecular collisions per unit volume per unit time is written as

$$f = \frac{NkT_l}{h} \tag{8.15}$$

In Eq. (8.15) N is the number of molecules per m^3 and h is Planck's constant. Substituting Eq. (8.15) in (8.14), dividing by J and taking logarithm of both sides gives

$$\frac{\Delta A_{max}}{kT_l} = \ln\left(\frac{NkT_l}{Jh}\right) \tag{8.16}$$

With ΔA_{max} given by Eq. (8.13), the critical cluster radius is obtained from Eq. (8.16) as

$$r_c = \left[\frac{3kT_l}{4\pi\sigma}\ln\left(\frac{NkT_l}{Jh}\right)\right]^{1/2} \tag{8.17}$$

The critical radius as given by Eq. (8.17) is not very sensitive to the value of J which is not known. However, variation in J from 10^{-6} to 10^6 will yield critical radii which will differ only by a few percent from each other. In the above expression the surface tension σ and the number density N are evaluated at the liquid temperature T_l. Using the critical radius in the condition for the mechanical equilibrium, the pressure in the cluster which now appears as a bubble can be written as

$$p_b - p_l = \frac{2\sigma}{\left[\frac{3kT_l}{4\pi\sigma}\ln\left(\frac{NkT_l}{Jh}\right)\right]^{1/2}} \tag{8.18}$$

Equation(8.18) has two unknowns p_b and T_l. However thermal equilibrium requires T_l to be the saturation temperature corresponding to pressure p_b. Thus employing the equation of state, both p_b and T_l can be determined.

Example 2: Calculate the superheat necessary for homogeneous nucleation in water at one atmosphere pressure. $N_A = 6.022 \times 10^{26}$ molecules/ kg-mol, $h = 0.626 \times 10^{-34} Js$, $k = 1.3806 \times 10^{-23} J/molK$.

Let us carry out the calculations by assuming J = 1. We devise a table (see Table 8.1) by choosing different values of T_l and calculating P_b from Eq. (8.18) and comparing it with that obtained from the steam tables.

Table 8.1

Temperature K	Equation (8.17) $Pb, N/m^2$	Steam Table $Pb, N/m^2$
520	27.3×10^6	3.78×10^6
560	12.77×10^6	7.13×10^6
572	8.46×10^6	8.46×10^6

Thus water can sustain a superheat of almost 200 K before homogeneous nucleation occurs.

8.2.2 Heterogeneous Nucleation

As mentioned earlier heterogeneous nucleation is a process in which bubbles nucleate at imperfections on the heater surface submerged in liquid. These imperfections are the scratches, pits and grooves commonly called cavities which form on the surface during its preparation. The early work of Bankoff [3] indicated that the superheats associated with heterogeneous nucleation were much lower than those for homogeneous nucleation. In the experiments, the test surface was submerged in a pool of liquid maintained at a constant temperature while the system pressure was gradually decreased until a bubble nucleated at the surface. In these experiments the liquid superheat corresponding to the system pressure at which the first bubble nucleated was noted.

The reason for lower inception superheat can be easily rationalized if we consider that the cavities generally trap air or other noncondensibles and have radii which are much larger than those given by Eq. (8.17). As we will see shortly, the amount of air that is trapped in the cavity depends on the surface tension, the contact angle and the shape of the cavity. With heating, the gas expands and pushes the interface outwards. During this process, the radius of curvature decreases (Fig. 8.4) and is smallest when the bubble nucleus just covers the mouth of the cavity ($\beta = 90°$). As the bubble grows further the radius of curvature will begin to increase. Since the liquid superheat required for thermal equilibrium increases with reduction in

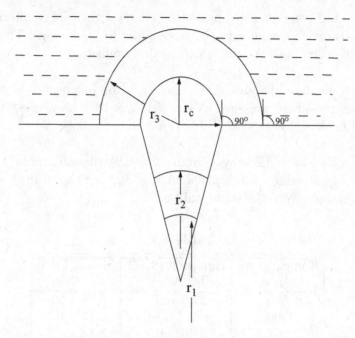

Figure 8.4 Bubble formation in a non-uniform temperature field ($\beta = 90°$)

the radius of the bubble, the highest superheat will be needed when the bubble radius is equal to the cavity radius. Thus the bubble inception superheat will be controlled by the size of the cavity. For commercial surfaces the size of the cavities is several orders of magnitudes larger than the critical cluster radius, hence lower is the observed superheat for heterogeneous nucleation. The precise value of the nucleation superheat strongly depends on the availability of unflooded cavities. As these cavities become fewer and fewer and their size decreases, the observed nucleation temperatures will approach the homogeneous nucleation temperature.

The trapped gas not only promotes nucleation but it also reduces the vapor pressure in the bubble. Using Dalton's law of partial pressures, the pressure in the nucleus can be written as

$$p_b = p_v + p_g \tag{8.19}$$

Since the vapor pressure difference across the interface is reduced by p_g, the liquid superheat for thermal equilibrium is also reduced accordingly, such that

$$T_l - T_{sat} = \frac{\left(\frac{2\sigma}{R} - p_g\right)}{\rho_v h_{fg}} T_{sat} \tag{8.20}$$

Next let us look at the conditions under which cavities will trap gas. Figure 8.5 shows the advance of liquid and gas fronts over wedge shaped cavities. According to Bankoff [3], if the contact angle, β, is greater than the cavity angle, ϕ, the advancing liquid front will fill the upper portion of the cavity while trapping gas underneath. Such a cavity will trigger early and should serve as a favored nucleation site. An advancing gas or vapor front however will not be able to displace the liquid if $\pi - \beta > \phi$. A liquid filled cavity will not be conducive to nucleation. Table 8.2 lists the trapping abilities of various wadge-shaped cavities.

Unable to displace gas   Unable to displace liquid

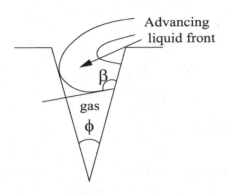

If $\beta > \phi$, the advancing liquid front will hit the farther wall of the cavity before it flows completely down the nearer wall

If $\pi - \beta > \phi$, the advancing gas front will strike the farther wall of the cavity before it has advanced completely down the nearer wall

Figure 8.5 Vapor or gas trapping behavior of cavities

Table 8.2
Trapping ability of cavities

Type of Cavity	Condition	Contact Angle	Trapping Ability
Steep Shallow	Poorly wetted	$\beta > \pi/2$	Traps gas or vapor only Traps gas or vapor and liquid
Steep Shallow	Well wetted	$\beta < \pi/2$	Traps liquid only Traps neither

The extent to which an advancing liquid front will penetrate into a cavity not only depends on the contact angle and size and shape of the cavity but also on the pressure in the liquid and the temperature. The higher the liquid pressure and lower the temperature, deeper will be the penetration of the liquid. The sketch on the left in Fig. 8.6 shows the shape of the liquid gas/vapor interface after the penetration is over. The interface is concave upward indicating that the pressure in the gas or vapor is lower than that in the liquid. After heating of the wall is initiated the interface turns over and becomes convex upward. This is shown on the right side of Fig. 8.6.

For prediction of heterogeneous nucleation, the criterion developed by Hsu [4] has been found to be quite successful. This criterion does a fairly good job in predicting the wall superheat necessary for nucleation. The criterion requires that

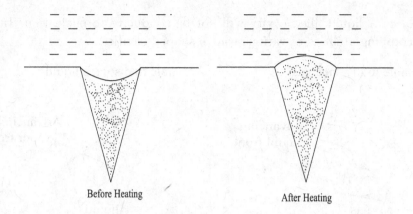

Figure 8.6   Liquid-gas interface shape before heating and after heating

the top of a bubble embryo be covered with warm liquid before it can grow. Since the bubble embryo must be at saturation temperature corresponding to the vapor pressure which is higher than the pool pressure by $2\sigma/r_b$, the liquid surrounding the bubble must be superheated. If the required superheat does not exist, the heat transfer into colder liquid will cause the bubble to collapse. Because the heat is transferred from the wall, the liquid temperature decreases with distance from the wall and Hsu's criterion is satisfied everywhere around the embryo if the liquid at the tip of the bubble has the required superheat.

Figure 8.7 shows how Hsu's criterion could be satisfied for a bubble embryo when the wall heat flux is gradually increased. In plotting the temperature a fictitious film thickness, δ, is defined which with a linear temperature profile gives the same heat flux at the wall as the actual thermal layer, δ_{th}. Also, it is assumed that δ is independent of the temperature difference across the layer. During natural convection this is not a correct assumption. However in natural convection, the dependence of film thickness on temperature difference is weak. The temperature profile traced as the solid line indicates the wall heat flux at which Hsu's criterion is just satisfied. If the cavity for which the criterion is satisfied is available on the surface (i.e. is not completely filled with liquid), this cavity will be the first one to nucleate. The lowest superheat at which nucleation is possible, can be determined from the expressions for the temperature distribution in the film and the liquid temperature dictated by the thermal equilibrium considerations. If the bulk liquid is saturated, the temperature profile in the liquid film is given by

$$T - T_{sat} = (T_w - T_{sat})(1 - y/\delta) \tag{8.21}$$

The vapor bubble thermal equilibrium condition equation (8.6) is written as

$$(T_l - T_{sat}) = \frac{2\sigma T_{sat}}{r_b \rho_v h_{fg}}$$

For a contact angle of $\pi/2$, the height of the bubble and the radius of the bubble embryo will be equal to the radius of the mouth of the cavity. For contact angles

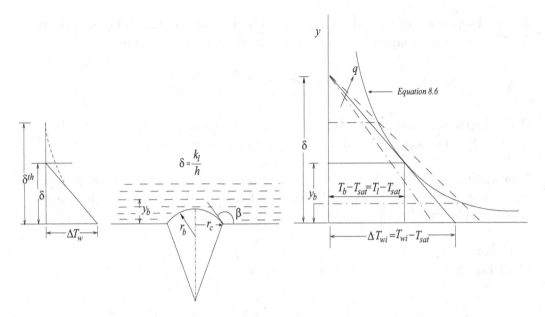

Figure 8.7 Inception of nucleation in a non-uniform temperature field

different than $\pi/2$, the radius and height of a spherical embryo are related to the cavity radius as

$$r_b = \frac{r_c}{C_1} \qquad \text{where} \quad C_1 = sin\beta \qquad (8.22)$$

$$y_b = \frac{r_c C_2}{C_1} \qquad \text{where} \quad C_2 = 1 + cos\beta \qquad (8.23)$$

Differentiating the liquid temperature in Eqs. (8.21) and (8.6) with respect to distance and equating the two derivatives we obtain

$$\frac{2C_2 \sigma T_{sat}}{y^2 \rho_v h_{fg}} = \frac{(T_w - T_{sat})}{\delta} \qquad (8.24)$$

In terms of the radius of the nucleating cavity, equation (8.24) yields

$$r_c = \left[\frac{2C_1^2 \sigma T_{sat} \delta}{C_2 \rho_v h_{fg}(T_w - T_{sat})} \right]^{1/2} \qquad (8.25)$$

If we replace the wall superheat with wall heat flux, the expression for r_c becomes

$$r_c = \left[\frac{2C_1^2 \sigma T_{sat} k_l}{C_2 \rho_v h_{fg} q} \right]^{1/2} \qquad (8.26)$$

Substituting for r_b into equation (8.6), after obtaining r_c from equation (8.26) the required liquid superheat at the tip of the bubble is

$$(T_l - T_{sat}) = \left[\frac{2C_2 \sigma T_{sat} q}{\rho_v h_{fg} k_l} \right]^{1/2} \qquad (8.27)$$

The wall superheat at nucleation is obtained from equation (8.21) after substituting for $(T_l - T_{sat})$ from equation (8.27) and y_b from equation (8.23) as

$$(T_w - T_{sat}) = \left[\frac{8C_2\sigma T_{sat}q}{\rho_v h_{fg}k_l}\right]^{1/2} \tag{8.28}$$

Wall superheat at nucleation is twice the liquid superheat at the tip of the bubble embryo. For a contact angle of $\pi/2$, both C_1 and C_2 are unity.

Example 3: Calculate the cavity size and the minimum wall superheat for nucleation from a surface submerged in a pool of saturated water at one atmosphere pressure, The wall heat flux is $10^5 W/m^2$. Assume the contact angle to be $\pi/2$.

Solution
From Eq. (8.26), the cavity radius is calculated as

$$r_c = \left[\frac{2(58.7)\ 10^{-3}\ (373.15)\ (0.681)}{(0.597)\ (2.257)\ 10^6\ (10^5)}\right]^{1/2} = 14.88 \times 10^{-6}m$$

The wall superheat for inception of boiling is found from Eq. (8.28) as

$$(T_w - T_{sat}) = \left[\frac{8(58.7)\ 10^{-3}\ (373.15)(10^5)}{(0.597)\ (2.257)\ 10^6\ (0.681)}\right]^{1/2} = 4.36°C$$

The cavity size obtained here is fairly large. It is possible that such a cavity will be filled with liquid and will not serve as a nucleation site. It is interesting to note that the size of the nucleating cavity is about 3 orders of magnitude larger than the equilibrium cluster size for homogeneous nucleation.

Next, let us determine the range of cavities that can nucleate if the wall superheat exceeds the minimum wall superheat for nucleation. Eliminating the liquid temperature from Eqs. (8.6) and (8.21) and substituting for bubble radius and bubble height from Eqs. (8.22) and (8.23), we obtain

$$(T_w - T_{sat})\left(1 - \frac{C_2 r_c}{C_1 \delta}\right) = \frac{2C_1\sigma T_{sat}}{\rho_v h_{fg}r_c} \tag{8.29}$$

or

$$r_c^2 - \frac{C_1}{C_2}\delta r_c + \frac{2C_1^2\sigma T_{sat}\delta}{C_2\rho_v h_{fg}(T_w - T_{sat})} = 0 \tag{8.30}$$

This is a quadratic equation in r_c and the two roots of the equation are

$$r_{c_{1,2}} = \frac{C_1\delta}{2C_2}\left[1 \pm \sqrt{1 - \frac{8C_2\sigma T_{sat}}{\rho_v h_{fg}(T_w - T_{sat})\delta}}\right] \tag{8.31}$$

In Fig. 8.7, the intercepts of the dotted line representing temperature distribution in the film with the liquid superheat predicted from Eq. (8.6) represent the cavity radii r_{c1} and r_{c2}. For a given wall heat flux and corresponding superheat any cavity

Figure 8.8 Thermal layer development after bubble departure

lying in the size range r_{c1} and r_{c2} can nucleate. If the heated surface is subjected to forced flow conditions, the thermal layer will thin. This in turn will reduce the film thickness δ. As can be seen from Eq. (8.31), the radius of the nucleating cavities will shrink as δ is reduced. We know from the earlier discussions that small cavities require higher superheat to nucleate. Thus the flow over the surface will tend to increase the inception superheat or suppress the nucleation process. Another point that should be made here is that in developing flows, the thermal layer thickness is a function of distance from inlet or the leading edge. For example, for laminar flow over a flat plate, the thermal layer thickness increases as square root of distance from the leading edge. Therefore, the farther the location, the larger will be the size of the nucleating cavities and smaller the required wall superheat. Theoretically, the effect of flow velocity and distance from the leading edge or inlet could be included in the film thickness (Fig. 8.7). However, as mentioned earlier, the predictions will be sensitive to the size and shape of the available cavities and the contact angle.

8.2.3 Waiting Period

After inception a bubble will grow until it encounters cold liquid such that the condensation rate at the top of the bubble equals the evaporation rate or the bubble departs as determined by forces acting on it. During the growth period, the bubble will push the surrounding liquid outwards causing some mixing of the cold liquid with the warm liquid in the thermal layer. After the bubble departs, colder-liquid will tend to fill the space vacated by the departing bubble. Figure 8.8 shows this process schematically. A new bubble at this location will not form until the superheated liquid layer is re-established and the inception criterion is satisfied. The time taken by the thermal layer to redevelop prior to the inception is termed as the waiting period. Hsu and Graham [5] proposed a model to calculate this period. Their model assumed that an ambient liquid layer of thickness δ and extending to infinity in the plane of the heater comes in contact with the wall and is transiently heated. During the heating period the temperature of the wall and the thickness of the thermal layer remain constant.

The transient heat diffusion equation for an infinite slab of thickness δ with appropriate initial and boundary conditions can be written as

$$\frac{\partial^2 T}{\partial y^2} = \frac{1}{\alpha_l} \frac{\partial T}{\partial t} \tag{8.32}$$

The initial and boundary conditions are:

Initial Condition:
$$T = T_{sat} \quad \text{at} \quad t = 0 \quad \text{for all } y \tag{8.33}$$

Boundary Conditions:
$$T = T_w \quad \text{at} \quad y = 0 \quad \text{for all } t \geq 0 \tag{8.34}$$

and
$$T = T_{sat} \quad \text{at} \quad y = \delta \quad \text{for all } t \geq 0 \tag{8.35}$$

The solution of the problem can easily be written as

$$\frac{T - T_{sat}}{T_w - T_{sat}} = \left(1 - \frac{y}{\delta}\right) + \frac{2}{\pi} \sum_{n=1}^{\infty} \frac{(-1)^n}{n} \sin\left[n\pi\left(1 - \frac{y}{\delta}\right)\right] e^{-n^2 \pi^2 \tau} \tag{8.36}$$

where
$$\tau = \frac{\alpha_l t}{\delta^2} \tag{8.37}$$

The temperature profiles from equation (8.36) for various times, τ, are plotted in Fig. 8.9. The required liquid superheat determined from Eq. (8.6) is also plotted. The intersection of this curve with the temperature profiles gives the waiting time for a particular size cavity. Since with increase in wall superheat the size of the nucleating cavity decreases, the effect of increased wall superheat will be to reduce the waiting time for a particular size cavity. However, as can be seen from Fig. 8.9, for a given wall superheat certain very small and very large cavities may never nucleate.

An analytical expression for the waiting time can be obtained if the cold liquid layer contacting the heated wall is assumed to be semi-infinite. This is exactly the approach which was taken by Han and Griffith [6]. During transient diffusion of heat in a semi-infinite slab, the film thickness can be written as

$$\delta = \sqrt{\pi \alpha_l t} \tag{8.38}$$

Elimination of liquid temperature between Eqs. (8.21) and (8.6) and upon substituting for bubble height and bubble radius in terms of cavity size we obtain,

$$\frac{C_2 r_c}{C_1 \delta} = \left[1 - \frac{2C_1 \sigma T_{sat}}{(T_w - T_{sat})\rho_v h_{fg} r_c}\right] \tag{8.39}$$

or

$$\delta^2 = \left[\frac{C_2 r_c}{C_1\left[1 - \frac{2C_1 \sigma T_{sat}}{(T_w - T_{sat})\rho_v h_{fg} r_c}\right]}\right]^2 \tag{8.40}$$

Substituting for δ^2 from Eq. (8.38), the waiting time is obtained from Eq. (8.40) as

$$t_w = \left[\frac{C_2 r_c}{\sqrt{\pi \alpha_l} C_1\left[1 - \frac{2C_1 \sigma T_{sat}}{(T_w - T_{sat})\rho_v h_{fg} r_c}\right]}\right]^2 \tag{8.41}$$

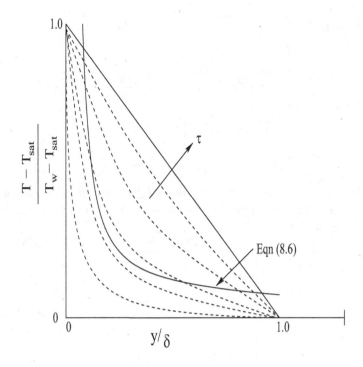

Figure 8.9 Waiting period for activation of various sized cavities

It can be seen from Eq. (8.41) that the waiting time will first decrease and then increase with cavity size but it will continue to decrease as the wall superheat is increased.

The waiting times obtained from Eq. (8.41) or from Fig. 8.9 are based on the assumption that during heating of the liquid, the heater wall temperature remains constant. This assumption corresponds to a wall heat flux that varies with time. Generally in laboratory experiments and in applications it is the heat flux that is held constant. This condition in turn implies that during transient heating, the wall temperature will vary. Also, it is possible that during the bubble growth period the heat flux associated with evaporation of the microlayer, as we will discuss in more detail later, can exceed the wall heat flux and thereby reduce the wall temperature. Thus during transient heating of the liquid adjacent to the wall, the wall temperature will vary. These factors leading to variation of wall temperature need to be included in the model for a more realistic prediction of waiting time. However Fig. 8.9 or Eq. (8.41) give trends which are consistent with the experimental data.

Example 4: Calculate and plot the waiting time as a function of cavity size for a surface submerged in a pool of saturated water at one atmosphere pressure. Assume the wall superheat to be fixed at 10°C and the contact angle to be $\frac{\pi}{2}$.

Solution: For $\beta = \frac{\pi}{2}$ both the constant C_1 and C_2 are unity. Evaluating the various physical properties at the mean temperature of 105°C, the expression for waiting time from Eq. (8.41) is

$$
\begin{aligned}
t_w &= \left[\frac{r_c}{\sqrt{\pi(1.68)10^{-7}} \left[1 - \frac{2(57.9)10^{-3}(373)}{(10)(0.695)(2.245)10^6} \right]} \right]^2 \\
&= \left[\frac{1.38 \times 10^{+3}(r_c)}{1 - \frac{2.77 \times 10^{-6}}{r_c}} \right]^2 \\
&= 364 \times 10^{-6}\text{s} \quad \text{for } r_c = 10 \times 10^{-6}
\end{aligned}
$$

It can be seen from the above equation that cavities equal to or smaller than 2.77 microns will never nucleate. Figure 8.10 shows the variation of the waiting time as a function of radius of a cavity.

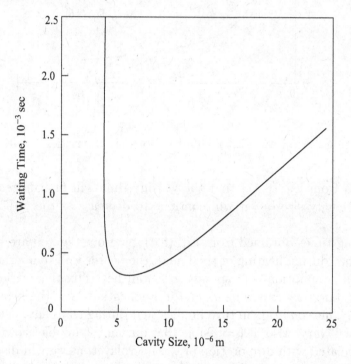

Figure 8.10 Dependence of waiting time on cavity size

The lifespan of a bubble can be divided into two periods: the waiting period during which conditions conductive to bubble formation are created near the heating surface, and the growth period during which the bubble actually grows to a certain size before departure. The growth of a bubble at the heated wall is influenced by several factors such as the thermal and flow field around the bubble and the volume and thickness of the microlayer underneath the bubble. The bubble shape is influenced by the contact angle, the direction and magnitude of gravitational acceleration and the flow around the bubble. The description of growth of such a bubble is complex. We can hope to address analytically some features of this process only after developing an understanding of the growth of a totally spherical bubble in an infinite medium under gravity free conditions.

8.2.4 Bubble Growth without Heat or Mass Transfer

The growth of a high pressure gas bubble in an infinite liquid medium was first described by Lord Rayleigh [7]. For an inviscid incompressible liquid, the potential function can be written as

$$\Phi = \frac{R^2 \dot{R}}{r} \tag{8.42}$$

where R is the instantaneous radius of the bubble, $\dot{R}$ is first time derivative of the radius and r is any radius measured from the center of the bubble. From Eq. (8.42), the radial velocity is obtained as:

$$u = -\frac{\partial \Phi}{\partial r} = \frac{R^2 \dot{R}}{r^2} \tag{8.43}$$

For a uniform pressure in the bubble, the momentum equation describing the radial velocity of the inviscid liquid can be written as

$$\frac{\partial u}{\partial t} + u \frac{\partial u}{\partial r} = -\frac{1}{\rho_l} \frac{\partial p}{\partial r} \tag{8.44}$$

After substitution for u from Eq. (8.43), Eq. (8.44) becomes

$$\frac{R^2 \ddot{R}}{r^2} + \frac{2R\dot{R}^2}{r^2} + \frac{1}{2} \frac{\partial}{\partial r}\left(\frac{R^4 \dot{R}^2}{r^4}\right) = -\frac{1}{\rho_l} \frac{\partial p}{\partial r} \tag{8.45}$$

Integration of Eq. (8.45) from far away from the bubble ($r = \infty$) to the outer surface of the bubble ($r = R$), yields

$$-\left[\frac{R^2 \ddot{R}}{r} + \frac{2R\dot{R}^2}{r}\right]_\infty^R + \left[\frac{R^4 \dot{R}^2}{2r^4}\right]_\infty^R = -\frac{1}{\rho_l}[p]_\infty^R \tag{8.46}$$

or

$$R\ddot{R} + \frac{3}{2}\dot{R}^2 = \frac{1}{\rho_l}[p_{bo} - p_l] \tag{8.47}$$

where

$$p_{bo} = p_b - \frac{2\sigma}{R} \tag{8.48}$$

and p_l is the pressure in the liquid far away from the bubble. Equation (8.47) is known as Rayleigh's Equation.

8.2.5 Dynamics of Bubble Growth on Wall

The size to which a bubble attached to the wall will grow before detaching depends on balance between several forces acting on the bubble. In this section we identify these forces.

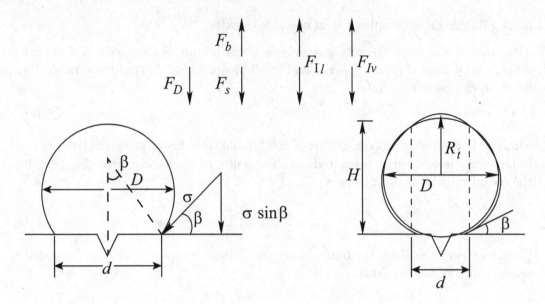

Figure 8.11   Forces acting on a bubble attached to a horizontal wall

Forces Acting on Bubbles on Walls

In 1935, Fritz [8] proposed that main forces acting on a bubble were the surface tension and buoyancy. The surface tension, tends to hold the bubble to the wall while buoyancy tries to lift it away. Referring to the spherical bubble shown in the left hand sketch in Fig. 8.11, the surface tension force is

$$F_s = \pi D \sigma \sin^2 \beta \tag{8.49}$$

The buoyancy force is

$$F_B = \frac{\pi D^3}{6}(\rho_l - \rho_v)g \left[\frac{1 + \cos\beta}{2}\right]^2 [2\cos\beta] \tag{8.50}$$

Equating the surface tension and buoyancy force from equations (8.49) and (8.50) we get

$$D_d = f(\beta)\sqrt{\frac{\sigma}{g(\rho_l - \rho_v)}} \tag{8.51}$$

where

$$f(\beta) = \left[\frac{24\sin^2\beta}{(2 + \cos\beta)\,(2 - \cos\beta)}\right]^{1/2} \tag{8.52}$$

Fritz [8] correlated his experimentally observed bubble diameters at departure as

$$D = 0.0208\beta\sqrt{\frac{\sigma}{g(\rho_l - \rho_v)}} \tag{8.53}$$

where β now is the contact angle measured in degrees. Figure 8.12 shows the dependence on β of function $f(\beta)$ obtained from Eq. (8.52) and the equivalent function obtained by Fritz and given in Eq. (8.53). The difference between the two expressions is probably reflective of the non-sphericity of the bubbles and non-inclusion of other forces which act on the bubble. However, for contact angles smaller than $\pi/2$, the difference between the two expressions is rather small. As pointed out earlier, for wettable surfaces the contact angle is smaller than $\pi/2$.

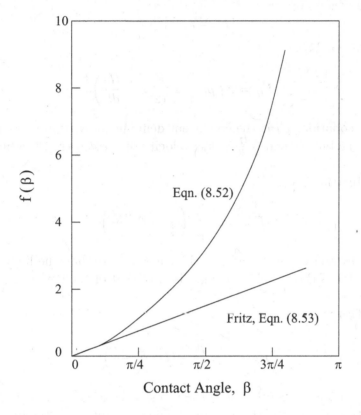

Figure 8.12 Computation of prediction of function $f(\beta)$ from Fritz's expression and Eq. (8.52)

Bubbles are generally not totally spherical and as many investigations such as those of Siegel and Keshock [9], and Cochran and Aydelott [10] have shown, that the bubbles not only experience surface tension and buoyancy forces during their growth but several other forces come into play. These additional forces are inertia and drag. Since these forces act both on volume and surface, it is difficult to determine a common equivalent diameter of a non-spherical bubble. However an equivalent diameter based on bubble volume can be defined as

$$D = \left(\frac{6}{\pi}V\right)^{1/3}$$

Various forces that act on a bubble are shown on the right side of Fig. 8.11. Depending on the functional dependence of bubble growth rate on time, the direction of liquid and vapor inertia force can change direction during bubble growth.

Expressions for various forces are:

(i) Buoyancy Force:

$$F_b = \frac{\pi}{6} D^3 (\rho_l - \rho_v) g \tag{8.54}$$

(ii) Surface Tension Force:

$$F_s = \pi d\sigma \sin \beta \tag{8.55}$$

(iii) Liquid Inertia Force:

$$F_{Il} = K_l \, \rho_l \, \frac{d}{dt} \left(\frac{\pi}{12} \, D^3 \, \frac{dD}{dt} \right) \tag{8.56}$$

In the above equation, K_l is the coefficient defining the virtual mass of the liquid surrounding the bubble and $\frac{1}{2}\frac{dD}{dt}$ is the velocity of the center of the bubble.

(iv) Bubble Inertia Force:

$$F_{Iv} = \rho_v \, \frac{d}{dt} \left(\frac{\pi}{12} \, D^3 \, \frac{dD}{dt} \right) \tag{8.57}$$

At low pressure the vapor density, ρ_v, is much smaller than the liquid density ρ_l. Thus the inertia of the bubble in comparison to that of the liquid can be ignored.

(v) Viscous Drag Force:

$$F_D = C_D \, \frac{\rho_l}{2} \, \frac{\pi}{4} \, D^2 \, \left(\frac{dD}{dt} \right)^2 \tag{8.58}$$

In writing Eq. (8.58), the velocity of the top surface of the bubble is used instead of the velocity of the center of the bubble. The constant C_D is the drag coefficient.

The buoyancy force written as Eq. (8.54) is for a body fully submerged in a liquid. Since the bubble is attached to the wall, the buoyancy force as has been suggested by Cochran and Aydelott [10] should be corrected such that

$$F_B = (\rho_l - \rho_v) \, g \, \left(\frac{\pi}{6} \, D^3 - \frac{\pi}{4} \, d^2 H \right) + \frac{\sigma}{R_t} \frac{\pi}{4} d^2 \tag{8.59}$$

In the above equation the second term accounts for excess pressure in the bubble due to surface tension, R_t is the radius of curvature at the top and the expression is obtained by assuming that the bubble top has no curvature in a plane normal to the plane of the figure.

8.2.6 Bubble Growth with Heat and Mass Transfer

The analysis for the growth of a bubble attached to a heated wall can be extremely involved because of the continuous change in the shape of the bubble, the convective motion around the bubble, microlayer evaporation underneath the bubble and the effect of the neighboring bubbles. In literature several attempts to model the growth of bubbles attached to a wall have been made. All of these efforts have tended to be modifications of the model for growth of a bubble in a uniform temperature field. Correction factors are generally incorporated to account for the shape of a bubble being different than that of a sphere. One such attempt is by Mikic, Rohsenow and Griffith [11] and will be followed here. Complete numerical simulations of bubble growth on a heated surface including microlayer have been carried out in the recent past by Son, Dhir and Ramanujapu [12].

We begin by first analyzing the growth of a single bubble in a pool of super-heated liquid.

Simple Analysis of Bubble Growth in a Uniform Temperature Field

Assume that bubble growth rate is slow enough and bubble radius is small such that the momentum equation Eq. (8.47) can be written as

$$p_{v_0} - p_l = \frac{3}{2}\rho_l \dot{R}^2 \tag{8.60}$$

If we assume that the contribution of surface tension to the pressure in the bubble is small, the above equation can be modified as

$$p_v - p_l = \frac{3}{2}\rho_l \dot{R}^2 \tag{8.61}$$

where p_v is the pressure of vapor in the bubble. By introducing a constant to account for the difference in shape between a bubble and a sphere, we can rewrite Eq. (8.61) as

$$\dot{R}^2 = \frac{b(p_v - p_l)}{\rho_l} \tag{8.62}$$

where $b = 2/3$ for a spherical bubble. A value of $1/3$ for b has been suggested by Mikic, Rohsenow and Griffith [11] for a bubble attached to a wall. Using Clausius–Clapeyron relationship, we can rewrite Eq. (8.62) as

$$\dot{R}^2 = \frac{b(T_v - T_{sat})h_{fg}\rho_v}{\rho_l T_{sat}} \tag{8.63}$$

By defining the temperature difference ΔT as

$$\Delta T = T_\infty - T_{sat} \tag{8.64}$$

where T_∞ is the pool temperature, Eq. (8.63) is obtained as

$$\dot{R}^2 = \frac{A^2(T_v - T_{sat})}{\Delta T} \tag{8.65}$$

where

$$A^2 = \left(\frac{b\rho_v h_{fg}\Delta T}{\rho_l T_{sat}}\right) \tag{8.66}$$

and A^2 has dimensions of square of velocity.

In a thermally controlled situation, the growth rate of the bubble is limited by the rate at which heat can diffuse to the interface from the superheated liquid. The expression for the growth rate of a bubble using a plane interface analysis is obtained as

$$\frac{d}{dt}\left(4/3\,\pi\,R^3\,\rho_v\,h_{fg}\right) = \frac{k_l}{\sqrt{\pi\alpha_l t}}\,(T_\infty - T_v)\,4\pi R^2 \tag{8.67}$$

$$\frac{dR}{dt} = \frac{k_l}{\sqrt{\pi\alpha_l t}}\,\frac{(T_\infty - T_v)}{\rho_v h_{fg}} \tag{8.68}$$

$$\frac{dR}{dt} = \left(\frac{\alpha_l}{\pi t}\right)^{1/2}\frac{\rho_l c_{pl}(T_\infty - T_v)}{\rho_v h_{fg}} \tag{8.69}$$

Nearly exact solution of Plesset and Zwick [13] suggests that to account for sphericity of the interface the right-hand side of Eq. (8.69) should be multiplied by $\sqrt{3}$. After this correction is applied, Eq. (8.69) can be written as

$$\frac{dR}{dt} = \frac{1}{2}\frac{B}{\sqrt{t}}\frac{(T_\infty - T_v)}{\Delta T} \tag{8.70}$$

where

$$B = \left(\frac{12\alpha_l}{\pi}\right)^{1/2}\left(\frac{c_{pl}\,\Delta T\,\rho_l}{h_{fg}\,\rho_v}\right) \tag{8.71}$$

The second term on the right-hand side of Eq. (8.71) is called Jakob number

$$Ja \equiv \left(\frac{c_{pl}\,\Delta T\,\rho_l}{h_{fg}\,\rho_v}\right) \tag{8.72}$$

and parameter B has dimensions of $m/(s)^{1/2}$. Equation (8.70) can be written as

$$\frac{dR}{dt} = \frac{B}{2\sqrt{t}}\left(1 - \frac{(T_v - T_{sat})}{\Delta T}\right) \tag{8.73}$$

Eliminating $\frac{T_v - T_{sat}}{\Delta T}$ between Eqs. (8.65) and (8.73) yields

$$\frac{dR}{dt} = \frac{B}{2\sqrt{t}}\left(1 - \frac{\dot{R}^2}{A^2}\right) \tag{8.74}$$

or

$$\frac{1}{A^2}\left(\dot{R}^2\right) + \frac{2\sqrt{t}}{B}\dot{R} - 1 = 0 \tag{8.75}$$

With dimensionless radius and dimensionless time defined as

$$R^* = \frac{AR}{B^2} \tag{8.76}$$

and

$$t^* = \frac{A^2 t}{B^2} \tag{8.77}$$

Eq. (8.75) becomes

$$\left(\frac{dR^*}{dt^*}\right)^2 + 2\sqrt{t^*}\frac{dR^*}{dt^*} - 1 = 0 \tag{8.78}$$

or

$$\left(\frac{dR^*}{dt^*}\right) = \frac{-2\sqrt{t^*} + \sqrt{4t^* + 4}}{2} = (t^* + 1)^{1/2} - (t^*)^{1/2} \tag{8.79}$$

If we use the initial condition that $R^* = 0$ at $t^* = 0$, the integration of the above equation yields

$$R^* = \frac{2}{3}\left[(t^* + 1)^{3/2} - (t^*)^{3/2} - 1\right] \tag{8.80}$$

If $t^* << 1$, or when growth is inertia controlled the above equation reduces to the form

$$R^* = t^* \tag{8.81}$$

On the other hand, for $t^* >> 1$, Eq. (8.80) yields

$$R^* = t^{*^{1/2}} \tag{8.82}$$

Thus for early times, the bubble will grow linearly with time and for longer times, i.e. in the limit of thermally controlled growth the radius of the bubble will increase as square root of time.

Simple Analysis of Bubble Growth on a Wall

After bubble departure from a given site, one must wait for a certain period of time before a new bubble is formed at the same site. The time that elapses between departure of a bubble and inception of a new bubble is called the waiting period. During the waiting period, the liquid next to the wall is heated and a bubble embryo develops into a bubble when inception criterion (e.g., Hsu's criterion) is satisfied. Thereafter as the bubble grows, the superheated liquid around the bubble is also pushed outward along with the bubble. Now the growth rate of the bubble neglecting contribution of microlayer is dictated by the rate at which thermal energy can be transferred from the superheated liquid to the interface. Mikic, Rohsenow and Griffith [11] also obtained a solution for growth of such a bubble. In carrying out the analysis, they assumed that

1. Liquid layer adjacent to the wall is semi-infinite.

2. During bubble growth no relative motion exists between the bubble and the surrounding liquid.

3. Liquid-vapor interface is plane.

4. Interfacial heat flux can be obtained by assuming that vapor temperature remains constant. In reality, vapor temperature changes during early period of growth.

The transient diffusion of heat into the liquid, initially at T_l, during the waiting period and out of the superheated liquid during the growth of the bubble could be broken up into two one-dimensional transient conduction problems as

$$\frac{\partial T}{\partial t} = \alpha_l \frac{\partial^2 T}{\partial y^2} \tag{8.83}$$

for $-t_w \leq t \leq 0$ for $0 \leq t$

$T = T_l$ for all y and $t \leq -t_w$ $T = T_l$ for all y at $t = 0$

$T = T_w$ at y = 0 for $0 \geq t \geq -t_w$ $T = T_l - (T_w - T_v)$ at y = 0 for $t \geq 0$

$T = T_l$ as $y \to \infty$ for $t \geq -t_w$ $T = T_l$ as $y \to \infty$ for $t \geq 0$

Superimposing the solutions for the two cases, we obtain

$$T - T_l = (T_w - T_l) \ \text{ertc} \left[\frac{y}{2\sqrt{\alpha_l(t + t_w)}} \right] - (T_w - T_v) \ \text{ertc} \left[\frac{y}{2\sqrt{\alpha_l t}} \right] \tag{8.84}$$

$$q_i = -\left[-k_l \frac{\partial T}{\partial y}\Big|_{y=0} \right] = -k_l(T_w - T_l)\frac{2}{\sqrt{\pi}}\frac{1}{2}\frac{1}{\sqrt{\alpha_l(t + t_w)}} + (T_w - T_v)\frac{2}{\sqrt{\pi}}\frac{k_l}{2\sqrt{\alpha_l t}} \tag{8.85}$$

$$= k_l \left[\frac{(T_w - T_v)}{\sqrt{\pi \alpha_l t}} - \frac{(T_w - T_l)}{\sqrt{\pi \alpha_l (t + t_w)}} \right] \tag{8.86}$$

To account for the geometry of the bubble, we multiply q_i is multiplied by $\sqrt{3}$. The energy and mass balance at the bubble liquid interface yields

$$\frac{d}{dt}(\rho_v h_{fg} (4/3)\pi R^3) = \frac{\sqrt{3}k_l}{\sqrt{\pi \alpha_l t}} \left[(T_w - T_v) - (T_w - T_l)\sqrt{\frac{t}{t + t_w}} \right] 4\pi R^2$$

$$\frac{dR}{dt} = \frac{\sqrt{3}\,k_l}{\rho_v\,h_{fg}\,\sqrt{\pi\,\alpha_l\,t}} \left[(T_w - T_v) - (T_w - T_l)\sqrt{\frac{t}{t + t_w}} \right] \tag{8.87}$$

$$= \frac{\rho_l c_{pl}}{\rho_v h_{fg}} \Delta T \sqrt{\frac{\alpha_l}{\pi t}} \sqrt{3} \left[\frac{T_w - T_v}{\Delta T} - \frac{T_w - T_l}{\Delta T}\sqrt{\frac{t}{t + t_w}} \right]$$

$$= \left(\frac{12\alpha_l}{\pi t} \right)^{1/2} \frac{Ja}{2} \left[\frac{T_w - T_v}{\Delta T} - \theta\sqrt{\frac{t}{t + t_w}} \right] \tag{8.88}$$

where $\Delta T = T_w - T_{sat}$ and $\theta = \frac{T_w - T_l}{\Delta T}$.

In terms of the parameter B defined earlier equation (8.88) can be written as

$$\frac{dR}{dt} = \frac{B}{2\sqrt{t}} \left[\frac{T_w - T_v}{\Delta T} - \theta\sqrt{\frac{t}{t + t_w}} \right] \tag{8.89}$$

or

$$1 - \frac{2\sqrt{t}}{B}\frac{dR}{dt} = 1 - \frac{T_w - T_v}{\Delta T} + \theta\sqrt{\frac{t}{t + t_w}}$$

$$= \frac{T_w - T_{sat} - T_w + T_v}{T_w - T_{sat}} + \theta\sqrt{\frac{t}{t + t_w}} \tag{8.90}$$

or

$$1 - \frac{2\sqrt{t}}{B}\frac{dR}{dt} = \frac{T_v - T_{sat}}{\Delta T} + \theta\sqrt{\frac{t}{t + t_w}} \tag{8.91}$$

Rayleigh equation in an approximate form is written as

$$\left(\frac{dR}{dt}\right)^2 = b\left(\frac{p_v - p_l}{\rho_l}\right)$$

or after using Clausius Clapeyron relation as

$$= \frac{b(T_v - T_{sat})h_{fg}\rho_v}{T_{sat}\rho_l \Delta T}\Delta T \tag{8.92}$$

or in terms of parameter A defined earlier, Eq. (8.92) can be written as

$$\left(\frac{dR}{dt}\right)^2 = A^2\frac{T_v - T_{sat}}{\Delta T} \tag{8.93}$$

Substituting for $\frac{T_v - T_{sat}}{\Delta T}$ from Eq. (8.93) into Eq. (8.91) we get

$$1 - \frac{2\sqrt{t}}{B}\frac{dR}{dt} = \frac{1}{A^2}\left(\frac{dR}{dt}\right)^2 + \theta\sqrt{\frac{t}{t + t_w}} \tag{8.94}$$

Or in dimensionless form, the above equation becomes

$$\left(\frac{dR^*}{dt^*}\right)^2 + 2\sqrt{t^*}\frac{dR^*}{dt^*} - \left[1 - \theta\left(\frac{t^*}{t^* + t_w^*}\right)^{1/2}\right] = 0 \tag{8.95}$$

Solution of (8.95) yields

$$\frac{dR^*}{dt^*} = \left[t^* + 1 - \theta\left(\frac{t^*}{t^* + t_w^*}\right)^{1/2}\right]^{1/2} - (t^*)^{1/2} \tag{8.96}$$

where we have used the + sign in the solution of the quadratic Eq. (8.95). From Eq. (8.96) we see that as $t_w^* \to \infty$, the growth rate approaches that for a bubble in a superheated pool. Effect of liquid subcooling ($\theta > 1$) will be to slow down the bubble growth. Knowing the bubble diameter at departure Eq. (8.96) can be integrated to determine the growth period.

8.3   BUBBLE DEPARTURE DIAMETER AND RELEASE FREQUENCY

In principle, one could use the expressions developed in the previous sections with respect to bubble growth rate and forces that act on the bubble to calculate its diameter at departure. However, in a real boiling system such an approach meets with little success. Some of the reasons for this are:

1. Cavities of different sizes and shapes exist on the surface. Waiting times between cavities may differ significantly.

2. Because of the differences in waiting times, thickness of thermal layers may differ from cavity to cavity. This is true, however, only as long as sufficient distance exists between cavities.

3. Evaporation from liquid film (micro/macro layers) underneath a bubble can contribute significantly to growth. This process has not been included in the bubble growth model described earlier.

4. Because of the development of flow field and interaction between neighboring sites, thermal layers around bubbles may be distorted.

5. Merger of bubbles at neighboring sites can alter their shapes and sizes and thereby significantly influence their departure diameters and frequencies.

Several correlations have been suggested in the literature for bubble diameter at departure and bubble release frequency. Based on bubble departure diameters observed with several liquids, Rohsenow [14] has given a correlation of the type:

For water

$$
D_d \sqrt{\frac{g(\rho_l - \rho_v)}{\sigma}} = 1.5 \times 10^{-4} Ja^{*5/4} \tag{8.97}
$$

For other liquids

$$
D_d \sqrt{\frac{g(\rho_l - \rho_v)}{\sigma}} = 4.65 \times 10^{-4} Ja^{*5/4} \tag{8.98}
$$

where

$$
Ja^* = \frac{\rho_l c_{pl} T_{sat}}{\rho_v h_{fg}} \tag{8.99}
$$

Unlike Fritz's correlation, the correlation of Rohsenow includes the Jakob number based on the saturation temperature of the liquid but does not account for the effect of the contact angle.

Since the bubble release frequency is influenced by the diameter at departure (i.e., larger bubbles taking longer to grow), generally correlations combining departure diameter and frequency have been reported in the literature. General form of such correlations is

$$
fD_d^n \equiv C \tag{8.100}
$$

where C is a constant.

Values of n between 1/2 and 3 have been suggested. Jakob [15] suggested a value of unity, while Ivey [16] suggested that $n = 2$ when bubble growth is inertia controlled and 1/2 when it is thermally controlled. For inertia controlled bubble growth Cole [17] has suggested the constant C to be

$$
C = 4/3 \frac{g(\rho_l - \rho_v)}{\rho_l} \tag{8.101}
$$

Zuber [18] on the other hand assumed that bubble growth rate was proportional to terminal bubble rise velocity and that waiting time was much smaller than the growth time. Under these assumptions he found that exponent n should have a value of unity and the constant C should be

$$C = 0.59 \left[\frac{\sigma g (\rho_l - \rho_v)}{\rho_v^2} \right]^{1/4} \tag{8.102}$$

In the above equation, the multiplier 0.59 was obtained by matching the predictions with the data.

Malenkov [19] has suggested a correlation of the type

$$fD_d = \frac{V_d}{\pi \left[1 - \frac{1}{1 + (V_d \, \rho_v \, h_{fg})/q} \right]} \tag{8.103}$$

where

$$V_d = \left[\frac{g D_d (\rho_l - \rho_v)}{2 (\rho_l + \rho_v)} + \frac{2\sigma}{D_d (\rho_l + \rho_v)} \right]^{1/2} \tag{8.104}$$

For bubbles with large diameter at departure

$$V_d \sim D_d^{1/2} \tag{8.105}$$

whereas for small bubbles

$$V_d \sim D_d^{-1/2} \tag{8.106}$$

Thus for bubbles with thermally controlled growth rates, (large bubbles), Malenkov's expression reduces to a form similar to that of Ivcy.

8.4   BUBBLE MERGER AND TRANSITION FROM PARTIAL TO FULLY DEVELOPED NUCLEATE BOILING

With increase in wall superheat, the number density of nucleation sites and bubble release frequency increase. If at a given site, bubble growth rate exceeds the rate at which a preceding bubble moves away from the surface, bubbles start to merge in the vertical direction. Now vapor appears to leave the heater in the form of vapor columns attached to the solid surface. With increase in the nucleation site density bubbles also start to merge in the lateral direction. This in turn leads to formation of large mushroom-type bubbles with several vapor stems. The formation of vapor columns and mushroom-type bubbles on the heater surface represents a transition from partial to fully developed nucleate boiling. Photographs in Fig. 8.13 obtained from Dhir [20] show merger of vapor bubbles in the vertical and lateral directions in the vicinity of the heater.

Moissis and Berenson [21] identified from experiments on a flat plate that vapor columns start to appear on the heater surface when

$$\frac{\dot{V}_v}{A_v V_t} = 9 \tag{8.107}$$

where $\dot{V}_v$ is the vapor volume flow rate over a vapor producing area A_v and V_t is the terminal velocity of a bubble rising in a pool of liquid.

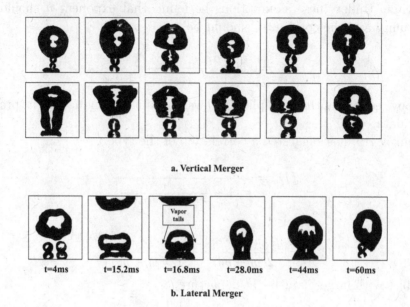

a. Vertical Merger

| t=4ms | t=15.2ms | t=16.8ms | t=28.0ms | t=44ms | t=60ms |

b. Lateral Merger

Figure 8.13 Photographs of bubble merger in the vertical and lateral directions on a heated surface

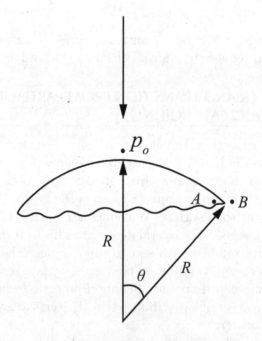

Figure 8.14 A large bubble rising in a pool and liquid

The terminal velocity of a large spherical cap shaped bubble as shown in Fig. 8.14 can be obtained from hydrodynamic considerations. Vapor bubble merger in the vertical direction as noted by Moissis and Berenson [21] was far away from the surface whereas photographs in Fig. 8.13 depict bubble merger in the vertical direction near the surface. Distant fluid will appear to move downwards at velocity V_t. For an inviscid fluid flow over a sphere, local velocity of liquid can be written as

$$V = \frac{3}{2} V_t \; \sin\theta \tag{8.108}$$

where θ is the angular position measured from the forward stagnation point. Since surface tension contribution for a large bubble is small, the static pressure inside and outside the bubble can be considered to be equal. The pressure at a point A inside the bubble can be written as

$$p_A = p_0 + \rho_v g \Delta H \tag{8.109}$$

Similarly, an expression for the static pressure at point B on the liquid side is obtained as

$$p_B = p_0 + \rho_l g \Delta H - \frac{1}{2}\rho_l V^2 \tag{8.110}$$

Equating p_A and p_B from Eqs. (8.109) and (8.110) and substituting for V from Eq. (8.108), we obtain

$$(\rho_l - \rho_v)g\Delta H = \frac{1}{2}\rho_l \left(\frac{3}{2}V_t \; \sin\theta\right)^2 \tag{8.111}$$

The vertical height difference, ΔH, between the upper stagnation point and points A or B is obtained as

$$\Delta H = R(1 - \cos\theta) \tag{8.112}$$

Substitution of Eq. (8.112) into (8.111) yields in terms of bubble diameter

$$(\rho_l - \rho_v)gD(1 - \cos\theta) = \frac{9}{4}\rho_l V_t^2 \sin^2\theta \tag{8.113}$$

For small values of θ, Eq. (8.113) after expansion of $\cos\theta$ and $\sin\theta$ in terms of θ reduces to

$$V_t = \frac{1}{3}\left(\frac{2g\,(\rho_l - \rho_v)D}{\rho_l}\right)^{1/2} \tag{8.114}$$

Use of Eq. (8.114) in Eq. (8.107) yields

$$\dot{V}_v = 3A_v \left(\frac{2g\,(\rho_l - \rho_v)D}{\rho_l}\right)^{1/2} \tag{8.115}$$

If all of the energy from the heater is assumed to be utilized to produce vapor and bubble diameter at departure is assumed to be given by Fritz's expression, the above equation becomes

$$q = 0.61 \; \rho_v h_{fg} \frac{A_v}{A_t}\sqrt{\beta}\left[\frac{\sigma g(\rho_l - \rho_v)}{\rho_l^2}\right]^{1/4} \tag{8.116}$$

If contact angle is taken to be 90° and the ratio occupied area of the vapor column to the total heater area is assumed to be $\pi/16$, Eq. (8.116) becomes

$$q = 1.14 \, \rho_v h_{fg} \left[\frac{\sigma g(\rho_l - \rho_v)}{\rho_l^2} \right]^{1/4} \qquad (8.117)$$

Thus according to Moissis and Berenson [21] transition from partial to fully developed nucleate boiling is expected to occur at wall heat flux given by Eq. (8.117).

8.5   BUBBLE SITE DENSITY

As wall superheat or heat flux is increased, the number density of nucleation sites that become active increases. Gaertner and Westwater [22], using a novel technique in which nickel from nickel salts dissolved in water deposited on the heater surface, obtained the number density of active nucleation sites as

$$N_a \approx q^{2.1} \qquad (8.118)$$

Hsu and Graham [5] have presented a summary of the earlier observations of several investigators including the work of Gaertner [23] with respect to dependence of site density on wall heat flux. Figure 8.15 shows the active nucleation site density data of Gaertner and Westwater along with the data of Sultan and Judd [24]. Both sets of data were taken with water boiling at 1 atm. on horizontal copper surfaces.

The data of Sultan and Judd show a weaker dependence of nucleation site density on heat flux than the data of Gaertner and Westwater. However, the number of active sites in Sultan and Judd's experiments is several times higher than that reported by Gaertner and Westwater. In Sultan and Judd's experiments the exponent on q generally varied between 1 and 2. The proportionality constant and the magnitude of the exponent depend on several parameters such as surface wettability, surface preparation procedure, and liquid properties and experimental conditions. Cornwell and Brown [25] made a systematic study of active nucleation sites on copper surfaces during boiling of water at 1 atm. Their study was limited to low heat fluxes, and the surface condition ranged from a smooth to a scratched rough surface. From their work, it was concluded that the active site density varied with wall superheat as

$$N_a \sim \Delta T^{4.5} \qquad (8.119)$$

The proportionality constant in Eq. (8.119) increased with surface roughness but the exponent on ΔT was independent of roughness. From an electron microscope measurement of cavity size distribution, they observed that the number density of cavities, N_S, present on the surface was related to cavity size such that

$$N_s \sim \frac{1}{D_c^2} \qquad (8.120)$$

By assuming that only conical cavities existed on the surface and that a minimum volume of trapped gas was needed for nucleation, they justified the observed

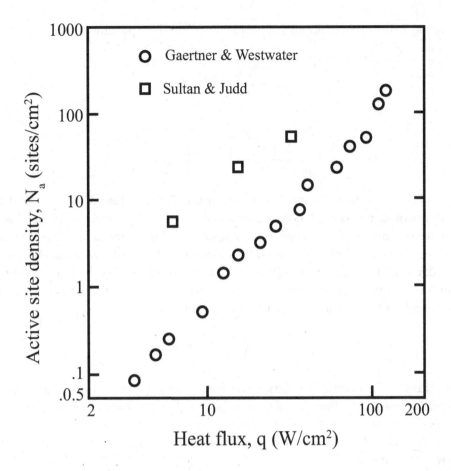

Figure 8.15 Comparison of cumulative nucleation site density observed by Gaertner and Westwater [22] and Sultan and Judd [24]

functional dependence of active site density on wall superheat. Singh et al. [26] compared nucleation behavior of water with organic liquids on four surfaces of different roughness. It was concluded that for a given boiling heat flux, the ratio of surface superheats required for the two fluids remained constant and was unaffected by the value of surface roughness. Kocamustafaogullari and Ishii [27], have correlated the cumulative nucleation site density reported by various investigators for water boilling on a variety of surfaces at pressures varying from 1 to 198 atm as

$$N_a^* = \left[D_c^{*-4.4} \ F(\rho^*) \right]^{1/4.4} \tag{8.121}$$

where

$$N_a^* = N_a D_d^2; \quad D_c^* = D_c/D_d \tag{8.122}$$

and

$$F(\rho^*) = 2.157 \times 10^{-7} \ \times \ \rho^{*-3.2} (1 + 0.0049\rho^*)^{4.13} \tag{8.123}$$

In the above equation, D_d is the bubble diameter at departure and is obtained by multiplying Fritz's equation with $0.0012 \times \rho^{*0.9}$. The parameters ρ^* and D_c are defined as

$$\rho^* = \frac{(\rho_l - \rho_v)}{\rho_v} \tag{8.124}$$

$$D_c = \frac{4\sigma\left[1 + \frac{\rho_l}{\rho_v}\right]}{p_l} \cdot exp\left[\frac{h_{fg}(T_v - T_{sat})}{R_v T_v T_{sat}}\right] - 1 \tag{8.125}$$

In Eq. (8.125), T_v is the temperature of vapor and p_l is the liquid pressure and R_v is the gas constant. For certain data an order of magnitude deviation in the observed active site density was observed with respect to the correlation. One reason for the scatter may be that the correlation does not take into account the surface wettability and the surface roughness. Mikic and Rohsenow [28] have proposed that on commercial surfaces the cumulative number of active sites per unit area can be assumed to vary in partial nucleate boiling as

$$N_a \sim \left(\frac{D_s}{D_c}\right)^m \tag{8.126}$$

where D_s is the diameter of the largest active cavity present on the surface and m is an empirical constant. The size, D_c, of a cavity that nucleates at a wall superheat ΔT is obtained from relation between D_c and ΔT developed earlier. Bier et al. [29], on the other hand, have deduced an expression for active site density from heat transfer data as

$$lnN_a = lnN_{max}\left[1 - \left(\frac{D_c}{D_s}\right)^m\right] \tag{8.127}$$

In Eq. (8.127) N_{max} is the maximum value of N_a which occurs at $D_c = 0$. The value of m was found to depend on the manner in which a surface was prepared. With Freon 115 and Freon 11 boiling on a chemically etched copper surface and on a turned surface, values of 0.42 and 0.26, respectively, were noted for m. In the heat transfer experiments the reduced pressure was varied from 0.0037 to 0.9. It was found that to correlate the data at low and high saturation pressures some changes in the functional form of Eq. (8.127) were necessary.

The studies on nucleation site density as described previously can be divided into two groups. In the first group are the studies in which the density of active nucleation sites as a function of wall superheat or heat flux has been obtained from experiments. In the second set of studies, the functional dependence of site density on wall superheat has been obtained by matching model predictions of heat flux with the data or vice versa. With the exception of Cornwell and Brown [25], no attempt has been made in these studies to relate the cavities that exist on the surface to those that actually nucleate. Though Cornwell and Brown tried to determine the functional dependence of cumulative site density from the local distribution of cavities that exist on the surface, the attempt was qualitative in nature. Also, no attention was given to the surface wettability and shape of the cavities.

By comparing gas and vapor bubble nucleation behavior, Eddington et al. [30] concluded that the number of active nucleation sites in boiling was much smaller than that in gas diffusion. The cause of this difference was thought to be the thermal interference between sites. From sub cooled flow boiling experiments, Eddington et al. [30] provided evidence that thermal interference inhibits nucleation within a region one diameter around an active site. This behavior in turn may distort random distribution of sites. Sultan and Judd [24] studied the bubble growth pattern at neighboring sites during nucleate pool boiling of water on a copper surface. They found that elapsed time between the start of bubble growth at two neighboring active sites increased as the distance separating the two sites increased. It was proposed that thermal diffusion in the substrate in the immediate vicinity of the boiling surface may be responsible for this behavior. Their work suggested that some relation may exist between distribution of active nucleation sites and bubble nucleation phenomenon.

Judd [31] in his summary of the results of nucleation site interactions notes that for dimensionless separation distances of $0.5 < S/D_d < 1$ between nucleation sites, the formation of a bubble at the initiating site promotes the formation of bubbles at the adjacent sites (site seeding). For separation distances $1 < S/D_d < 3$, formation of a bubble at the initiating site inhibits the formation of bubbles at the adjacent site (deactivation of sites). However, these observations are subject to the number density of active sites. A detailed discussion of site interactions has also been given by Fujita [32] in his review article.

Wang and Dhir [33],[34] have developed a criterion for entrapment of gas in cavities of spherical type on copper surface. Using the gas entrapment criterion and systematically varying the surface wettability, they were able to determine the fraction of cavities present on the surface that became active at a particular wall superheat. Subsequently, Basu et al. [35] have correlated number density of active nucleation sites as a function of wall superheat for water at one atmosphere pressure and contact angle as

$$N_a = 0.34(1 - cos\beta)\Delta T^2 \; sites/cm^2 \Delta T < 15°C$$

$$N_a = 3.4 \times 10^{-5}(1 - cos\beta)\Delta T^{5.3} \; sites/cm^2 \Delta T \geq 15°C$$

8.6 HEAT TRANSFER MECHANISMS

In the literature four mechanisms have been identified that contribute to total boiling heat flux under pool boiling conditions. These are transient conduction at the area of influence of a bubble growing on a nucleation site; evaporation (a fraction of which may be included in the transient conduction) at the vapor-liquid interface and from microlayer; enhanced natural convection on the region in the immediate vicinity of a growing bubble; and natural convection over the area that has no active nucleation sites and is totally free of the influence of the former three mechanisms. However, the importance of these mechanisms depends strongly on the magnitude of the wall superheat and other system variables such as heater geometry, orientation with respect to gravitational acceleration, magnitude of gravitational acceleration, etc. Contributions of various mechanisms vary both spatially and temporally.

The effect of orientation of a surface on the relative importance of the various mechanisms of boiling heat transfer can be seen from the data of Nishikawa et al. [36] plotted in Fig. 8.16. The data were taken on a polished copper plate oriented at different angles, θ, to the horizonal. Saturated water at 1 atm was used as the test liquid.

The data plotted in Fig. 8.16 clearly show that in partial nucleate boiling there is a strong effect of orientation of plate with respect to gravitational acceleration. However, in fully developed nucleate boiling (the dotted line in the figure corresponds to prediction for formation of columns with $\beta = 90°$), no such effect is discernible. The contributions of transient conduction and enhanced natural convection are affected by bubble dynamics, which in turn depends on the magnitude and direction of components of gravitational acceleration. The nondependence of fully developed nucleate boiling (post first transition) heat fluxes on plate orientation suggests that the mechanisms (transient conduction and enhanced convection) that are associated with bubble movement are of little consequence. The contribution of natural convection is generally small. Therefore, in fully developed nucleate boiling, evaporation appears to be the most dominant mode of heat transfer.

8.7 PARTIAL NUCLEATE BOILING

In the isolated bubble regime, transient conduction into liquid adjacent to the wall is probably the most important mechanism for heat removal from the wall. After bubble inception, the superheated liquid layer is pushed outward and it mixes with the bulk liquid. The bubble acts like a pump in removing hot liquid from the surface and replacing it with cold liquid. The mechanism was originally proposed by Forster and Greif [37]. Mikic and Rohsenow [28] were the first to formalize the derivation of the functional dependence of partial nucleate boiling heat flux on wall superheat. Assuming that the contact area of bubbles in relation to area of influence was small, and the contribution of evaporation to total heat removal rate is included in transient conduction they obtained an expression for the partial nucleate boiling heat flux by combining contributions of transient conduction and steady state natural convection as

$$q = \frac{K^2}{2}\sqrt{\pi(k\rho c_p)_l f} D_d^2 N_a \Delta T + \left(1 - \frac{K^2}{4} N_a \pi D_d^2\right) h_{nc}\Delta T \qquad (8.128)$$

In Eq. (8.128), the parameter K^2 is reflective of the area of influence of a bubble, and a value of about 2 is assigned to it. The site density N_a is obtained from an equation such as (8.126) whereas dependence on wall superheat ΔT of the size D_c, of a nucleating cavity is obtained from the inception criterion. For the bubble

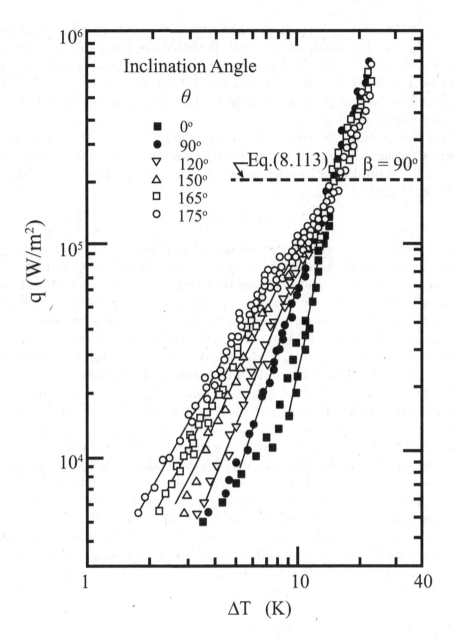

Figure 8.16 Nucleate boiling data of Nishikawa et al. [36] on plates oriented at different angles to the horizontal

diameter at departure, D_d, they used Eqs. (8.97) and (8.98). The product of bubble departure diameter and frequency, f, is obtained from the following correlation:

$$fD_d = 0.6 \left[\frac{\sigma g (\rho_L - \rho_G)}{\rho_L^2} \right]^{1/4} \tag{8.129}$$

For natural convection heat transfer coefficient, any of the correlations available in the literature can be used. It has been noted by Rohsenow [14] that predictions from Eq. (8.128) show a better agreement with data when observed values for D_d and f rather than correlations are used. It should be noted that a quantitative prediction from Eq. (8.128) of dependence of heat flux on wall superheat requires a knowledge of several empirical constants. Though Eq. (8.128) was derived for partial nucleate boiling, it has been suggested that it could be extrapolated to fully developed nucleate boiling.

Judd and Hwang [38] employed an approach similar to that of Mikic and Rohsenow but included micro and macrolayer evaporation at the base of the bubble as well. Thus, a third term for microlayer contribution was added to the right-hand side of Eq. (8.128) as

$$q_e = V_e N_a \rho_l h_{fg} f \tag{8.130}$$

where V_e is the volume of the micro/microlayer associated with each bubble. Using the microlayer volume discerned from experiments in which dichloromethane was boiled on a glass surface, and assuming that parameter K in equation (8.128) has value of $\sqrt{1.8}$, they were able to match the predictions with the data. Experimentally measured values of active nucleation site density and bubble release frequency were also used in the model. Figure 8.17 shows their data and predictions. At the highest measured heat flux, q_m, of about 6 W/cm^2 about a third of the energy is dissipated through evaporation at the bubble base and boundary.

Equation (8.128) is not in a form that can be readily used to determine the dependence of nucleate boiling heat flux on wall superheat. To circumvent this difficulty most of the text books in heat transfer use the correlation developed by Rohsenow [39]. The physics underlying this correlation is not correct and is not in agreement with the mechanistic approach described above. However, the empirical constants used in this correlation generally yield correct dependence of wall heat flux on wall superheat. Hence it has found a wide acceptance in literature.

In developing the correlation, Rohsenow assumed that heat transfer with flow induced by cyclic release of bubbles from the surface was similar to that during forced convection. Hence a formulation similar to that for forced convection could be used. The superficial velocity of liquid was obtained by knowing the mass of liquid needed to compensate for the evaporation rate. In terms of bubble diameter at departure, number density and frequency, the vapor mass flow rate per unit area of the heater surface can be written as

$$\dot{m}_v = \rho_v \frac{\pi}{6} D_d^3 f N_a \tag{8.131}$$

Assuming that all of the energy lost from the heater is utilized in evaporation, the vapor mass flow rate can be related to the heat flux as

$$\dot{m}_v = q/h_{fg} \tag{8.132}$$

Equating the mass flow rate of liquid with that of vapor, the superficial velocity of the liquid moving towards the surface can be written as

$$V_l = q/\rho_l h_{fg} \tag{8.133}$$

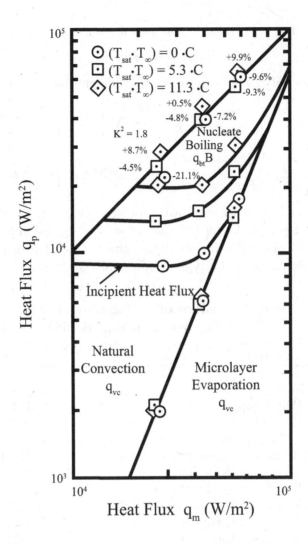

Figure 8.17 Relative contributions of various mechanisms to nucleate boiling heat flux

or Reynolds number characterizing the flow of liquid towards the surface is obtained as

$$Re = \frac{\rho_l V_l D_d}{\mu_l} = \frac{qD_d}{\mu_l h_{fg}} \tag{8.134}$$

From Fritz's expression, bubble diameter at departure is written as

$$D_d \sim \sqrt{\frac{\sigma}{g(\rho_l - \rho_v)}} \tag{8.135}$$

Substituting Eq. (8.135) in (8.134) yields

$$Re = \frac{q\sqrt{\frac{\sigma}{g(\rho_l - \rho_v)}}}{\mu_l h_{fg}} \tag{8.136}$$

For forced convention heat transfer, the data are correlated in the form

$$Nu = f(Re, Pr) \tag{8.137}$$

Using the definition of Nu, the above relationship can be written as

$$Re = f_1\left(Pr, \frac{cp_l \Delta T}{h_{fg}}\right)$$

or

$$\left[\frac{q}{\mu_l h_{fg}}\sqrt{\frac{\sigma}{g(\rho_l - \rho_v)}}\right] = \left[\frac{1}{C_s}\frac{c_{pl}\Delta T}{h_{fg}}\right]^n Pr^{-m} \tag{8.138}$$

In Eq. (8.138), C_s, n and m are empirical constants. On commercial surfaces, values of n generally lie between 3 and 4. Constant m was suggested to have a value of 3 for water and 5.1 for other liquids. The values of C_s depends on fluid-surface combination. Table 8.3 gives the typical values for C_s.

Table 8.3
Selected values of surface correction factor for use with Eq. (8.138)

Surface-Fluid Combination	C_s
Water-nickel	0.006
Water-platinum	0.013
Water-copper	0.013
Water-brass	0.006
CCl_4-copper	0.013
Benzene-chromium	0.010
n-Pentane-chromium	0.015
Ethyl alcohol-chromium	0.0027
Isopropyl alcohol-copper	0.0025
35% K_2CO_3-copper	0.0054
50% K_2CO_3-copper	0.0027
n-Butyl alcohol-copper	0.0030

It should be noted that magnitude of C_s depends on the surface wettability and roughness. Liaw and Dhir [40] systematically studied the effect of surface wettability on vertical copper surfaces placed in a pool of saturated water. They used static contact angle as a measure of wettability. Table 8.4 lists their reported values of C_s for a vertical surface. Note that C_s increases as heater wettability improves.

Table 8.4
Values of C_s for different contact angles

deg	C_s
14	0.0209
27	0.0202
38	0.0194
69	0.0186
90	0.0172

Before proceeding further with heat transfer under fully developed nucleate boiling conditions, it is imperative to determine the resistance to evaporation at a liquid vapor interface. Knowledge of such a resistance is needed to quantify the evaporation rate at the liquid-vapor interface that may form at the bubble base. The thickness of the microlayer intervening liquid layer between solid and vapor-liquid interface may vary in space and with time or remain constant with time if liquid can feed into the layer to compensate for the evaporation rate. If the liquid inflow into the film falls short of the evaporation rate the film will dry out. Equation (8.130) developed earlier does not account for liquid feeding and represents a global mass and energy balance over the growth period of a bubble.

For transfer of heat across a stationary (or a quasi-stationary) liquid layer formed on a heater, the conductive resistance of the liquid layer and the evaporative resistances at the liquid-vapor interface need to be considered. Conductive resistance can be simply calculated using Fourier's law. To determine the evaporative resistance, we assume that a thin layer known as Knudsen layer exists just outside of the interface. In the layer the exchange of molecules from the liquid side and vapor side occurs. The number density of molecules in the vapor space can be written by invoking the ideal gas law as

$$n_v = \frac{p_v N_A}{MRT_v} \tag{8.139}$$

where R is the gas constant, M is molecular weight and N_A is Avogadro's number. Similarly the number density of molecules in the Knudsen layer from the liquid side, assuming their temperature to be equal to that of the liquid is obtained as

$$n_l = \frac{p_l N_A}{MRT_l} \tag{8.140}$$

The average speed of molecules in terms of the vapor and liquid temperatures can be respectively written as

$$\bar{C}_v = \left(\frac{8\,RT_v}{\pi}\right)^{1/2} \tag{8.141}$$

and

$$\bar{C}_l = \left(\frac{8\,RT_l}{\pi}\right)^{1/2} \tag{8.142}$$

The net efflux of molecules leaving from the liquid side is

$$J_m = \frac{n_l \bar{C}_l}{4} - \frac{n_v \bar{C}_v}{4} = \frac{p_l N_A}{4MRT_l}\left(\frac{8RT_l}{\pi}\right)^{1/2} - \frac{p_v N_A}{4MRT_v}\left(\frac{8RT_v}{\pi}\right)^{1/2} \tag{8.143}$$

$$J_m = \frac{p_l N_A}{M}\left(\frac{1}{2\pi RT_l}\right)^{1/2} - \frac{p_v N_A}{M}\left(\frac{1}{2\pi RT_v}\right)^{1/2} \tag{8.144}$$

The difference between absolute temperatures T_v and T_l is generally small. As such we can assume $T_v \simeq T_l \simeq T_{sat}$ or

$$J_m = \frac{(p_l - pv)N_A}{M}\left(\frac{1}{2\pi RT_{sat}}\right)^{1/2} \tag{8.145}$$

Equation (8.145) can be written in terms of the heat flux as

$$q = \frac{J_m M}{N_A} \cdot h_{fg} = (p_l - pv)h_{fg}\left(\frac{1}{2\pi RT_{sat}}\right)^{1/2} \tag{8.146}$$

The pressure difference can be replaced by corresponding saturation temperature difference by using Clausius-Clapeyron equation. As such

$$q \simeq (T_l - T_v)\frac{h_{fg}^2}{T_{sat}v_{fg}}\left(\frac{1}{2\pi RT_{sat}}\right)^{1/2} \tag{8.147}$$

From Eq. (8.147), the evaporative resistance can be simply evaluated as

$$(R)_{evap} = \frac{T_{sat}(2\pi RT_{sat})^{1/2}\,v_{fg}}{h_{fg}^2} \tag{8.148}$$

This resistance is generally much smaller than the liquid film resistance unless the film is of the order of few nanometers.

8.8 FULLY DEVELOPED NUCLEATE BOILING

Transition from partial to fully developed nucleate boiling occurs when bubbles start to merge both in the vertical and horizontal directions. Merger of bubbles in the vertical direction leads to formation of vapor columns or jets whereas merger of bubbles in the horizontal directions leads to development of vapor mushrooms. Figure 8.18 originally proposed by Gaertner [23] shows vapor structures in nucleate boiling. According to Gaertner, after the transition from partial to fully developed nucleate boiling, evaporation is the dominant mode of heat transfer. Evaporation occurs at the peripheries of vapor stems. Energy for the phase change is supplied by the superheat liquid layer in which the stems are implanted. Thus the boiling heat flux can be calculated if the fractional area occupied by the stems and the thickness of the thermal layer are known. The heater area fraction occupied by stems (vapor) is equal to the product of the member density of stems and the area occupied by one stem. Experimental observations of Iida and Kobayasi [41] and Liaw and Dhir [42] show that vapor occupied fractional area of the heater increases with wall superheat or heat flux. Gaertner and Westwater [22], however, found the stem diameter to decrease with wall heat flux. Because of the rapid increase in number density of nucleation sites, reduction in stem diameter is accompanied by net increase in heater fractional area occupied by vapor. Gaertner and Westwater [22] and Iida and Kobayasi [41] determined that the thickness of the thermal layer varied spatially and the average thermal layer thickness decreased with wall heat

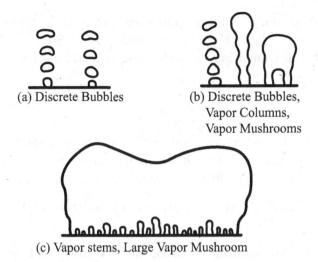

(a) Discrete Bubbles

(b) Discrete Bubbles,
Vapor Columns,
Vapor Mushrooms

(c) Vapor stems, Large Vapor Mushroom

Figure 8.18 Gaertner's identification of vapor structures in nucleate boiling. (a) Discrete bubble region, (b) first transition, (c) vapor mushroom region

flux. The functional dependence of film thickness on wall heat flux was found to be about the same as of the diameter of the vapor stems. Bobst and Colver [43] also measured the thickness of the thermal layer adjacent to the wall and found the thickness δ_{th}, to vary with wall superheat, ΔT, as

$$\delta_{th} \approx \Delta T^{-6} \tag{8.149}$$

The only mechanistic model for fully developed nucleate boiling is that of Dhir and Liaw [44]. Their model assumes that vapor stems of diameter, D, locate themselves on a square grid with spacing L, being equal to $Na^{-1/2}$. The vapor stem diameter was deduced from measured void fractions and was found to vary as $\Delta T^{-1/4}$. Energy from the wall was conducted into the liquid micro/macro layers and was used in evaporation at the stationary liquid-vapor interface of vapor stems. The heat transfer rate into the thermal layer and the temperature distribution in it were determined by solving a two-dimensional steady state conduction equation in the liquid occupied region. The conduction equation was solved for each ray while including the evaporative resistance at the liquid vapor interface. The heater surface area over which stems existed was assumed to be dry. Using average heat transfer coefficients over the liquid and vapor occupied regions, an expression for the wall heat flux was obtained as

$$q = \left\{ \bar{h}_l \left(1 - \frac{\pi/4 D^2}{L^2} \right) + \bar{h}_v \frac{\pi/4 D^2}{L^2} \right\} \Delta T \tag{8.150}$$

In nucleate boiling the contribution of the second term in the curly brackets is much smaller than the first term and hence may be neglected. Equation (8.150) is also valid in transition boiling. At high wall superheat the second term starts to dominate and cannot be neglected. Also, it should be noted that the model was quasi-static and it did not account for reformation of stems as large bubbles depart and new mergers take place. More recent studies (e.g., Son and Dhir [45]) have shown that numerical simulations provide more realistic modelling of the boiling process.

REFERENCES

1. S, Nukiyama, "The Maximum and Minimum values of the Heat Transmitted from Metals to Boiling Water under Atmospheric Pressure", J. Japan Soc. Mech Engrs, Vol. 37, pp. 367-374, 1934.

2. R. Cole, "Boiling Nucleation", Advances in Heat Transfer, Vol. 10, pp. 85-166, 1974.

3. S. G. Bankoff, "Entrainment of Gas in the Spreading of a Liquid over a Rough Surface", AIChE J., Vol. 4, pp. 24-26, 1958.

4. Y. Y. Hsu, "On the Size Range of Active Nucleation Cavities on a Heating Surfaces", J. Heat Transfer, Vol. 84, pp. 207-216, 1962.

5. Y. Y. Hsu, and R. W. Graham, "Transport Processes in Boiling and Two-Phase Systems", Hemisphere Publishing, Washington, 1976.

6. C. Y. Han and P. Griffith, "The Mechanism of Heat Transfer in Nucleate Pool Boiling-Part I: Bubble Initiation, Growth and Departure", Int. J. Heat and Mass Transfer, Vol. 8, pp. 887-904, 1965.

7. J. W. S. Rayleigh, "Philosphical Magazine", Vol. 94, p. 122, 1917.

8. W. Fritz, "Maximum Volume of Vapor Bubbles", Physik Zeitschr., Vol. 36, pp. 379-384, 1935.

9. R. Siegel, and E. G. Keshock, "Effects of Reduced Gravity on Nucleate Bubble Dynamics in Water", AIChE J., Vol. 10, No. 4, pp. 509-516, 1964.

10. T. K. Cochran, and J. C. Aydelott, "Effects of Subcooling and Gravity Level on Boiling in the Discrete Bubble Region", NASA TN-D-3449, 1966.

11. B. B. Mikic, W. M. Rohsenow, and P. Griffith, "On Bubble Growth Rates", Int. J. Heat and Mass Transfer, Vol. 13, pp. 657-666, 1970.

12. G. Son, V. K. Dhir and N. Ramanujapu, "Single Bubble During Nucleate Boiling on a Horizontal Surface", J. Heat Transfer, Vol. 121, pp. 623-631, 1999.

13. M. S. Plesset and S. A. Zwick, "Growth of Vapor Bubbles in Superheated Liquids", J. App. Phys., Vol. 25, pp. 493-500, 1954.

14. W. M. Rohsenow, Boiling. In Handbook of Heat Transfer Fundamentals, 2nd ed., W. M. Rohsenow, J. P. Hartnett, and E. N. Ganic, Eds., McGraw Hill, New York, pp. 12-1-12-4, 1985.

15. M. Jakob, "Heat Transfer", Vol. 1, Wiley, New York, 1949.

16. H. J. Ivey, "Relationships between bubble frequency, departure diameter and rise velocity in nucleate boiling", Int. J. Heat Mass Transfer, Vol. 10, pp. 1023-1040, 1967.

17. R. Cole, "Frequency and departure diameter at sub-atmospheric pressures", AIChE Journal, Vol. 13, pp. 779-783, 1967.

18. N. Zuber, "The Dynamics of Vapour Bubbles in Non-Uniform Temperature Field", Int. J. Heat Mass Transfer, Vol. 2, pp. 83-98, 1961.

19. F. G. Malenkov, "Detachment Frequency as a Function of Size of Vapor Bubbles", (translated), Inzh. Fiz. Zhur., Vol. 20, p. 99, 1971.

20. V. K. Dhir, "Nucleate Pool Boiling", Handbook of Thermal Science and Engineering, Vol. 3, pp. 1645-1694, 2018.

21. R. Moissis, and R. J. Berenson, "On the Hydrodynamic Transitions in Nucleate Boiling," J. Heat Transfer, Vol. 85, No. 3, pp. 221-229, 1963.

22. R. F. Gaertner, and J. W. Westwater, "Population of Active Sites in Nucleate Boiling Heat Transfer," Chem. Eng. Prog. Symp. Ser., Vol. 56, No. 30, pp. 39-48, 1960.

23. R. F. Gaertner, "Distribution of Active sites in the Nucleate Boiling of Liquids," Chem. Engr. Prog. Symp. Ser. Vol. 59, pp. 52-61, 1963.

24. M. Sultan, and R. L. Judd, "Interaction of the Nucleation Phenomena at Adjacent Sites in Nucleate Boiling," J. Heat Transfer, Vol. 105, pp. 3-9, 1983.

25. K. Cornwell, and R. D. Brown, "Boiling Surface Topography" Proc. 6th Int. Heat Transfer Conf., Toronto, Canada, Vol. 1, pp. 157-161, 1978.

26. A. Singh, B. B. Mikic, andW. M. Rohsenow, "Relative Behavior ofWater and Organics in Boiling," Proc. 6th Int. Heat Transfer Conf., Toronto, Canada, Vol. 1, pp. 163-168, 1978.

27. G. Kocamustafaogullari, and M. Ishii, "Interfacial Area and Nucleation Site Density in Boiling Systems," Int. J. Heat and Mass Transfer, Vol. 26, pp. 1377-1387, 1983.

28. B. B. Mikic, and W. M. Rohsenow, "A New Correlation of Pool Boiling Data Including the Effect of Heating Surface Characteristics," J. Heat Transfer, Vol. 9, pp. 245-250, 1969.

29. K. Bier, D. Gorenflo, M. Salem, and Y. Tanes, "Pool Boiling Heat Transfer and Size of Active Nucleation Centres for Horizontal Plates with Different Surface Roughness", Proceedings of 6th International Heat Transfer Conference, Toronto, Vol. 1, pp. 151-156, 1978.

30. R. I. Eddington, D. B. R. Kenning, and A. I. Korneichev, "Comparison of Gas and Vapor Bubble Nucleation on a Brass Surface in Water," Int. J. Heat and Mass Transfer, Vol. 21, pp. 855-862, 1978.

31. R. L. Judd, "On the Nucleation Site Interaction," J. Heat Transfer, Vol. 116, pp. 475-478, 1988.

32. Y. Fujita, "Recent Developments in Pool Boiling Transfer," Proc. 4th Int. Topical Meeting on Nuclear Rector Thermal Hydraulics, Karlsruhe, Germany, Vol. 2, pp. 1068-1086, 1989.

33. C. H. Wang and V. K. Dhir, "On the Gas Entrapment and Nucleation Site Density During Pool Boiling of Saturated Water", J. Heat Transfer, Vol. 115, pp. 670-679, 1993a.

34. C. H. Wang and V. K. Dhir, "Effect of Surface Wettability on Active Nucleation Site Density During Pool Boiling of Surface Water", J. Heat Transfer, Vol. 115, pp. 659-669, 1993b.

35. N. Basu, G. R.Warrier, and V. K. Dhir, "Onset of Nucleate Boiling and Active Nucleation Site Density During Subcooled Flow Boiling", J. Heat Transfer, Vol. 124(4), pp. 717-728, 2002.

36. K. Nishikawa, Y. Fujita, and H. Ohta, "Effect of Surface Configuration on Nucleate Boiling Heat Transfer," Int. J. Heat and Mass Transfer, Vol. 27, pp. 1559-1571, 1974.

37. D. E. Forster, and R. Greif, "Heat Transfer to a Boiling Liquid Mechanism and Correlation," J. Heat Trans., Vol. 81, No. 1, pp. 43-53, 1959.

38. R. L. Judd, and K. S. Hwang, "A Comprehensive Model for Nucleate Pool Boiling Heat Transfer Including Microlayer Evaporation," J. Heat Transfer, Vol. 98, pp. 623-629, 1976.

39. W. M. Rohsenow, "A Method of Correlation Heat Transfer Data for Surface Boiling of Liquids," Trans. ASME, Vol. 74, No. 3, pp. 969-976, 1952.

40. S. P. Liaw, and V. K. Dhir, "Effect of Surface Wettability on Transition Boiling Heat Transfer from a Vertical Surface," Proc. 8th Int. Heat Transfer Conf., San Francisco, CA, Vol. 4, pp. 2031-2036, 1986.

41. Y. Iida, and K. Kobayashi, "Distribution of Void Fraction above a Horizontal Heating Suface in Pool Boiling," Bull. JSME, Vol. 12, pp. 283-290, 1969.

42. S. P. Liaw, and V. K. Dhir, "Void Fraction Measurements During Saturated Pool Boiling of Water on Partially Wetted Vertical Surfaces," J. Heat Transfer, Vol. 111, pp. 731-738, 1989.

43. R.W. Bobst, and C. P. Colver, "Temperature Profiles up to Burnout Adjacent to a Horizontal Heating Surface in Nucleate Pool Boiling of Water," Chem. Eng. Prog. Symp. Ser., Vol. 64, pp. 26-32, 1968.

44. Dhir, V.K, and Liaw, S.P, "Frame work for a Unified Model for Nucleate amd Transition Pool Boiling",J. Heat Transfer, vol. 111, pp. 203-210, 1989.

45. G. Son and V. K. Dhir, "Numerical Simulation of Nucleate Boiling on a Horizontal Surface at High Heat Fluxes", Int. J. Heat Mass Transfer, Vol. 51, pp. 2566-2582, 2008.

EXERCISES

1. Calculate equilibrium radius of a vapor bubble in water and R-113 for liquid superheats of 3°C, 5°C, 7°C, 9°C, and 11°C. Assume the pressure to be one atmosphere.

2. Calculate equilibrium radius of a vapor bubble in water for system pressures of 10, 50, and 100 atmospheres at a superheat of 5°C.

3. Plot from Exercises 1 and 2 above, equilibrium bubble size for water as a function of superheat at fixed pressure, and as a function of pressure for a given superheat and explain your results.

4. Calculate the critical radius and homogeneous nucleation temperature of water at one atmosphere pressure. Assume $J = 1$ cluster/cm^3 sec. How would your answer change if $J = 1 \times 10^{-2}$ cluster/cm^3 sec or if system pressure was 100 bars? What is the ratio of T_{hom}/T_{crit} for these cases?

5. Calculate the minimum wall superheat and the corresponding cavity size for boiling of saturated water at one atmosphere pressure on an upward facing horizontal plate. Wall heat flux is $10^3 W/m^2$. If laminar natural connection existed on the plate prior to nucleation, is the predicted wall superheat consistent with the heat flux? Assume contact angle to be 90°.

6. Calculate the wall superheat and the corresponding size of nucleating cavities for saturated water at one atmosphere pressure. The wall heat flux is $10^4 W/m^2$ and turbulent natural connection exists on the heater prior to nucleation. Assume the contact angle to be 60°.

7. Calculate the waiting time for a cavity of radius $10^{-5}m$ to nucleate in saturated water at one atmosphere pressure. The wall superheat is 8°C and thermal layer has a thickness of $6 \times 10^{-4}m$. Assume contact angle to be 90°.

8. A spherical vapor bubble is growing in a pool of water maintained at 110°C at one atmosphere pressure. Calculate and plot the bubble radius as a function of time.

9. A heater is placed in a pool of saturated water at one atmosphere pressure. The contact angle is 20°. If at a wall superheat of 8K, the wall heat flux is $6W/cm^2$, calculate the size range of cavities that will nucleate and the corresponding waiting time. Choose two size cavities—one the smallest and one the largest. For the largest cavity, use the model developed in the chapter for a bubble growing in a non-uniform temperature field to calculate the bubble radius as a function of time. Plot your result.

10. For the largest cavity chosen in Exercise 9, plot the forces acting on the bubble as a function of time. Determine the time at which a bubble will detach and the corresponding diameter of the bubble at departure. What will be the bubble release frequency for this case?

11. Calculate the evaporative resistance for water at 1, 10, and 100 atmosphere pressures, respectively. Compare the evaporative resistance with conductive resistance of a 10^{-6} m thick layer of water.

12. Working with the mechanistic model, plot for saturated water at one atmosphere pressure the nucleate boiling heat flux as a function of wall superheat. In the model, use Fritz' expression for bubble diameter at departure and Malenkov's correlation for the product of bubble diameter at departure and frequency. For number density of active sites use Basu's correlation. Assume contact angle to be 45°. Discuss your results.

13. Employing Rohsenow's correlation, plot the wall heat flux as a function of wall superheat for saturated water and PF-5060 at one atmosphere pressure. Heater surface material is copper. C_s for water and PF-5060 has a value of 0.013 and 0.006, respectively.

## CHAPTER 9

Maximum and Minimum Heat Fluxes, Film and Transition Boiling

Before proceeding with the above topics, we develop background in the interfacial instabilities.

9.1   TAYLOR INSTABILITY

Taylor instability is the name given to instability of an interface between two superposed fluid layers when the heavier fluid lies over a lighter fluid and acceleration is directed from the heavier to the lighter fluid. This instability between two fluids in the absence of interfacial tension and gravity was first analyzed by Rayleigh [1]. Subsequently Taylor [2] considered this instability between two incompressible fluid layers of infinite extent in the manner in which we have come to recognize it but without surface tension. Bellman and Pennington [3] extended Taylor's analysis by incorporating the effect of surface tension and viscosity. Figure 9.1 shows a typical

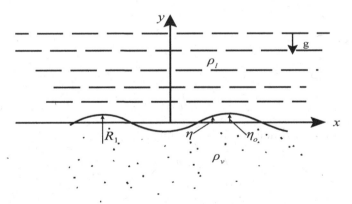

Figure 9.1   Configuration for Taylor instability analysis

configuration for a two-dimensional interface. Under the assumptions that fluids are inviscid, incompressible, fluid layers are of infinite depth, and the non-linear terms are small, the continuity and momentum equations in x and y directions for the two fluids can be written as

$$\frac{\partial u}{\partial x} + \frac{\partial v}{\partial y} = 0 \tag{9.1}$$

$$\frac{\partial u}{\partial t} = -\frac{1}{\rho}\frac{\partial p}{\partial x} \tag{9.2}$$

$$\frac{\partial v}{\partial t} = -\frac{1}{\rho}\frac{\partial p}{\partial y} - g \tag{9.3}$$

Let the plane interface be perturbed such that

$$\eta = \eta_0 e^{i(w_r + iw_i)t} \cos \kappa x \tag{9.4}$$

where w_r and w_i are the real and imaginary components of the wave frequency and κ is the wave number. If the wave is assumed to be stationary, $w_r = 0$ or

$$\eta = \eta_0 e^{-w_i t} \cos \kappa x \tag{9.5}$$

If w_i is a negative number, the interface will grow with time and become unstable. If on the other hand w_i is positive, the perturbation will die with time. For $w_i = 0$, magnitude of perturbation will not change with time. Such a disturbance is called neutral. Effect of perturbation also dies far-away from the interface.

The kinematic condition at the vapor-liquid interface can be written as

$$\frac{d\eta}{dt} \sim \frac{\partial n}{\partial t} = v \tag{9.6}$$

or

$$-w_i \eta_0 e^{-w_i t} \cos \kappa x = v \tag{9.7}$$

or

$$v = -\eta w_i \tag{9.8}$$

if we define a potential function, Φ, such that

$$u = \frac{\partial \Phi}{\partial x} \tag{9.9}$$

$$v = \frac{\partial \Phi}{\partial y} \tag{9.10}$$

Noting that far away from the interface, perturbation components of velocity should vanish and eliminating v between Eqs. (9.8) and (9.10) we find, for the liquid phase

$$\Phi_l = +\eta_0 \frac{w_i}{\kappa} e^{-\kappa y} e^{-w_{it}} \cos \kappa x \tag{9.11}$$

and for the vapor phase

$$\Phi_v = -\eta_0 \frac{w_i}{\kappa} e^{+\kappa y} e^{-w_{it}} \cos \kappa x \tag{9.12}$$

In terms of potential function for the liquid phase, the momentum equations in the x and y directions become

$$\frac{\partial^2 \Phi_l}{\partial x \partial t} = -\frac{1}{\rho_l} \frac{\partial p_l}{\partial x} \tag{9.13}$$

and

$$\frac{\partial^2 \Phi_l}{\partial y \partial t} = -\frac{1}{\rho_l} \frac{\partial p_l}{\partial y} - g \tag{9.14}$$

Integrating Eq. (9.14) with respect to y we find

$$\frac{\partial \Phi_l}{\partial t} = -\frac{1}{\rho_l} p_l - gy + f(x) \tag{9.15}$$

The constant of integration $f(x)$ can without any loss of generality be assumed to be a constant. Equation (9.15) will thus satisfy Eq. (9.13) as well. The constant of integration can be determined if we evaluate Eq. (9.15) in the limits 0 and y while recognizing that at the unperturbed interface $\frac{\partial \Phi}{\partial t} = 0$ and therefore

$$f(x) = C = \frac{p_{ol}}{\rho_l} \tag{9.16}$$

where p_{ol} is the pressure on the liquid side of the unperturbed interface.

Or Eq. (9.15) can be written as

$$\frac{\partial \Phi_l}{\partial t} = \frac{(p_{ol} - p_l)}{\rho_l} - gy \tag{9.17}$$

An equation similar to (9.17) for the vapor phase is obtained as

$$\frac{\partial \Phi_v}{\partial t} = \frac{(p_{ov} - p_v)}{\rho_v} - gy \tag{9.18}$$

where p_{ov} is the pressure on the vapor side of the unperturbed interface.

At the undisturbed interface $p_{ol} = p_{ov}$, Eliminating p_{ol} and p_{ov}, between Eqs. (9.17) and (9.18), we obtain

$$\rho_l \frac{\partial \Phi_l}{\partial t} - \rho_v \frac{\partial \Phi_v}{\partial t} = (p_v - p_l) - (\rho_l - \rho_v)gy \tag{9.19}$$

Substituting for $\frac{\partial \Phi_l}{\partial t}$ and $\frac{\partial \Phi_v}{\partial t}$ from Eqs. (9.11) and (9.12), respectively, and evaluating Eq. (9.19) at $y = \eta$, we obtain

$$-\left[\rho_l \frac{(w_i)^2}{\kappa}\eta + \rho_v \frac{(w_i)^2}{\kappa}\eta\right] = (p_v - p_l)|_\eta - (\rho_l - \rho_v)g\eta \tag{9.20}$$

For an interface with curvature only in the $x - y$ plane, the pressure difference between vapor and liquid can be written as

$$(p_v - p_l)|_\eta = \frac{\sigma}{R_1} \tag{9.21}$$

The curvature, $\frac{1}{R_1}$, is related to the interface shape as

$$\frac{1}{R_1} = -\frac{\frac{\partial^2 \eta}{\partial x^2}}{\left[1 + \left(\frac{\partial \eta}{\partial x}\right)^2\right]^{3/2}} \simeq -\frac{\partial^2 \eta}{\partial x^2} = \kappa^2 \eta \tag{9.22}$$

Substituting for $(p_v - p_l)$ from Eq. (9.21), Eq. (9.20) becomes

$$(\rho_l + \rho_v)w_i^2 \frac{\eta}{\kappa} = -\kappa^2 \eta \sigma + (\rho_l - \rho_v)g\eta \tag{9.23}$$

or

$$w_i^2 = \left[-\frac{\sigma\kappa^3}{(\rho_l + \rho_v)} + \frac{(\rho_l - \rho_v)g\kappa}{(\rho_l + \rho_v)}\right]$$

$$w_i = \pm\left[-\frac{\sigma\kappa^3}{(\rho_l + \rho_v)} + \frac{(\rho_l - \rho_v)g\kappa}{(\rho_l + \rho_v)}\right]^{1/2}$$

Choosing the (-ve) sign, we find

$$-w_i = \left[\frac{(\rho_l - \rho_v)g\kappa}{(\rho_l + \rho_v)} - \frac{\sigma\kappa^3}{(\rho_l + \rho_v)}\right]^{1/2} \tag{9.24}$$

From Eq. (9.24) we see that buoyancy tends to destabilize the interface whereas surface tension stabilizes it.

For a neutral wave $w_i = o$, or

$$\frac{(\rho_l - \rho_v)g\kappa}{(\rho_l + \rho_v)} = \frac{\sigma\kappa^3}{(\rho_l + \rho_v)}$$

or

$$\kappa_c^2 = \frac{(\rho_l - \rho_v)g}{\sigma}$$

or critical wavelength,

$$\lambda_c = 2\pi\sqrt{\frac{\sigma}{g(\rho_l - \rho_v)}} \tag{9.25}$$

The "most dangerous" wavelength λ_d is the one with maximum growth rate. Differentiating w_i with respect to κ and setting $\frac{dw_i}{d\kappa} = 0$, we obtain

$$\frac{(\rho_l - \rho_v)}{(\rho_l + \rho_v)} - \frac{3\sigma\kappa^2}{(\rho_l + \rho_v)} = 0$$

or

$$\kappa_d^2 = \frac{(\rho_l - \rho_v)g}{3\sigma} \tag{9.26}$$

or

$$\lambda_d = 2\pi\sqrt{3}\sqrt{\frac{\sigma}{(\rho_l - \rho_v)g}} \tag{9.27}$$

Substituting for κ_d from Eq. (9.26) into Eq. (9.24), we calculate the wave frequency corresponding to the fastest growth rate as

$$(-w_i)_{max} = \sqrt{\frac{2}{3\sqrt{3}}\frac{g(\rho_l - \rho_v)}{(\rho_l + \rho_v)}}\sqrt{\frac{g(\rho_l - \rho_v)}{\sigma}} \tag{9.28}$$

These results are based on the assumption of inviscid fluids and with wave amplitude much smaller than the wavelength. We also see that the "most dangerous" wavelength is $\sqrt{3}$ times longer than the critical wavelength.

9.2 INSTABILITY OF PLANE INTERFACE BETWEEN TWO PARALLEL FLOWING STREAMS

Figure 9.2 shows the typical configuration in which streams of liquid and vapor at velocities U_l and U_v, respectively move parallel to each other.

In carrying out the instability analysis of the interface in the absence of heat and mass transfer, we assume that:

1. Fluids are inviscid and incompressible and no interfacial heat/mass transfer occurs.

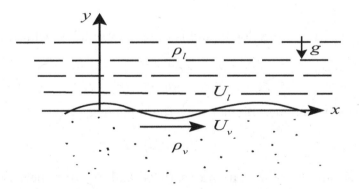

Figure 9.2 Configuration for instability between two parallel streams

2. Second order non-linear terms can be neglected.

3. Wave amplitude is much smaller than the wavelength.

4. Fluid layers are of infinite extent.

5. Liquid and gas/vapor layers move at a free stream velocity of U_l and U_v, respectively.

6. Only two dimensional waves are considered.

If an initially plane interface between two streams is disturbed such that

$$\eta = \eta_0 e^{i(\omega t - \kappa x)} \tag{9.29}$$

where ω is a complex frequency defined as

$$\omega = \omega_r + i\omega_i \tag{9.30}$$

Potential functions satisfying the continuity equations for the perturbed liquid and vapor streams can be written as

$$\Phi_l = U_l x + A_l \ e^{-\kappa y} \ e^{i(wt - \kappa x)} \tag{9.31}$$

$$\Phi_v = U_v x + A_v \ e^{\kappa y} \ e^{i(wt - \kappa x)} \tag{9.32}$$

A relation between constants A_l, A_v and η_0 is obtained by using the kinematic condition at the interface as

$$v = \frac{\partial \Phi}{\partial y} = \frac{d\eta}{dt} = \frac{\partial \eta}{\partial t} + U \frac{\partial \eta}{\partial x} \tag{9.33}$$

Substitution in Eq. (9.33) for Φ_l from Eq. (9.31) and for η from Eq. (9.29) yields for the liquid phase

$$A_l = -i\eta_0 (c - U_l) \tag{9.34}$$

In Eq. (9.34), c is the wave velocity and is a complex number defined as

$$c = \frac{\omega}{\kappa} \equiv \frac{\omega_r + i\omega_i}{\kappa} \tag{9.35}$$

where κ is the wave number. Similarly for the vapor/gas phase

$$A_v = i\eta_0 (c - U_v) \tag{9.36}$$

After A_l and A_v are substituted from Eqs. (9.34) and (9.36), respectively, into Eqs. (9.31) and (9.32), the expressions for the potential functions become

$$\Phi_l = U_l x - i\eta_0 (c - U_l) e^{-ky} \ e^{i(wt - kx)} \tag{9.31a}$$

$$\Phi_v = U_v x - i\eta_0(c - U_v)e^{ky} e^{i(wt-kx)} \tag{9.32a}$$

Neglecting the second order terms, the momentum equations in the x and y direction for either of the phases can be written as

$$\frac{\partial u}{\partial t} + U\frac{\partial u}{\partial x} = -\frac{1}{\rho}\frac{\partial p}{\partial x} \tag{9.37}$$

$$\frac{\partial v}{\partial t} + U\frac{\partial v}{\partial x} = -\frac{1}{\rho}\frac{\partial p}{\partial y} - g \tag{9.38}$$

In terms of the potential function, Eqs. (9.37) and (9.38) are written as

$$\frac{\partial^2\Phi}{\partial x\partial t} + U\frac{\partial^2\Phi}{\partial x^2} = -\frac{1}{\rho}\frac{\partial p}{\partial x} \tag{9.39}$$

$$\frac{\partial^2\Phi}{\partial y\partial t} + U\frac{\partial^2\Phi}{\partial x\partial y} = -\frac{1}{\rho}\frac{\partial p}{\partial y} - g \tag{9.40}$$

Integration of Eq. (9.40) with respect to y, yields

$$\frac{\partial\Phi}{\partial t} + U\frac{\partial\Phi}{\partial x} = -\frac{p}{\rho} - gy + f(x) \tag{9.41}$$

Without any loss of generality, the function $f(x)$, can be assumed to be a constant. Equation (9.41) can be shown to satisfy Eq. (9.39) as well. The constant of integration can be evaluated by setting the limits of integration on Eq. (9.41) as

$$\left[\frac{\partial\Phi}{\partial t}\right]_0^y + \left[U\frac{\partial\Phi}{\partial x}\right]_0^y = -\left[\frac{p}{\rho}\right]_0^y - gy \tag{9.42}$$

At the plane undisturbed interface ($y = 0$), $\frac{\partial\Phi}{\partial t} = 0$ and $\frac{\partial\Phi}{\partial x} = U$. With these substitutions Eq. (9.42) becomes for any position of the interface

$$\frac{\partial\Phi}{\partial t} + U\frac{\partial\Phi}{\partial x} = -\frac{1}{\rho}[p - p_0] - gy + U^2 \tag{9.43}$$

In Eq. (9.43), p_0 is the static pressure at the undisturbed interface. A comparison of Eq. (9.43) with (9.41) shows that the constant of integration in Eq. (9.41) is

$$\frac{p_0}{\rho} + U^2 \tag{9.44}$$

In Eqs. (9.43) and (9.44), p_0 is the static pressure at the undisturbed interface. After substitution for Φ from Eqs. (9.31) and (9.43) when evaluated at $y = \eta$ yields for the liquid phase with $k\eta \simeq 0$;

$$\rho_l\left[\eta_0(c - U_l)w - U_l\eta_0(c - U_l)k\right]e^{i(wt-\kappa x)} = p_0 - p_l - \rho_l g\eta \tag{9.45}$$

Similarly for the vapor phase

$$\rho_v \left[-\eta_0(c - U_v)\omega + U_v \right] (c - U_v)\eta_0 k e^{i(wt - \kappa x)} = p_0 - p_v - \rho_v g \eta \tag{9.46}$$

Elimination of p_0 between Eq. (9.45) and (9.46) yields

$$(\rho_l + \rho_v)c^2 - 2(\rho_v U_v + \rho_l U_l)c + \rho_l U_l^2 + \rho_v U_v^2 = \frac{p_v - p_l}{\eta k} - \frac{(\rho_l - \rho_v)g}{\kappa}$$

$$= \frac{\sigma}{R_1 \eta k} - \frac{(\rho_l - \rho_v)g}{\kappa} \tag{9.47}$$

where R_1 is the radius of curvature of the interface. For a two dimensional interface, R_1^{-1} can be written as

$$R_1^{-1} \simeq -\frac{\partial^2 \eta}{\partial x^2} = k^2 \eta \tag{9.48}$$

After substitution for R_1^{-1} from Eq. (9.48), Eq. (9.47) becomes

$$(\rho_l + \rho_v)c^2 - 2(\rho_v U_v + \rho_l U_l)c + \rho_l U_l^2 + \rho_v U_v^2 = \sigma k - \frac{(\rho_l - \rho_v)g}{k} \tag{9.49}$$

Equation (9.49) is a quadratic equation in c. The solution for this equation is written as

$$c = \left[\frac{\rho_l U_l + \rho_v U_v}{(\rho_l + \rho_v)} \right] \pm i \sqrt{ \frac{\rho_v \rho_l (U_v - U_l)^2}{(\rho_l + \rho_v)^2} - \frac{\sigma k}{(\rho_l + \rho_v)} + \frac{(\rho_l - \rho_v)g}{(\rho_l + \rho_v)k} } \tag{9.50}$$

For $U_v = U_l = 0$, Eq. (9.50) reduces to the dispersion relation derived earlier for stationary fluid layers. The first term under the square root sign represents the contribution of inertia, the second term indicates surface tension and the last term represents the contribution of buoyancy. We can see that inertia and buoyancy tend to destabilize the interface whereas surface tension stabilizes the interface. From instability considerations, our interest turns to the imaginary part of the wave velocity. Critical condition will occur when the growth rate given by the imaginary component is zero. For such a case we find

$$\frac{\rho_v \rho_l (U_v - U_l)^2}{(\rho_l + \rho_v)} = \sigma \kappa - \frac{(\rho_l - \rho_v)g}{\kappa}$$

If gravity is zero or the streams move parallel to the gravitational acceleration vector, we find the critical velocity difference between the two streams for instability is

$$|(U_v - U_l)| = \sqrt{ \frac{\sigma \kappa (\rho_l + \rho_v)}{\rho_v \rho_l} } \tag{9.51}$$

This instability is generally known as Kelvin-Helmholtz instability.

9.3 MAXIMUM HEAT FLUX

Kutateladze [4] was one of the first researchers to suggest that the maximum heat flux condition was purely a hydrodynamic phenomenon. Based on similarity considerations related to the equations of motion and energy, he identified two dimensionless groups

$$\frac{q_{max}}{\rho_v h_{fg} U_v} \quad \text{and} \quad \frac{\sigma(\rho_l - \rho_v)g}{U_v^4 \rho_v^2}$$

where q_{max} is the maximum heat flux and U_v is the superficial vapor velocity in the vapor columns that exist in fully developed nucleate boiling. By eliminating U_v contained in the two dimensionless groups, he obtained an expression for the maximum heat fluxes as

$$q_{max} = 0.16 \, \rho_v h_{fg} \left[\frac{\sigma g(\rho_l - \rho_v)}{\rho_v^2} \right]^{1/4} \tag{9.52}$$

The constant in Eq. (9.52) was obtained empirically by comparing the predictions with the data available at that time. Subsequently Zuber [5] argued that onset of maximum heat flux was a consequence of hydrodynamic instability of vapor columns. The maximum heat flux occurred when vapor velocity in the jets reached a critical velocity at which vapor jets became unstable. The instability of the jets thus placed an upper limit on the vapor removal rate from the heater surface. If the vapor generation rate exceeded the vapor removal rate, vapor accumulation occurred near the surface. This in turn led to a rapid increase in the dry fraction of the heater area. The insulating effect of vapor causes wall superheat to increase while at the same time leading to a reduction in heat transfer rate from the surface. The difference in the heat input rate and the heat transfer rate from the surface further amplifies the increase in wall superheat.

If U_{vc} is the vapor velocity at which vapor jets become unstable, an expression for the maximum heat flux is obtained by calculating the energy balance at the heater surface as

$$q_{max} = \rho_v h_{fg} U_{vc} A_v / A \tag{9.53}$$

In Eq. (9.53), A_v/A is the fraction of the heater area that is occupied by vapor jets. In writing Eq. (9.53) it is assumed that all of the energy released from the heater is utilized in production of vapor at and near the heater surface.

The critical velocity of the jets was assumed to be given by Helmholtz instability of two parallel flowing streams. For inviscid fluid streams flowing parallel to each other, an expression for the relative velocity between two streams was obtained in the previous section in terms of dominant wavelength as

$$(U_v - U_l)_c = \sqrt{\frac{2\pi\sigma(\rho_l + \rho_v)}{\lambda_{HC}\rho_v\rho_l}} \tag{9.54}$$

In Eq. (9.54), λ_{HC} is the dominant wavelength on the interface. Equation (9.54)

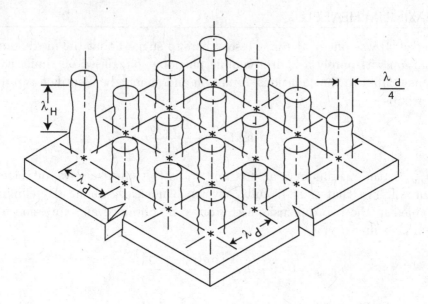

Figure 9.3 Zuber's vapor removal configuration on "infinite" flat plate (Lienhard and Dhir [6])

as shown earlier is obtained from plane interface analysis, but is being applied to a vapour column or jet.

Zuber argued that on a large horizontal plate, the jets locate themselves on a square grid with spacing based on a two-dimensional Taylor unstable wavelength. The diameter of the jets is equal of the jet spacing. The support for this assumption was based on the spacing and bubble diameter relationship observed in film boiling. The vapor removal configuration is shown in Fig. 9.3. Since liquid must flow in the downward direction to keep up with the vapor outflow, a mass balance over one cell with spacing, λ, yields

$$\rho_v U_v \left[\pi \left(\frac{\lambda}{4} \right)^2 \right] = -\rho_l U_l \left[\lambda^2 - \pi \left(\frac{\lambda}{4} \right)^2 \right] \tag{9.55}$$

or

$$-U_1 = \frac{\rho_v}{\rho_l} U_v \left[\frac{\frac{\pi}{16}}{1 - \frac{\pi}{16}} \right] \tag{9.56}$$

Substituting for U_l from Eq. (9.56) into Eq. (9.54), an expression of the critical velocity of vapor is obtained as

$$U_{vc} = \sqrt{\frac{2\pi\sigma(\rho_l + \rho_v)}{\lambda_{HC}\rho_v\rho_l}} \left[1 + \frac{\rho_v}{\rho_l} \frac{\pi}{16 - \pi} \right]^{-1} \tag{9.57}$$

Zuber assumed that Helmholtz unstable wavelength was the critical Rayleigh unstable wavelength on a jet in vacuum. The wavelength is equal to the circumference of the jet. Upon substitution for A_v/A and U_{vc}, Eq. (9.53) becomes

$$q_{max} = \rho_v h_{fg} \frac{\pi}{16} \sqrt{\frac{4\sigma(\rho_l + \rho_v)}{\lambda \rho_v \rho_l}} \left[1 + \frac{\rho_v}{\rho_l} \frac{\pi}{16 - \pi}\right]^{-1} \qquad (9.58)$$

Zuber could not decide whether the jet spacing λ should be equal to λ_c, the two-dimensional critical Taylor wavelength or, λ_d, the "most dangerous" Taylor wavelength. Substituting in Eq. (9.58) for $\lambda = \lambda_c$ or $\lambda = \lambda_d$, from the previous section an expression for the maximum heat flux is obtained as

$$q_{max} = \frac{\pi}{24} \rho_v h_{fg} \left[\frac{\sigma g(\rho_l - \rho_v)}{\rho_v^2}\right]^{1/4} \left[\frac{\rho_l + \rho_v}{\rho_l}\right]^{1/2} \left[\frac{\rho_l(16 - \pi)}{\rho_l(16 - \pi) + \pi \rho_v}\right][1.196 \text{ or } 0.909] \qquad (9.59)$$

where the constant in the last square bracket on the extreme right of Eq. (9.59) is 1.196 when $\lambda = \lambda_c$ and 0.909 when $\lambda = \lambda_d$. Zuber took the constant in the square bracket to be unity which is a good mean value. At low pressures the components of second and third terms in square brackets on the right-hand side are nearly equal to unity. Thus Zuber was able to obtain an elegant analytical solution for a very complex process. It should be noted that the above expression is derived for a large flat plate. Deviations will occur when the width is less than about two Taylor wave lengths (Lienhard and Dhir [6]). They also suggested that lead constant in Eq. (9.59) would be 0.15 when λ_{HC} was taken to be equal to λ_d. Over the years a number of competing mechanisms of maximum heat flux have been proposed in the literature.

9.4 MINIMUM HEAT FLUX

Zuber was also successful in obtaining an expression for the minimum heat flux purely from hydrodynamic considerations. He rationalized that vapor film collapse occurred when vapor production rate on a horizontal plate fell just short of that required to sustain a cyclic growth of the interface as determined by Taylor instability. Thus an expression for the minimum heat flux in a saturated liquid was obtained as

$$q_{min} = \rho_v h_{fg} N_b V_b f_b \qquad (9.60)$$

where N_b is the number of bubbles that are released per unit area, V_b is the volume of a bubble and f_b is the bubble release frequency. The bubble release frequency was determined by knowing the time taken by the interface to travel a distance equal to a Taylor wavelength. This distance is calculated by assuming that bubbles are released alternatively from the nodes and antinodes of a Taylor wave. The bubbles grow to a height 0.4λ and time to travel a distance of 0.1λ is lost during bubble break off at each site. The average interface speed was obtained by averaging the growth rate over the distance travelled by the interface. The average growth rate of the interface was obtained as

$$\frac{d\bar{\eta}}{dt} = \frac{1}{0.4\lambda} \int_0^{0.4\lambda} \frac{d\eta}{dt} \cdot d\eta. \qquad (9.61)$$

$$= 0.2\lambda(-w_i) \tag{9.62}$$

or the expression for the bubble frequency, f_b, is written as

$$f_b = \frac{1}{\lambda}\frac{d\bar{\eta}}{dt} = 0.2(-w_i) \tag{9.63}$$

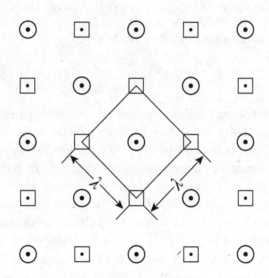

Figure 9.4 Taylor wave pattern on infinite flat plate

If the bubbles locate on a square grid as shown in Fig. 9.4, two bubbles will be released per λ^2 area of the heater. One bubble is released from the corner and the second bubble from the center. If bubble diameter is assumed to be equal to half the wavelength, Eq. (9.60) becomes

$$q_{min} = \rho_v h_{fg} \cdot 2\left(\frac{4}{3}\pi\left(\frac{\lambda}{4}\right)^3\right)0.2(-w_i)/\lambda^2 \tag{9.64}$$

$$= \frac{\pi}{24}\cdot\rho_v h_{fg}\cdot(0.2)(-w_i)\lambda$$

If the dominant wavelength is assumed to be the "most dangerous" two-dimensional Taylor wavelength and $(-w_i)$ is taken to be the frequency corresponding to the fastest growing wave, Eq. (9.64) becomes,

$$q_{min} = 0.177\rho_v h_{fg}\sqrt[4]{\frac{\sigma g(\rho_l - \rho_v)}{(\rho_l + \rho_v)^2}} \tag{9.65}$$

Equation (9.65) over-predicts by about a factor of two the minimum heat fluxes observed on a large horizontal plate. The derivation of Eq. (9.65) is flawed for several reasons:

1. The growth rate, as noted by Berenson [7], should have been averaged over time rather than over distance.

2. Growth rate obtained from linear analysis was used even when interface amplitude was comparable to the wavelength.

3. Two-dimensional Taylor wave configuration was used for a three-dimensional situation.

Although Zuber used two-dimensional Taylor wavelength for the spacing between adjacent nodes, interestingly as can be deduced from Fig. 9.4, three-dimensional wavelength would have given the same spacing as shown in Fig. 9.4.

Berenson [8] suggested that because of these deficiencies in the model and unavailability of data on growth rate, it was more appropriate to determine the lead constant by comparing predictions from Eq. (9.65) with the data. He suggested a value of 0.09 for the lead constant.

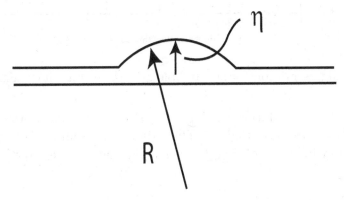

Figure 9.5 Assumed bubble shape during film boiling

Subsequently, Lienhard and Dhir [6] used the interface growth data during film boiling in a cylinder to develop an expression for the minimum heat flux on a flat plate. They assumed that at any interface position, the bubble as shown in Fig. 9.5 could be assumed as a segment of sphere. The volume of such a segment could be written as

$$V = \pi R \eta^2 (1 - \eta/3R) \qquad (9.66)$$

where R is the radius of curvature and η is the displacement of the interface above the mean position of the film. During the growth of the interface both η and R change, thus the rate of change of bubble volume with time can be written as

$$\frac{dV}{dt} = \pi(2R\eta - \eta^2)\frac{d\eta}{dt} + \pi\eta^2\frac{dR}{dt} \qquad (9.67)$$

Visual observations show that at the time of detachment, the bubble is larger than a hemisphere. They assumed that bubble radius does not change after it becomes a hemisphere.

With $\frac{dR}{dt}$ set to zero, Eq. (9.67) becomes

$$\frac{dV}{dt} = \pi(2R\eta - \eta^2)\frac{d\eta}{dt} \qquad (9.68)$$

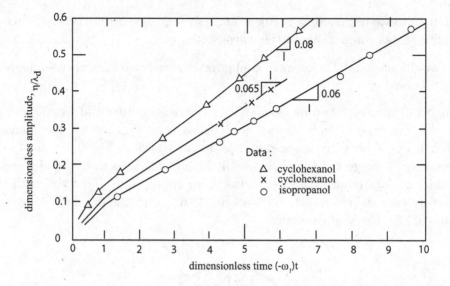

Figure 9.6 Typical observations of late bubble growth during film boiling [6]

The bubble growth rate during the late stages of growth was obtained from the data such as those plotted in Fig. 9.6. These data were from movies of film boiling on a horizontal cylinder. From Fig. 9.6 it is seen that bubble growth rate can be written as

$$\frac{d\eta}{dt} = (-w_i)(c\lambda) \tag{9.69}$$

The substitution of $\frac{d\eta}{dt}$ from Eq. (9.69) in Eq. (9.68) yields

$$\frac{dV}{dt} = \pi(2R\eta - \eta^2)(-w_i)(c\lambda) \tag{9.70}$$

The rate of change of bubble volume as given by Eq. (9.70) is maximum when $\eta = R$. Lienhard and Dhir argued that since bubbles grow larger than a hemisphere, the vapor production rate per cell should at least be equal to the maximum rate demanded by the growth process. With $\eta = R$,

$$\frac{dV}{dt} = \pi c\lambda(-w_i)R^2 \tag{9.71}$$

The visual observations both on cylinders and flat surfaces show that bubbles indeed are spaced a distance approximately equal to the "most dangerous" Taylor wavelength. Substitution in Eq. (9.71) for $\lambda = \lambda_d, R = \lambda_d/4$ and $(-w_i) = (-wi)_{max}$, based on two-dimensional analysis, an expression for the minimum heat flux is obtained as

$$q_{min} = 0.091\rho_v h_{fg} \sqrt[4]{\frac{\sigma g(\rho_l - \rho_v)}{(\rho_l + \rho_v)^2}} \tag{9.72}$$

In Eq. (9.72), a mean value of $c = 0.069$ as given in Fig. (9.6) was used. It is interesting to note that this expression is the same as Zuber–Berenson Eq. (9.65) when the lead constant in it was chosen by matching the predictions with the data.

9.5   FILM BOILING HEAT TRANSFER COEFFICIENT

Berenson [8] was probably the first to suggest a model for heat transfer coefficient during laminar film boiling on a horizontal plate. Berenson's model was simple but it reflected a significant insight of the phenomenon. A key assumption made in Berenson's model is that energy transferred across the vapor film is utilized in evaporation at the vapor liquid interface. The vapor flows through the film and into the protruding interface which acts as a vapor reservoir. The friction resistance encountered by vapor as it flows through the film is balanced by the hydrostatic head difference between points 1 and 2 as shown in Fig. 9.7. Various other assumptions made in Berenson's analysis are:

1. Vapor film thickness between adjacent bubbles is constant and flow through the film is modeled as one dimensional flow.

2. Flow is laminar and fluids are incompressible.

3. Inertia and convective terms in the momentum and energy equation can be ignored.

4. Interface between vapor and liquid is free of any ripples.

5. Properties are evaluated at the mean temperature between wall and film interface.

6. Bubbles are located on a square grid a distance λ_d apart. Two bubbles are supported per λ_d^2 area.

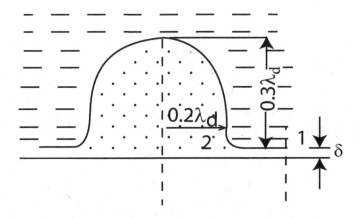

Figure 9.7   Bubble shape assumed in Berenson's analysis

7. Mean bubble radius is 0.2 λ_d whereas the mean bubble height is $0.3\lambda_d$.

8. Although rectangular coordinates are used to locate the bubbles, an axisymmetric flow configuration as shown in Fig. 9.8 was used. For this configuration the plane of symmetry is located at $r_1 = \lambda/\sqrt{2\pi}$.

9. Heat transfer by radiation is neglected.

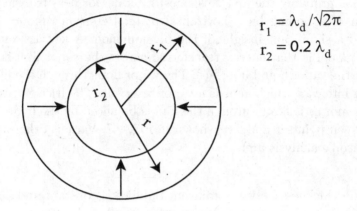

$$r_1 = \lambda_d/\sqrt{2\pi}$$
$$r_2 = 0.2\,\lambda_d$$

Figure 9.8   Vapor flow configuration

The downward velocity of vapor produced at the interface is written by making an energy balance at the interface as

$$v = \frac{k_v \Delta T}{\delta \rho_v h_{fg}} \tag{9.73}$$

The average radial velocity, $\bar{u}$, in the film at any location r, can be written as

$$\bar{u} \cdot 2\pi r \delta = \pi(r_1^2 - r^2)v \tag{9.74}$$

$$\bar{u} = \frac{r_1^2 - r^2}{2r\delta}v \tag{9.75}$$

Substituting for v in Eq. (9.75) from Eq. (9.73), an expression for $\bar{u}$ is obtained as

$$\bar{u} = -\frac{(r_1^2 - r^2)}{2r\delta^2} \cdot \frac{k_v \Delta T}{\rho_v h_{fg}} \tag{9.76}$$

In Eq. (9.76), $-ve$ sign indicates vapor flows in a direction opposite to increasing r. In the absence of inertia, vapor momentum equation in the radial direction can be written as

$$\frac{dp}{dr} = \mu_v \frac{\partial^2 u}{\partial y^2} \tag{9.77}$$

It should be noted that Eq. (9.77) is only approximate as the effect of downward velocity of the vapor generated at the interface is not included. The solution of Eq. (9.77) requires two boundary conditions: The first condition is no slip at the wall, i.e., $u = 0$ at $y = 0$. The second condition at the interface is more difficult to specify since no attempt is being made here to solve for the complex liquid notion adjacent to film. Two limiting boundary conditions can be rigid interface, i.e., $u = 0$ at $y = \delta$, or dynamic interface, i.e., $\frac{\partial u}{\partial y} = 0$ at $y = \delta$. Integration of Eq. (9.77) yields

$$u = \frac{1}{\mu_v}\frac{dp}{dr}\left(\frac{y^2}{2} - \frac{\delta y}{2}\right) \ or \ \frac{1}{\mu_v}\frac{dp}{dr}\left(\frac{y^2}{2} - \delta y\right) \tag{9.78}$$

the first expression is for a rigid interface whereas the second expression results when the interface is assumed to be dynamic (slip condition). The average vapor velocity in the film is obtained as

$$\bar{u} = \frac{1}{\delta}\int_0^\delta udy = -\frac{1}{\mu_v}\frac{dp}{dr}\cdot\frac{\delta^2}{\beta} \tag{9.79}$$

where $\beta = 12$ for a rigid interface and $\beta = 3$ for a dynamic interface.

Substitution of $\bar{u}$ from Eq. (9.76) into Eq. (9.79) yields

$$\frac{dp}{dr} = \frac{\beta\cdot\mu_v k_v \Delta T}{\rho_v h_{fg}\delta^4}\frac{(r_1^2 - r^2)}{2r} \tag{9.80}$$

Integration of Eq. (9.80) between points 1 and 2 shown in Fig. 9.8 gives

$$(p_1 - p_2) = \frac{\beta\mu_v k_v \Delta T}{2\rho_v h_{fg}\delta^4}\left[r_1^2 ln\frac{r_1}{r_2} - \frac{r_1^2}{2} + \frac{r_2^2}{2}\right] \tag{9.81}$$

Since $r_1 = \frac{\lambda_d}{\sqrt{2\pi}}$ and $r_2 = 0.2\lambda_d$, Eq. (9.81) becomes

$$p_1 - p_2 = \frac{0.079\,\mu_v k_v \Delta T}{\rho_v h_{fg}\delta^4}\frac{\lambda_d^2\beta}{\pi} \tag{9.82}$$

For a bubble of radius $0.2\lambda_d$, and height $0.3\lambda_d$, the pressure difference between points 1 and 2 can be written as

$$p_1 - p_2 = (\rho_l - \rho_v)g(0.3\lambda_d) - \frac{2\sigma}{0.2\lambda_d} \tag{9.83}$$

Eliminating $(p_1 - p_2)$ between Eqs. (9.82) and (9.83) and substituting for λ_d, the two dimensional "most dangerous" Taylor wavelength, we obtain

$$\sqrt{\sigma_g(\rho_l - \rho_v)}[3.26 - 0.919] = 2.98\,\frac{\beta\mu_v k_v \Delta T\sigma}{\rho_v h_{fg}\delta^4\cdot g(\rho_l - \rho_v)} \tag{9.84}$$

or

$$\delta^4 = \frac{1.273\beta\mu_v k_v \Delta T}{g(\rho_l - \rho_v)\rho_v h_{fg}}\sqrt{\frac{\sigma}{g(\rho_l - \rho_v)}} \tag{9.85}$$

If a mean value of $\beta(= 7.5)$ between 12 and 3 is used in Eq. (9.85), an expression for the vapor film thickness becomes

$$\delta = 1.76 \left[\frac{\mu_v k_v \Delta T}{\rho_v h_{fg} g(\rho_l - \rho_v)} \sqrt{\frac{\sigma}{g(\rho_l - \rho_v)}} \right]^{1/4} \tag{9.86}$$

Since the bubble height is much larger than the film thickness, heat transfer across the bubble can be neglected. Thus the total heat loss from the cell will be

$$Q = \frac{k_v \Delta T}{\delta} \left\{ \frac{\lambda_d^2}{2} - \pi (0.2 \lambda_d)^2 \right\} \tag{9.87}$$

Thus if $\bar{\delta}$ is the effective film thickness over total area of the cell, the heat loss from the heater per unit cell is

$$Q = \frac{k_v \Delta T}{\bar{\delta}} \cdot \frac{\lambda_d^2}{2} \tag{9.88}$$

Eliminating Q between Eqs. (9.87) and (9.88) yields an expression for the effective film thickness as

$$\bar{\delta} = 1.34 \delta \tag{9.89}$$

Substituting for δ from Eq. (9.86) into Eq. (9.89) gives

$$\bar{\delta} = 2.36 \left[\frac{\mu_v k_v \Delta T}{\rho_v h_{fg} g(\rho_l - \rho_v)} \sqrt{\frac{\sigma}{g(\rho_l - \rho_v)}} \right]^{1/4} \tag{9.90}$$

or the average film boiling heat transfer coefficient on an infinite horizontal plate is given by

$$\bar{h} = 0.42 \left[\frac{k_v^3 \rho_v g(\rho_l - \rho_v) h_{fg}}{\mu_v \Delta T \sqrt{\frac{\sigma}{g(\rho_l - \rho_v)}}} \right]^{1/4} \tag{9.91}$$

Combining the expression for the minimum heat flux developed earlier and Eq. (9.91) for film boiling heat transfer coefficient, an expression for the wall superheat at film collapse is obtained as

$$\Delta T_{min} = 0.13 \frac{\rho_v h_{fg}}{k_v} \left[\frac{g(\rho_l - \rho_v)}{(\rho_l + \rho_v)} \right]^{2/3} \left[\frac{\mu_v}{g(\rho_l - \rho_v)} \right]^{1/3} \left[\frac{\sigma}{g(\rho_l - \rho_v)} \right]^{1/2} \tag{9.92}$$

In evaluating ΔT_{min}, the vapor properties are to be evaluated at the mean temperature. As such an iterative procedure will be required. Also, h_{fg} should be corrected for sensible heating of vapor.

9.6 TRANSITION BOILING

Transition boiling is a mixed model process that combines features of both film and nucleate boiling. At any instance a large fraction of the heater area is covered by vapor. Discrete spots exist on the surface where liquid makes contact with the surface. A given location on the heater surface may experience short periods of liquid contacts followed by long periods during which vapor is present. Transition boiling is characterized by a reduction in heat flux with increase in wall superheat. This feature makes it difficult to obtain steady state data in transition boiling. This fact has led to relatively fewer studies on transition boiling. Consequently, it is not surprising that transition boiling is one of the least understood processes.

Two types of steady state experiments have generally been performed in obtaining transition boiling data. In temperature controlled experiments vapor of a high boiling point liquid is condensed on the underside of the heater. The heater temperature is changed by controlling the pressure or thereby the saturation temperature of the vapor. Such experiments have been carried out by Berenson [7], Hesse [9], and more recently by Ramilison and Lienhard [10]. The difficulty with these experiments is that only few data points are accessible in transition boiling because of the limitations imposed by the sum of resistances between the heat source and the heat sink.

In the heat flux controlled experiments such as those performed by Bui and Dhir [11] it is the mismatch between the heat input and heat flux at the heated surface that causes the process to become time dependent. As such only in a limited range of heat fluxes, steady state transition boiling data have been obtained. Transient transition boiling data can be obtained either through the heat flux or the temperature controlled experiments. A convenient way of obtaining transient transition boiling data is through quenching experiments.

Berenson's [7] experiments were probably the first well-documented and carefully conducted experiments on transition boiling. He boiled n-pentane and carbon tetrachloride on a horizontal disc heated by condensation of steam underneath the disc. The data points from the nucleate boiling side were different from those obtained from the film boling side. Using maximum and minimum heat flux points as a guide, Berenson found a single curve through a limited data set available in transition boiling.

Witte and Lienhard [12] questioned the uniqueness of the transition boiling curve. They argued that two different transition boiling curves are obtained when accessed from nucleate and film boiling sides. They called the curve accessed from nucleate boiling sides as transitional nucleate boiling and that from film boiling as transitional film boiling curves. Subsequent experiments of Bui and Dhir [11], Dhir and Liaw [13] and Maracy and Winterton [14] show unequivocally the existence of two transition boiling curves. The data of Liaw and Dhir showed that the difference in the transition boiling curves accessed from nucleate and film boiling sides was dependent on the wettability of the surface. The difference decreased as the surface became more wettable. It has been argued by Auracher [15] that the differences in the two transition boiling curves obtained by Bui and Dhir and Maracy and

Winterton were due to the transient nature of the data. Had the data been obtained under steady state conditions, a unique transition boiling curve would have been observed. Until a significant number of steady state transition boiling data are taken under different conditions, the debate on the existence or non existence of a unique transition boiling curve will continue.

At present no mechanistic model for dependence of transition boiling heat flux on wall superheat exists in the literature. Most of the empirical correlations rely on the fact that the upper limit of the transition boiling heat flux is the maximum heat flux whereas the lower limit corresponds to the minimum heat flux. Nucleate boiling is assumed to prevail over the liquid occupied region whereas conditions similar to film boiling exist on the dry area. An expression for the transition boiling heat flux is written as

$$q = F_l q_l + (1 - F_l) q_v \tag{9.93}$$

In Eq. (9.93), F_i is the fractional area of the heater occupied by liquid, q_l is the time and area averaged heat flux on the liquid occupied area, and q_v is the heat flux on the area occupied by vapor. The fractional area, F_l, is related to the wall temperature as

$$F_l = \left[\frac{T_{min} - T}{T_{min} - T_{max}} \right]^n \tag{9.94}$$

In Eq. (9.94) n is an empirical constant. Equation (9.94) inplies that the heater is totally covered with liquid at q_{max} and is totally dry at q_{min}. This is not really correct as in nucleate boiling a certain fraction of the heater area is covered by vapor and some liquid-solid contacts may exist near the minimum heat flux condition. The heat fluxes q_l and q_v are evaluated using correlations for nucleate and film boiling respectively. There is a danger in this type of approach. Forcing the correlations to match the data, dependence of dry and wet fractional areas on wall superheat may be skewed. Bjonard and Griffith [16] somewhat alleviated this difficulty by assuming q_l and q_v to be equal to q_{max} and q_{min}, respectively.

REFERENCES

1. Rayleigh, Lord, Theory of Sound, Dover Publications, New York, 1945.

2. G. I. Taylor, "The instability of liquid surfaces when accelerated in a direction perpendicular to their planes," Proc. Roy Soc. (London), Vol. 201, pp. 192-196, 1950.

3. R. Bellman and R. H. Pennington, "Effects of surface tension and viscosity on Taylor instability", Q. App. Math., 12, 151-162, 1954.

4. S. S. Kutateladze, "A hydrodynamic theory of changes in a boiling process under free convection," Izv. Akad. Nauk SSSR, Otdel. Tekh. Nauk. 4, 529, 1951.

5. N. Zuber, "Hydrodynamic aspects of boiling heat transfer," U.S. AEC Report No. AECU 4439 (Technical Information Service, Oak Ridge, TN), 1959.

6. J. H. Lienhard and V. K. Dhir, "Extended hydrodynamic theory of peak and minimum pool boiling heat fluxes," NASA CR-2270, Contract No. NGL 18-001-035, 1973.

7. P. J. Berenson, "Experiments on pool boiling heat transfer," Int. J. Heat Mass Transfer, 5, 985-999, 1962.

8. P. J. Berenson, "Film boiling heat transfer from horizontal surface," ASME Trans. J. Heat Transfer 83, 351, 1961.

9. G. Hesse, "Heat transfer in nucleate boiling, maximum heat flux and transition boiling," Int. J. Heat Mass Transfer, 16, 1611, 1973.

10. J. M. Ramilison and J. H. Lienhard, "Transition boiling heat transfer and the film transition regime", J. Heat Transfer 109, 746, 1987.

11. T. D. Bui and V. K. Dhir, "Transition boiling heat transfer on a vertical surface", J. Heat Transfer, 107 (4), 756, 1985.

12. L. C. Witte and J. H. Lienhard , "On the existence of two transition boiling curves", Int. J. Heat Mass Transfer, 25, 771, 1982.

13. V. K. Dhir and S. P. Liaw, "Framework for a unified model for nucleate and transition pool boiling", J. Heat Transfer, 111(3), 739, 1989.

14. M. Maracy and R. H. S. Winterton, "Hysteresis and contact angle effects in transition pool boiling of water", 31(7), 1443, 1988.

15. H. Auracher, "Transition boiling". Proc. 9th Int. Heat Transfer Conference. Jerusalem 1, 69-90, 1990.

16. T. A. Bjornard and P. Griffith, "PWR blowdown heat transfer", Light Water React, 1, 17, 1977.

EXERCISES

1. Calculate and plot for saturated water maximum heat flux using Zuber's model on an infinite flat plate for pressures of 1. 61, 100, and 221.2 bars. Discuss your result.

2. For two dimensional Taylor wave, calculate the most dominant wavelength and corresponding frequency for gravity levels 1, 10^{-2}, and 10^{-6}ge. Assume test liquid to be saturated water and system pressure to be one atmosphere.

3. Calculate the minimum heat flux for saturated water boiling at one atmosphere pressure on an infinite flat plate. In obtaining your result, assume the wall superheat to be 100°C.

4. Calculate minimum wall superheat and minimum heat flux for water on a flat plate at pressures of 1, 61, 100, and 221.2 bars. Using your earlier results for qmax, plot the ratio of q_{max}/q_{min} as function of $p/p_{critical}$. Comment on your result.

5. Using Berenson's expression, calculate the film boiling heat transfer coefficient for saturated water at one atmosphere pressure. Assume a wall superheat of 100°C. How would you modify Berenson's model to include contribution of radiative heat transfer?

6. Calculate minimum film boiling temperature for PF-5060 at one atmosphere pressure.

## CHAPTER 10

Laminar Film Condensation

10.1 CONDENSATION ON PLANE AND AXISYMMETRIC BODIES

Nusselt [1] was the first researcher to provide a model for film condensation on a vertical plate and on a horizontal cylinder. Figure 10.1 shows the configuration of an axisymmetric surface. Here we will carry out a generalized analysis which following the approach of Dhir and Lienhard [2] is valid for plane (e.g., a vertical wall) and for axi-symmetric bodies (e.g., cylinders and spheres). Various assumptions made in carrying out the steady state analysis are:

(i) Surface is maintained at a constant temperature below the vapor temperature. Vapor is at its saturation temperature.

(ii) Condensate film thickness is much smaller than the radius of curvature.

(iii) Fluids are incompressible.

(iv) Flow in the film is laminar.

(v) Inertia and convective terms can be neglected in the momentum and energy equations, respectively.

(vi) The condensate properties are evaluated at the mean film temperature.

(vii) Interface between vapor and liquid is free of any ripples and vapor imposes no shear on the film.

(viii) No non-condensables are present in the vapor.

Since large velocity gradients occur normal to the surface, and inertia terms are neglected, the momentum equation for the condensate film can be written as

$$0 = -\frac{\partial p}{\partial x} + \rho_l g(x) + \mu_l \frac{\partial^2 u}{\partial y^2} \qquad (10.1)$$

where $\frac{\partial p}{\partial x}$ represents the pressure gradient in the vapor and is equal to $+\rho_v g(x)$. As such the momentum equation becomes

$$0 = g(x)(\rho_l - \rho_v) + \mu_l \frac{\partial^2 u}{\partial y^2} \qquad (10.2)$$

or

$$\frac{\partial^2 u}{\partial y^2} = -\frac{g(x)(\rho_l - \rho_v)}{\mu_l} \tag{10.3}$$

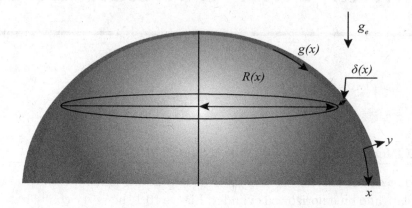

Figure 10.1 Laminar film condensation on an axi-symmetric surface

One boundary condition for Eq. (10.3) is zero velocity at the wall ($y = 0$). The other boundary condition is specified at the interface. Since we are not solving for the flow field in the vapor, it is difficult to specify an appropriate boundary condition. Two limiting possibilities are: slip or no slip boundary condition at the interface. For slip boundary condition $\frac{\partial u}{\partial y} = 0$ at $y = \delta$ while for no slip boundary condition $u = 0$ at $y = \delta$. Slip condition is appropriate for condensation at low pressures. With integration of Eq. (10.3) and use of the slip boundary condition, an expression of the condensate velocity is obtained by

$$u = \frac{g(x)(\rho_l - \rho_v)}{2\mu_l}(-y^2 + 2\delta y) \tag{10.3a}$$

Since convective terms in the energy equation are neglected and condensate temperature variation normal to wall is dominant, the energy equation can be written as

$$\frac{\partial^2 T}{\partial y^2} = 0 \tag{10.4}$$

The boundary conditions on the equation are: a specified wall temperature, T_w, and saturation temperature at the interface. An expression for the temperature distribution in the condensate film is obtained as

$$T - T_w = (y/\delta)(T_{sat} - T_w)$$
$$= (y/\delta)\Delta T$$

or

$$T_{sat} - T = (1 - (y/\delta))\Delta T \tag{10.5}$$

An energy balance at any location x on the wall yields

$$\frac{k_l(T_{sat} - T_w)}{\delta} 2\pi R(x) = \frac{\partial}{\partial x} \int_0^\delta \rho_l u 2\pi R(x) h_{fg} dy$$

$$+ \frac{\partial}{\partial x} \int_0^\delta \rho_l u 2\pi R(x) c_{pl}(T_{sat} - T) dy \quad (10.6)$$

The first term on the right-hand side results from the rate at which condensate is added to the film with distance. The second term represents the change with distance of the sensible heat content of the condensate flowing in the film. Substitution for u from Eq. (10.3a) and for $(T_{sat} - T)$ from Eq. (10.5) in Eq. (10.6) yields after integration

$$\frac{k_l(T_w - T_{sat})R(x)}{\delta} = \frac{h_{fg}\rho_l(\rho_l - \rho_v)}{3\mu_l} \frac{\partial}{\partial x}\left(R(x)g(x)\delta^3\left(1 + \frac{3}{8}\frac{c_{pl}\Delta T}{h_{fg}}\right)\right) \quad (10.7)$$

Defining

$$h'_{fg} = h_{fg}\left(1 + \frac{3}{8}\frac{c_{pl}\Delta T}{h_{fg}}\right)$$

and $\Delta = \delta R^{1/3}g^{1/3}$, Eq. (10.7) becomes

$$\frac{k_l\Delta T R^{4/3}g^{1/3}}{\Delta} = \frac{h'_{fg}\rho_l(\rho_l - \rho_v)}{3\mu_l} \frac{d}{dx}\Delta^3$$

or

$$k_l\Delta T R^{4/3}g^{1/3} = \frac{h'_{fg}\rho_l(\rho_l - \rho_v)}{4\mu_l} \frac{d}{dx}\Delta^4 \quad (10.8)$$

Integrating Eq. (10.8) with respect to x gives

$$\Delta^4 = \frac{4\mu_l k_l \Delta T}{h'_{fg}\rho_l(\rho_l - \rho_v)} \int_0^x R^{4/3}g^{1/3}dx \quad (10.9)$$

Since Δ is defined as a product of δ, $R^{1/3}$ and $g^{1/3}$, it will be zero when either one of the factors is 0. For a vertical plane wall, δ is 0 at $x = 0$, for a horizontal cylinder $g = 0$ at $x = 0$ and for a sphere g and R are both zero at $x = 0$.

For a vertical plane wall, $g(x) = g_e$ and $R(x) \to \infty$. As a result, Eq. (10.9) yields

$$\delta^4 = \frac{4\mu_l k_l \Delta T x}{\rho_l(\rho_l - \rho_v)g_e h'_{fg}}$$

or

$$\delta = \left[\frac{4\mu_l k_l \Delta T x}{\rho_l(\rho_l - \rho_v)g_e h'_{fg}}\right]^{1/4} \quad (10.10)$$

or local heat transfer coefficient is

$$h = \frac{k_l}{\delta} = \left[\frac{k_l^3 h_{fg}' \rho_l (\rho_l - \rho_v) g_e}{4 \mu_l \Delta T x} \right]^{1/4} \tag{10.11}$$

and the average heat transfer coefficient is

$$\bar{h} = \frac{4}{3} \left(\frac{1}{4} \right)^{1/4} \left[\frac{k_l^3 h_{fg}' \rho_l (\rho_l - \rho_v) g_e}{\mu_l k_l \Delta T x} \right]^{1/4} \tag{10.12}$$

The numerical constant in Eq. (10.12) is 0.943 for a dynamic interface where we have assumed slip condition at the interface.

For a long horizontal cylinder of radius, R, $g(x) = g_e \sin x/R$ and $R(x)$ is constant R. Substituting for $g(x)$ and R in Δ, Eq. (10.9) gives

$$\delta^4 = \frac{4 \mu_l k_l \Delta T}{\rho_l (\rho_l - \rho_v) g_e h_{fg}'} \frac{\int_0^x \left[\sin \left(\frac{x}{R} \right) \right]^{1/3} dx}{\left[\sin \frac{x}{R} \right]^{4/3}} \tag{10.13}$$

With local heat transfer coefficient being equal to k_l/δ, Eq. (10.13) leads to

$$h = \left[\frac{k_l^3 \rho_l (\rho_l - \rho_v) g_e h_{fg}'}{4 \mu_l \Delta T} \right]^{1/4} \left[\frac{(\sin \frac{x}{R})^{4/3}}{\int_0^x \sin \left(\frac{x}{R} \right)^{1/3} dx} \right]^{1/4}$$

or the average heat transfer coefficient is obtained as

$$\bar{h} = \frac{1}{\pi R} \left[\frac{k_l^3 \rho_l (\rho_l - \rho_v) g_e h_{fg}'}{4 \mu_l \Delta T} \right]^{1/4} \left[\int_0^{\pi R} \left[\frac{(\sin \frac{x}{R})^{4/3}}{\int_0^x \sin \left(\frac{x}{R} \right)^{1/3} dx} \right]^{1/4} dx \right]$$

After integration the second term is found to be $0.868 \pi R^{3/4}$. Or the expression for the average heat transfer coefficient becomes

$$\bar{h} = 0.73 \left[\frac{k_l^3 \rho_l (\rho_l - \rho_v) g_e h_{fg}'}{\mu_l \Delta T D} \right]^{1/4} \tag{10.14}$$

where D is the diameter of the cylinder. For a sphere, $R(x) = R \sin \frac{x}{R}$ and $g(x) = g_e \sin \frac{x}{R}$. By following a development similar to that for a cylinder, an expression for the average heat transfer coefficient on a sphere is found as

$$\bar{h} = 0.81 \left[\frac{k_l^3 \rho_l (\rho_l - \rho_v) g_e h_{fg}'}{\mu_l \Delta T D} \right]^{1/4} \tag{10.15}$$

Condensate flowing through the film accumulates in a droplet formed at the downstream stagnation point. This droplet releases cyclically from the surface. Since droplet height is much larger than the film thickness, the condensing surface area under the droplet is effectively insulated. The correction for this is relatively small.

Two issues concerning the approach used by Nusselt [1] and followed above in obtaining expressions for laminar film condensation on a vertical wall and axisymmetric bodies are to be addressed. These issues pertain to the two basic assumptions namely:

(i) The neglect of inertia even through the film thickness or the condensate velocity increases in the direction of the flow.

(ii) The neglect of convective terms in the energy equation even through the velocity in the film is nonzero.

Rohsenow [3] was the first to offer a correction to Nusselt's solution by eliminating the second assumption. Using an iterative scheme in an integral approach, Rohsenow accounted for the non-linearity of the temperature profile. The result was a new definition of the corrected latent of heat of vaporization

$$h'_{fg} = h_{fg} + 0.68c_{p\ell}\Delta T \tag{10.16}$$

The derivation of Eq. (10.16) is left to the reader as an exercise (e.g., Exercise 4). Eventually Sparrow and Greg [4] carried out full boundary layer type solutions for laminar film condensation on an isothermal vertical wall, thus eliminating both assumptions in Nusselt's original solution. Here we will carry out the complete boundary layer type solutions for a variety of surface shapes that generate similar solutions by reproducing the work of Dhir and Lienhard [5]. The component of gravitational acceleration along these surfaces will vary with distance x, from the leading edge or the stagnation point. The Sparrow and Gregg solution for a vertical wall will be one special case of this general formulation.

10.2   GENERALIZED SPARROW AND GREGG SOLUTION

Figure 10.2 shows the general shape of the two dimensional surface we will consider and also the coordinate system we will employ. In carrying out the analysis for steady state process we make the same assumptions used in Section 10.1 except that we relax the condition that inertia and convection terms are ignored.
We will discuss the implications of these assumptions later.

The continuity equation and the boundary layer type equation of motion for the film are:

$$\frac{\partial u}{\partial x} + \frac{\partial v}{\partial y} = 0 \tag{10.17}$$

and

$$u\frac{\partial u}{\partial x} + v\frac{\partial u}{\partial y} = \frac{g(x)(\rho_\ell - \rho_\nu)}{\rho_\ell} + \mu_\ell\frac{\partial^2 u}{\partial y^2} \tag{10.18}$$

The boundary conditions on Eq. (10.18) are:

$$u = 0 \quad at \quad y = 0 \tag{10.19}$$

$$v = 0 \quad at \quad y = 0$$

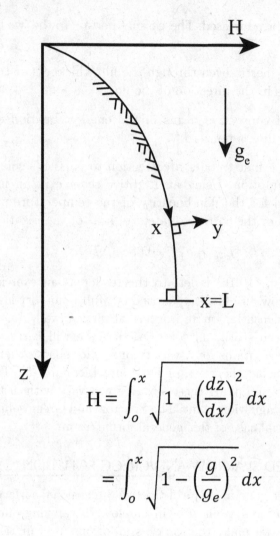

$$H = \int_o^x \sqrt{1 - \left(\frac{dz}{dx}\right)^2} \, dx$$

$$= \int_o^x \sqrt{1 - \left(\frac{g}{g_e}\right)^2} \, dx$$

Figure 10.2 Coordinates of two-dimensional surface

and

$$\frac{\partial u}{\partial y} = 0 \quad at \quad y = \delta$$

The energy equation for the film is

$$u\frac{\partial T}{\partial x} + v\frac{\partial T}{\partial y} = \alpha_\ell \frac{\partial^2 T}{\partial y^2} \qquad (10.20)$$

The boundary conditions are

$$T = T_w \quad at \quad y = 0$$

and

$$T = T_{sat} \quad at \quad y = \delta \qquad (10.21)$$

To satisfy the continuity equation, we define a stream function ψ such that

$$u = \frac{\partial \psi}{\partial y} \quad and \quad v = -\frac{\partial \psi}{\partial x} \tag{10.22}$$

We next seek a similarity variable and a similarity function in terms of an unknown dimensionless stretching function, $f(x)$ and a variable component, $g(x)$, of gravitational acceleration, g_e:

$$\eta = \frac{y\,\ell}{f(x)} \quad and \quad F(\eta) = \psi g_e/(\alpha_\ell g(x) f^3(x)) \tag{10.23}$$

where l has dimension of m^{-1} and is defined as

$$\ell = \left[\frac{g_e(\rho_l - \rho_\nu)}{\mu_\ell \alpha_\ell}\right]^{1/3} \tag{10.24}$$

The above formulation is similar to that employed in Falkner-Skan solutions (see Schichting [6]). Under these variables, the velocities and their derivatives become

$$u \equiv \frac{\partial \psi}{\partial y} = F'\,\alpha_\ell\,gf^2\ell/g_e$$

$$v \equiv -\frac{\partial \psi}{\partial x} = F'\,\alpha_\ell\,gf^2f'\,\frac{\eta}{g_e} - F\alpha_\ell\frac{d}{dx}(gf^3)/g_e$$

$$\frac{\partial u}{\partial y} \equiv \frac{\partial^2 \psi}{\partial y2} = F''\,\alpha_\ell\,gf\ell^2/g_e$$

$$\frac{\partial u}{\partial x} \equiv \frac{\partial^2 \psi}{\partial x \partial y} = -F''\,\alpha_\ell\,gff'\ell\eta/g_e - F'\,\alpha_\ell\,gff'\ell/g_e + F'\,\alpha_\ell\,\ell\frac{d}{dx}(gf^3)/(fg_e)$$

$$\frac{\partial^2 u}{\partial y^2} \equiv \frac{\partial^3 \psi}{\partial y^3} = F'''\,\alpha_\ell\,g\ell^3/g_e$$

In the above equations the prime symbols denote the derivatives of the functions with respect to the single independent variable. Writing Eq. (10.18) in terms of new variables and dividing throughout by $\alpha_\ell \nu_\ell g\ell^3/g_e$ we get

$$\frac{(F')^2}{Pr_\ell}\left[f^2\frac{d}{dx}(f^2g)/\ell g_e\right] - \frac{FF''}{Pr_\ell}\left[\frac{fd}{dx}(gf^3/\ell g_e)\right] = 1 + F''' \tag{10.25}$$

We will be successful in obtaining similar solutions only if the functions contained in the square parentheses are independent of the position coordinate x, i.e. constant or

$$f\frac{d}{dx}(f^3g)/(\ell g_e) = A \tag{10.26}$$

and

$$f^2\frac{d}{dx}(f^2g)/(\ell g_e) = B \tag{10.27}$$

We will soon see the constraint imposed by this requirement on the variation of $g(x)$ with x. Since the constants A and B are inter-related through $f(x)$ and $g(x)$, one of these constants can be set arbitrarily. We set constant $A = 3$ simply on the premise that our formulation should approach that of Sparrow and Gregg in the limit of constant g. Thus Eq. (10.25) can be rewritten as

$$F''' + \frac{A}{Pr_\ell}FF'' - \frac{B}{Pr_\ell}(F')^2 + 1 = 0 \tag{10.28}$$

and the boundary conditions in terms of the new variables are

$$F = F' = 0 \quad at \quad \eta = 0$$

and

$$F'' = 0 \quad at \quad \eta = \eta_\delta \tag{10.29}$$

where η_δ is the value of η at $y = \delta$. Similarly, the energy equation after introducing a similarity function, θ, for temperature as

$$\theta(\eta) = \frac{T - T_w}{T_{sat} - T_w} \tag{10.30}$$

becomes

$$\theta'' + AF\theta' = 0 \tag{10.31}$$

The boundary conditions on Eq. (10.31) are:

$$\theta = 0 \quad at \quad \eta = 0$$

and

$$\theta = 1 \quad at \quad \eta = \eta_\delta \tag{10.32}$$

Equations (10.28) and (10.31) are fully defined except that we do not know what value of η_δ to use while satisfying the boundary conditions. To relate η_δ to physical parameters of the problem, we make an energy balance at the vapor-liquid interface. The heat flux at the interface should be equal to the rate at which enthalpy is added during condensation at the interface. Such an energy balance gives

$$-k_\ell \frac{\partial T}{\partial y}\Big|_{y=\delta} = \rho_\ell\, h_{fg}\, [v|_{y=\delta} - u|_{y=\delta}\, \tan\phi] \tag{10.33}$$

where $\tan\phi = \frac{d\delta}{dx}$. In terms of the new variables Eq. (10.33) can be rewritten as

$$-k_\ell\theta'(\eta_\delta)\Delta T\ell/f(x)$$
$$= \rho_\ell h_{fg}\left[F'(\eta_\delta)\alpha_\ell g f^2 f'\eta_\delta/g_e - F(\eta_\delta)\alpha_\ell\frac{d}{dx}(fg^3)/g_e - F'\alpha_\ell g f^2\ell\frac{d\delta}{dx}/g_e\right] \tag{10.34}$$

We know from Eq. (10.23) that

$$\delta \sim \frac{f(x)}{\ell}$$

or

$$\frac{d\delta}{dx} \sim f'(x)/\ell \qquad (10.35)$$

and

$$\eta_\delta \sim 1 \qquad (10.36)$$

Substituting for $\frac{d\delta}{dx}$ and η_δ in Eq. (10.34) we find

$$\frac{c_{p\ell}\Delta T}{h_{fg}}\theta'(\eta_\delta) = f\frac{d}{dx}(gf^3)/(\ell g_e) \cdot F(\eta_\delta)$$

or

$$\frac{F(\eta_\delta)}{\theta'(\eta_\delta)} = \frac{1}{A}\frac{c_{p\ell}\Delta T}{h_{fg}} \qquad (10.37)$$

Equation (10.37) provides a relationship between η_δ, and the dimensionless group $c_{p\ell}\Delta T/h_{fg}$. The procedure for numerical integration of Eqs. (10.28) and (10.31) will be to assume a value of η_δ, and to complete the integration while satisfying the boundary conditions. Once $F(\eta_\delta)$ and $\theta'(\eta_\delta)$ are known the chosen value can be related to $c_{p\ell}\Delta T/h_{fg}$. The magnitude of heat flux at the wall can be simply written as

$$q_w = -k_\ell\frac{dT}{dy}|_{y=0} = k_\ell\Delta T\theta'(0) \cdot \frac{\ell}{f(x)} \qquad (10.38)$$

or the Nusselt number based on local heat transfer coefficient is

$$Nu \equiv \frac{hx}{k_\ell} = \frac{q_w x}{\Delta T \kappa_\ell} = \frac{\theta'(0) \cdot x\ell}{f(x)} \qquad (10.39)$$

Before going on to the discussion of results, let us determine the surface shapes which will admit similar solutions. We know from Eqs. (10.26) and (10.27) that

$$3 = \frac{1}{\ell g_e}[3f^3 f'g + f^4 g'] \qquad (10.40)$$

and

$$B = \frac{1}{\ell g_e}[2f^3 f'g + f^4 g'] \qquad (10.41)$$

Solving Eqs. (10.40) and (10.41) simultaneously we find that

$$\frac{df^4}{dx} = -\frac{4\ell g_e(B - 3)}{g} \qquad (10.42)$$

and

$$gg'' = \left[\frac{4(B-3)}{3(B-2)}\right](g')^2 \tag{10.43}$$

The solution of Eq. (10.43) can be written as

$$g = (C_1 x + C_2)^n \tag{10.44}$$

where

$$n = \frac{3B-6}{6-B} \tag{10.45}$$

or

$$B = \frac{6(n+1)}{n+3} \tag{10.46}$$

In Eq. (10.44), C_1 and C_2 are constants of integration. Since there is no flow at the leading edge, $x = 0$, the stream function ψ or gf^3 should be 0 there. This is only possible if $n \geq 0$, or C_2 is zero, or g is constant and say equal to g_e. Thus the appropriate form for g for which similar solutions are possible is

$$g(x) = C_3 x^n \tag{10.47}$$

It must be pointed out that negative values of n are not physically possible. Similarly integrating Eq. (10.42) and after substituting for B from Eq. (10.46) we get

$$f(x) = \left[\frac{12(n-1)g_e\ell}{(n+3)}\frac{x}{g(x)}\right]^{1/4} \tag{10.48}$$

Again noting that $gf^3 = 0$ at $x = 0$, the constant of integration has been set equal to zero.

Let us next consider the curved surfaces which at earth's gravity admit similar solutions. Referring to Fig. 10.2, the component of gravitational acceleration acting along the surface can be written as

$$\frac{g(x)}{g_e} = \frac{dz}{dx} \tag{10.49}$$

For such a surface with a forward stagnation point, the similar solutions are possible if

$$\frac{g(x)}{g_e} = \left(\frac{x}{L}\right)^n \tag{10.50}$$

where L is value of x at which the surface becomes vertical. Integrating Eq. (10.49) we get

$$\frac{z}{L} = \frac{\left(\frac{x}{L}\right)^{n+1}}{n+1} \tag{10.51}$$

From geometrical consideration we can also write

$$\frac{dH}{dx} = \sqrt{1 - \left(\frac{dz}{dx}\right)^2} \tag{10.52}$$

Integration of Eq. (10.52) after substitution for dz/dx yields

$$\frac{H}{L} = \int_0^{x/L} \sqrt{1 - \left(\frac{x}{L}\right)^{2n}} \, d\left(\frac{x}{L}\right) \tag{10.53}$$

Figure 10.3 displays the family of earth normal gravity surfaces corresponding to the shapes given by Eqs. (10.51) and (10.53). Next we consider the heat transfer results for a few such surfaces.

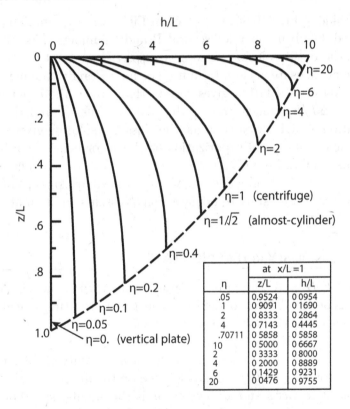

		at x/L =1	
	n	z/L	h/L
	.05	0.9524	0 0954
	1	0 9091	0 1690
	2	0 8333	0 2864
	4	0 7143	0 4445
	.70711	0 5858	0 5858
	10	0 5000	0 6667
	2	0 3333	0 8000
	4	0 2000	0 8889
	6	0 1429	0 9231
	20	0 0476	0 9755

Figure 10.3 The family of body shapes for two-dimensional stagnation point condensing flows which admit similar solutions (Dhir and Lienhard [5])

10.2.1 Vertical Plate

This is the surface which was treated in detail by Sparrow and Gregg. We see from Fig. 10.3 that for this surface $n = 0$ or $B = 2$ and $C_3 = g_e$. Thus for a vertical plate our governing equations reduce to exactly the same form as Sparrow and Gregg's [4] original equations. The function $f(x)$ for this case is obtained from Eq. (10.48) as

$$f(x) = (4\ell x)^{1/4} \tag{10.54}$$

Substituting Eq. (10.54) in Eq. (10.39) we get

$$Nu = \left[\frac{\rho_\ell g_e (\rho_\ell - \rho_v) h_{fg} x^3}{4\mu_\ell k_l \Delta T}\right]^{1/4} \theta'(0) \left[\frac{c_{p\ell}\Delta T}{h_{fg}}\right]^{1/4}$$

or

$$\frac{Nu}{\left[\frac{\rho_\ell g_e (\rho_\ell - \rho_v) h_{fg} x^3}{4\mu_\ell k_l \Delta T}\right]^{1/4}} = \theta'(0) \left[\frac{c_{p\ell}\Delta T}{h_{fg}}\right]^{1/4} \tag{10.55}$$

The left-hand side of Eq. (10.55) is plotted in Fig. 10.4 as a function of $c_{p\ell}\Delta T/h_{fg}$ or the modified Jakob number for several Prandtl numbers. Nusselt's solution is very close to the exact solution as long as $c_{p\ell}\Delta T/h_{fg} < 10^{-1}$ and the Prandtl number of the condensate is close to unity. For high Prandtl numbers, Rohsenow's correction to Nusselt's solution gives results that match the exact solution even for values of $c_{p\ell}\Delta T/h_{fg}$ much greater than 10^{-1}. Thus we find that for laminar film condensation of high Prandtl fluids, the Nusselt's solution gives quite accurate results after the latent heat of vaporization has been corrected according to the determination made by Rohsenow. Another point that can be made here is that $\theta'(0)$ varies as $[c_{p\ell}\Delta T/h_{fg}]^{-1/4}$ provided $cp_\ell\Delta T/h_{fg}$ values are small. For low Prandtl number fluids, the results start to deviate from Nusselt's solution at relatively low values of $cp_\ell\Delta T/h_{fg}$.

10.3 SYNCHONOUSLY ROTATING PLATE

When $n = 1$, or $B = 3$, the gravity varies linearly with the distance from the leading edge, as occurs during condensation on a plate rotating synchronously with the surrounding vapor (i.e., the plate and vapor rotate at the same speed as they would in a closed centrifuge). The component of gravitational acceleration will also vary linearly on a curved surface in an earth normal gravity such as that shown in Fig. 10.3. For a rotating plate with axis located on the line $x = 0$, the $g(x)$ and the g_e can be replaced by $w^2 x$ and $w^2 L$ where w is the angular speed of the plate. In this case any static component of external gravity field is considered to be much smaller than $w^2 x$. For this case the function f from Eq. (10.48) becomes

$$f = (3\ell L)^{1/4} \tag{10.56}$$

Thus we see that function f is independent of x. Substituting f in Eq. (10.35) we find that the thickness of the condensate film on the rotating plate will be constant. The Nusselt number from Eq. (10.39) is obtained as

$$Nu = \theta'(0) \left[\frac{\rho_\ell w^2 (\rho_\ell - \rho_v) h_{fg} x^4}{3\mu_\ell k_\ell \Delta T}\right]^{1/4} \left[\frac{c_{p\ell}\Delta T}{h_{fg}}\right]^{1/4}$$

or

$$\frac{Nu}{\left[\frac{\rho_\ell w^2 (\rho_\ell - \rho_v) h_{fg} x^4}{3\mu_\ell k_\ell \Delta T}\right]^{1/4}} = \theta'(0) \left[\frac{c_{p\ell}\Delta T}{h_{fg}}\right]^{1/4} \tag{10.57}$$

We would guess that for low values of $c_{p\ell}\Delta T/h_{fg}$, the right-hand side of Eq. (10.57) should be equal to unity. This indeed is the case when we carry out the integration. At large values of $c_{p\ell}\Delta T/h_{fg}$, the exact solutions start to deviate from the Nusselt's solution in the same manner as that for a vertical plate and plotted in Fig. 10.4. However, Nusselt's solution with Rhosonow's correction does quite well even for high values of $c_{p\ell}\Delta T/h_{fg}$ provided Pr is much greater than unity.

The temperature distribution in the film is plotted in Fig. 10.5 for $Pr = 1$ and for several values of $c_{p\ell}\Delta T/h_{fg}$. In accordance with the expectation, the profile is found to be nearly linear at low values of $c_{p\ell}\Delta T/h_{fg}$. The non-linearity in the profile increases as $c_{p\ell}\Delta T/h_{fg}$ becomes large. Larger values of $c_{p\ell}\Delta T/h_{fg}$ correspond to larger values of film thickness η_δ.

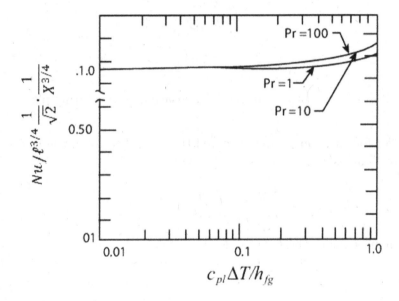

Figure 10.4 Heat transfer from a vertical plate

10.3.1 Limitations on Similarity Solution for Curved Surfaces Shown in Fig. 10.3

As n becomes large the surfaces tend to flatten out and gravitational acceleration acts normal to most of the surface. The limiting case of $n \to \infty$ or condensation on an upward facing horizontal surface was treated by Leppert and Nimmo [7]. Now the condensate is driven by the hydrostatic head. If we include the hydrostatic head term in the momentum equation we can write for the curved surfaces

$$u\frac{\partial u}{\partial x} + v\frac{\partial u}{\partial y} = \frac{g(x)(\rho_\ell - \rho_v)}{\rho_\ell} - g_e\frac{d\delta}{dx}\frac{dH}{dx} + u_\ell\frac{\partial^2 u}{\partial y^2} \tag{10.58}$$

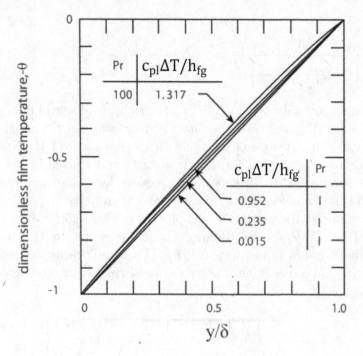

Temperature distribution in the condensate

Figure 10.5   Temperature profile during condensation on a rotating plate

In our earlier formulation we have ignored the second term on the right-hand side of Eq. 10.58. This is realistic if

$$g(x) >> \frac{g_e \rho_\ell}{(\rho_\ell - \rho_\nu)}\left(-\frac{d\delta}{dx}\right)\sqrt{1 - \left(\frac{x}{L}\right)^{2n}} \qquad (10.59)$$

The maximum value of the term under the square root sign can be unity. At low pressures, the value of $\frac{\rho_\ell}{\rho_\ell - \rho_\nu}$ will also be close to unity. If we ignore these terms we can write

$$\frac{g(x)}{ge} >> \left(-\frac{d\delta}{dx}\right) \qquad (10.60)$$

From Eq. (10.23) we know that $\delta \approx f/\ell$. If we substitute for f from Eq. (10.48) and for $g(x)/g_e$ from Eq. (10.50), we get

$$\left(\frac{x}{L}\right)^n >> -\frac{1}{\ell}\frac{d}{dx}\left[\frac{12x\ell}{(n+3)}\left(\frac{L}{x}\right)^n\right]^{1/4}$$

$$\left(\frac{x}{L}\right)^n >> \left[\frac{3(n-1)^4}{64(n+3)(L\ell)^3}\right]^{1/4}\left(\frac{x}{L}\right)^{-\frac{n+3}{4}} \qquad (10.61)$$

It is evident from Eq. (10.61) that for $n > 1$, the solution without the hydrostatic head term will not be valid over most of the surface. Thus the usefulness of the similarity solution for surfaces with $n > 1$ is limited.

Next we determine the reference temperature at which the properties should be evaluated. A numerical integration of the governing equation with variable properties has been carried by several investigators (e.g., Minkowyez and Sparrow [8] and Denny and Miles [9]). From their work, Minkowycz and Sparrow concluded that Nusselt's solution with Rohsenow's correction will give very accurate results if the properties are evaluated at a reference temperature defined as

$$T_{Reference} = T_w + 0.31\Delta T \tag{10.62}$$

Since in many practical situations of interest ΔT is generally much smaller than the wall temperature, little error is made if the wall temperature is used to calculate the properties. Also, we should note that, at higher values of ΔT which correspond to higher condensation rates, ripples appear at the film surface. The effect of the ripples on heat transfer is much more pronounced than that of the variable properties.

REFERENCES

1. W. Nusselt, Die Oberflachenkondensation des Wasserdamfes, Z. Ver. Deutsch. Ing., Vol. 60, pp. 541-546, 569-575, 1916.

2. V. Dhir and J. Lienhard, Laminer film condensation on plane and axisymmetric bodies in nonuniform gravity, J. Heat Transfer, Vol. 91, pp. 97-100, 1971.

3. W. M. Rohsenow, Heat transfer and temperature distribution in laminar film condensation, J. Heat Transfer, Trans. ASME, Vol. 78, pp. 1645-1648, 1958.

4. E. M. Sparrow, and J. L. Gregg, A boundary-layer treatment of laminar-film condensation, J. Heat Transfer, Trans. ASME, Vol. 81, pp. 13-18, 1959.

5. V. K. Dhir, and J. H. Lienhard, Similar solutions for film condensation with variable gravity or body shape, J. Heat Transfer, Vol. 95, pp. 483-486, 1973.

6. H. Schlichting, Boundary layer theory, McGraw Hill, New York, 1968.

7. G. Leppert and B. Nimmo, Laminar film condensation on surface normal to body or inertial forces, J. Heat Transfer, Trans. ASME, Vol. 90, pp. 178-179, 1968.

8. E. J. Minkowycz and E. M. Sparrow, Condensation heat transfer in the presence of non-condensibles, interfacial resistance, superheating, variable properties and diffusion, Int. J. Heat Mass Transfer, Vol. 9, pp. 1125-1144, 1966.

9. V. E. Denny, and A. F. Mills, Non similar solutions for laminar film condensation on a vertical surface, Int. J. Heat Mass Transfer, Vol. 12, pp. 965-979, 1969.

EXERCISES

1. Using the approach used by Rohsenow [3], obtain an expression for correction of the latent heat of vaporization to account for sensible cooling of the condensate resulting from non-linearity of the temperature profile.

2. Laminar film condensation occurs on a 20 cm tall vertical plate maintained at 90°C. Surface is exposed to pure stream at 100°C and one atmosphere pressure. Calculate the average condensation heat transfer coefficient. What is the mass flow rate of condensate from the plate?

3. Laminar film condensation occurs on a cone standing on its base. The cone angle is 30°. Obtain an expression for the average heat transfer coefficient. The base of the cone is 2.5 cm in diameter and the cone wall temperature is maintained at 90°C. What is the value of the average heat transfer coefficient if saturated water vapor, at a pressure of one atmosphere, condenses on the cone surface?

4. A circular disc of diameter D spins about its axis at N rpm. The disc temperature is maintained below the saturation temperature at the imposed pressure such that $T_w < T_{sat}$. The disc rotates synchronously with pure vapor at its saturation temperature and laminar film condensation occurs on the disc. Obtain an expression for the average heat transfer coefficient. Assume centrifugal acceleration is much larger than gravity.

5. Water at a temperature of 20°C flows at a velocity of 3 m/sec inside a circular aluminum tube with outer diameter of 19.1 mm and inner diameter of 16.1 mm. Saturated steam at a pressure of 0.1 atm condenses on the outside of the tube. Calculate the condensation rate of steam. Assume that there is no non-condensable present in the steam.

Index

Printed in the United States
by Baker & Taylor Publisher Services